CHEMICAL BIOPHYSICS
Quantitative Analysis of Cellular Systems

Simulation and analysis of biochemical systems is at the heart of computational and systems biology. This textbook covers mathematical and computational approaches to biochemical systems based on rigorous physical principles. Written with an interdisciplinary audience in mind, this book shows the natural connection between established disciplines of chemistry and physics and the emerging field of systems biology, enabling the reader to take an informed approach to quantitative biochemical systems analysis.

Organized into three parts, introducing the student to basic biophysical concepts before applying the theory to computational modeling and analysis through to advanced topics and current research, this book is a self-contained treatment of the subject.

- Background material – this part introduces kinetics and thermodynamics of biochemical networks, providing a strong foundation for understanding biological systems and applications to well-conceived biochemical models.
- Analysis and modeling of biochemical systems – topics covered include enzyme-mediated reactions, metabolic networks, signaling systems, biological transport processes, and electrophysiological systems.
- Special topics – explores spatially distributed systems, constraint-based analysis for large-scale networks, protein–protein interaction, and stochastic phenomena in biochemical networks.

Featuring end-of-chapter exercises, with problems ranging in scope from straightforward calculations to small computational simulation projects, this book will be suitable for advanced undergraduate or graduate level courses in systems biology, computational bioengineering, and molecular biophysics.

DANIEL A. BEARD is Associate Professor in the Department of Physiology and the Biotechnology and Bioengineering Center, Medical College of Wisconsin.

HONG QIAN is Professor of Applied Mathematics and Bioengineering at the University of Washington.

There is a growing number of physicists, engineers, mathematicians, and chemists who are interested in joining the post-genomics party and addressing cutting-edge problems in molecular and cell biology. The barrier to entry can be high and prohibitive. This marvelous new book opens the door for the quantitively inclined. Beard and Qian, in an accessible and clear style, present fundamental methods that can be used to model and analyze an array of biomolecular systems and processes, ranging from enzyme kinetics to gene regulatory networks to cellular transport. This book will appeal to autodidacts as well as professors looking for course texts.

J. J. Collins, Professor of Biomedical Engineering and MacArthur Fellow, Boston University

This is one of the most useful and readable accounts of biochemical thermodynamics that I have seen for a long time, if indeed ever. It is very definitely a book that I shall want to have on my shelves and to refer others to, because it contains a considerable amount of information not easy to find elsewhere.

Athel Cornish-Bowden, Directeur de Récherche, CNRS, Marseilles

Dan Beard and Hong Qian's *Chemical Biophysics: Quantitative Analysis of Cellular Systems* is a masterful portrayal of a critically important new area of science. The success of genomics now makes it imperative to understand the relationships between proteomics, biochemical systems behavior, and the physiology of the intact animal or human. This book provides the path. Its clarity of description, making the complexities seem simple by adhering to fundamental principles, avoiding cluttering detail while painting the broad picture to great depth, makes it a pleasure to read and a treasure to study. It's a must for scientists and scholars working to understand integrative biology.

James B. Bassingthwaighte, Professor of Bioengineering, Biomathematics and Radiology at the University of Washington, Seattle

This wonderful book will be indispensable to specialists in the fields of systems biology, biochemical kinetics, cell signaling, genetic circuits and quantitative aspects of biology, and also to undergraduate and graduate students. It presents a systematic approach to analyzing biochemical systems. The complex subjects are described in a clear style, with carefully crafted definitions and derivations. This unique book is an important step in the development and dissemination of systems biology approaches.

Aleksander S. Popel, Professor of Biomedical Engineering, Johns Hopkins University

As computational biology moves into a more integrative and multi-scale phase, to provide the quantitative framework for linking the mass of experimental data generated by molecular techniques at the subcellular level to tissue- and organ-scale physiology, it is vitally important that models are based on quantitative approaches that incorporate, wherever possible, thermodynamically constrained biophysical mechanisms. This new book *Chemical Biophysics: Quantitative Analysis of Cellular Systems* by Dan Beard and Hong Qian does a wonderful job of formulating models for metabolic pathways, gene regulatory networks, and protein interaction networks on the well-established principles of physical chemistry. Topics include enzyme-catalyzed reactions, reaction–diffusion modeling, membrane transport, the chemical master equation, and much more. This book will be of lasting value to computational biologists and bioengineers.

Professor Peter J. Hunter, Auckland Bioengineering Institute at the University of Auckland

Metabolic modeling often contains simplified assumptions to achieve convergence of equations and these sometimes violate principles of solution physical chemistry. Readers of this remarkable monograph will no longer find those approaches satisfactory because Beard and Qian elucidate the principles of kinetics and thermodynamics of electrolyte solutions relevant to metabolic modeling and computational biology. They show how these principles are essential for molecular modeling of cellular processes most of which involve ionized molecules and macromolecules in the cytoplasm. Their exposition is rigorous. The chapters have an enormous scope and depth that present clear derivations, explanations, and examples. Beard and Qian set the bar very high for future metabolic modeling yet show how the details involved can be managed well and correctly. Analyses at this level of detail are necessary before more complex concepts of molecular crowding and intracellular compartmentalization can be considered. I expect this monograph will become a landmark in computational and systems biology and will be read thoroughly by all scholars in these fields.

Martin J. Kushmerick, Professor of Radiology, Bioengineering, Physiology and
Biophysics at the University of Washington, Seattle

Chemical Biophysics: Quantitative Analysis of Cellular Systems by Daniel Beard and Hong Qian fills a significant niche. The text is a concise yet clear exposition of the fundamentals of chemical thermodynamics and kinetics, aimed specifically at practitioners of the new science of systems biology. It is marvelously illustrated with biochemical examples that will aid those who aim to analyze and model the workings of biological cells.

David Eisenberg, Director UCLA-DOE Institute for Genomics & Proteomics,
University of California, Los Angeles, Investigator, Howard Hughes Medical Institute

Cambridge Texts in Biomedical Engineering

Series Editors

W. Mark Saltzman, *Yale University*
Shu Chien, *University of California, San Diego*

Series Advisors

William Hendee, *Medical College of Wisconsin*
Roger Kamm, *Massachusetts Institute of Technology*
Robert Malkin, *Duke University*
Alison Noble, *Oxford University*
Bernhard Palsson, *University of California, San Diego*
Nicholas Peppas, *University of Texas at Austin*
Michael Sefton, *University of Toronto*
George Truskey, *Duke University*
Cheng Zhu, *Georgia Institute of Technology*

Cambridge Texts in Biomedical Engineering provides a forum for high-quality accessible textbooks targeted at undergraduate and graduate courses in biomedical engineering. It will cover a broad range of biomedical engineering topics from introductory texts to advanced topics including, but not limited to, biomechanics, physiology, biomedical instrumentation, imaging, signals and systems, cell engineering, and bioinformatics. The series will blend theory and practice, aimed primarily at biomedical engineering students but will be suitable for broader courses in engineering, the life sciences, and medicine.

CHEMICAL BIOPHYSICS

Quantitative Analysis of Cellular Systems

DANIEL A. BEARD

Department of Physiology
Medical College of Wisconsin

HONG QIAN

Department of Applied Mathematics
University of Washington

CAMBRIDGE
UNIVERSITY PRESS

CAMBRIDGE UNIVERSITY PRESS
Cambridge, New York, Melbourne, Madrid, Cape Town, Singapore,
São Paulo, Delhi, Dubai, Tokyo, Mexico City

Cambridge University Press
The Edinburgh Building, Cambridge CB2 8RU, UK

Published in the United States of America by Cambridge University Press, New York

www.cambridge.org
Information on this title: www.cambridge.org/9780521870702

First published 2008

A catalogue record for this publication is available from the British Library

ISBN 978-0-521-87070-2 Hardback
ISBN 978-0-521-15824-4 Paperback

10063376 10

To our wives
  Katie and Madeleine

and children
  Henry and Isabelle

Contents

Preface

The life sciences have strong traditions as quantitative disciplines. In several fields quantitatively minded research was at a zenith in the 1960s and 1970s. Flick through, for example, chapters of the American Physiological Society's *Handbook of Physiology* that were published in this era (and even into the 1980s) and one will see physiology revealed as an engineering science, applying the tools of chemical, mechanical, and electrical engineering to measure, analyze, and simulate biological systems. A great deal of biochemical research in the 1960s and 1970s was focused on the kinetics, thermodynamics, and generally physical chemistry of biochemical systems. From this work emerged an interdisciplinary field sometimes called *biophysical chemistry*, which encompasses a collection of physical and mathematical methods for analyzing molecular structure and dynamics.

This great era of quantitative physiology and biochemistry was temporarily sidetracked by revolutions in molecular biology and molecular genetics, which, at risk of oversimplification, are focused on the question of what is there (inside a cell) rather than how it works. In the 1980s and 1990s much of the physically oriented quantitative research in biology was similarly focused on isolated molecules. In the 1990s the term *molecular biophysics* arose as a popular name of new departments combining experimental techniques with theory and simulation, emphasizing physicochemical approaches to studying biological macromolecules.

Nowadays, with several genomes sequenced and large amounts of data available on what molecules are present in what quantities and inside what sorts of cells, attention is shifted to the question of how it all works. The new focus is sometimes called *systems biology*. Whatever we call it, although a number of recent publications would have the reader believe that *systems biology* is an entirely new endeavor, the basic idea of pursuing quantitative mechanistic-based understanding of how biological systems function is a shift back to the philosophy of a previous era.

Of course we should not imply that progress in biological systems analysis ever ceased or that the current trend calls for a wholesale abandonment of reductionist approaches in favor of integrative systems analysis. Yet it is obvious to even the casual browser of the headlines of the science magazines that, in some form or another, systems analysis in biology is in the spotlight for now and at least the foreseeable future. At the heart of a systems approach to biology is a recognition of the importance of dynamic behavior (and function) of a system (a cell, an organ, or an organism) emerging from the interaction of its components. Moreover, computational modeling and simulation is centrally important to analysis of such systems.

While it is in the context of this newfound attention on quantitative and computational biology that we hope this book is useful, some readers may find some of the content old fashioned. A student planning a career in systems biology may wonder whether our emphasis on the physical chemical basis of natural phenomena looks backward or forward. This text represents an attempt to do both in synthesizing a basic foundation in *chemical biophysics* for analysis and simulation of cellular systems. The title of the book, yet another permutation of phys-, chem-, bio-, and related syllables, arises from this desire to continue the rigorous tradition while at the same time define something new.

We are fortunate to have been mentored by a number of leading scientists, including James Bassingthwaigthe, Elliot Elson, Carl Frieden, John Hopfield, James Murray, John Schellman, and Tamar Schlick. In particular we have both benefited a great deal from our long-time association with Jim Bassingthwaigthe. His advice and inspiration is at the root of much of what we have endeavored to do, including writing this book. In addition, we owe a particular debt to Athel Cornish-Bowden who gave us advice, both specific and general, and encouragement on the text. His book on enzyme kinetics sets the standard for clarity that we can only strive for. These two books, perhaps together with one that emphasizes molecular biophysics, could provide appropriate material for a year-long sequence on biophysical chemistry, from macromolecules to biochemical systems. On its own, this book has been used for a semester-long course on computational biology.

Many others provided feedback on the text, discovered typos and errors, and suggested improvements. We are grateful particularly to Xuewen Chen, Ranjan Dash, Ed Lightfoot, Clark Miller, Luis Moux-Dominguez, Feng Qi, Rebecca Vanderpool, Kalyan Vinnakota, Fan Wu, and Feng Yang.

(v) To design optimal experiments. A computational model provides a means of designing experiments for which the hypotheses to be tested (see item (ii) in this list) are sensitive to the variables to be measured.

(vi) To transfer information obtained from one experimental regime to apply to another. For example, the potential impact of observations on enzyme expression levels on metabolic function can be explored using a computational model of the given enzymes and related biochemical pathways.

The key to being able to do all of the above is to be able to build physically realistic models of biological systems. Since basic physical principles circumscribe the behavior of biological systems, we place a special emphasis on physical realism in computational biology and simulation of biochemical systems in this text. Models developed this way attain certain *a priori* validity that is missing in models based on experimental data alone. Viewed from this perspective there is nothing fundamentally novel in the systems biology endeavor. In his Nobel lecture Eduard Buchner argued in 1907 that "the differences between the vitalistic view and the enzyme theory have been reconciled" [25]. By building biological theory on a foundation of physicochemical theory, we will ensure that vitalism does not creep back into the study of biology in the twenty-first century. As Buchner put it, "We are seeing the cells of plants and animals more and more clearly as chemical factories, where the various products are manufactured in separate workshops." This way of seeing cells (and organs and organisms) is at the philosophical foundation of this text.

Inherent in the study of biological systems is the notion of *emergent properties* – the idea that the functions of a complex system transcend the properties of all its individual parts. It is fair to say that the most rigorously understood emergent properties in nature are related to how the observed macroscopic properties of matter arise from the microscopic behavior of atoms and molecules. This is the domain of statistical thermodynamics. Thus the analysis of biochemical systems in terms of statistical thermodynamics provides a natural framework within which current applications in systems biology from electrical, chemical, and computer engineering (such as feedback in networks, optimization, statistical inference, and data mining) may be integrated into a consistent theory of biological systems [78]. One goal of this text is to build a bridge between physical chemical concepts and engineering-based analysis of biological systems.

Organization of this book

This book is organized into three parts. The first part introduces background material on physical chemistry and the treatment of kinetics and thermodynamics in biochemical reactions. While the concepts introduced in Chapter 1, *Concepts from physical chemistry*, are essential for assimilating material in the remainder of the text, detailed mathematical derivations related to this material are not essential and appear in Chapter 12. Key concepts that directly relate to material in the later chapters are introduced and/or reviewed. The second chapter of Part I, *Conventions and calculations for biochemical systems*, introduces the concepts of apparent equilibrium constants and apparent Gibbs free energy, and shows how these concepts are applied to in vivo and in vitro biochemical systems. The third chapter covers basic techniques in modeling and simulating chemical systems.

Part II of this book represents the bulk of the material on the analysis and modeling of biochemical systems. Concepts covered include biochemical reaction kinetics and kinetics of enzyme-mediated reactions; simulation and analysis of biochemical systems including non-equilibrium open systems, metabolic networks, and phosphorylation cascades; transport processes including membrane transport; and electrophysiological systems. Part III covers the specialized topics of spatially distributed transport modeling and blood–tissue solute exchange, constraint-based analysis of large-scale biochemical networks, protein–protein interactions, and stochastic systems.

Since the scope of this book is broad, one could write a whole book on the topics of several of the chapters herein. Indeed for many topics, such books exist. Therefore throughout this book, typically at the conclusion of a given chapter, we refer the reader to a number of excellent texts that we have found useful in studying and synthesizing the important concepts in the analysis of biochemical systems.

Part I

Background material

1

Concepts from physical chemistry

Overview

An essential property of all living systems is that they operate in states of flux, transporting and transforming mass, transducing free energy between chemical, electrical, and mechanical forms, and delivering signals and information in terms of biochemical activities. Consequently, the principles governing the behavior of biochemical systems are the principles of physical chemistry. As an introduction to background material necessary for describing and understanding the behavior of biochemical systems, this chapter covers the concepts of chemical thermodynamics including temperature, entropy, chemical potential, free energy, and Boltzmann statistics.

In the early nineteenth century Carnot gave birth to the field that came to be known as thermodynamics, with the first theoretical treatise on mechanical work and efficiency in heat engines [28]. Over the course of that century, a complete physical theory of how changes in heat, mechanical work, and internal energy of molecular systems are related – in short the theory of thermodynamics – was assembled by Clausius, Helmholtz, Boltzmann, Gibbs, and others. As part of the thermodynamic theory a number of familiar physical quantities were introduced, including entropy, enthalpy, and free energy. We shall see that understanding these quantities and how they are related is essential for building physically realistic simulations of biochemical systems.

1.1 Macroscopic thermodynamics

Our study begins with an enumeration of the laws of thermodynamics:

0. If two systems have the same temperature as a third, then they have the same temperature as one another.
1. The total energy of an isolated system is conserved.

7

2. The entropy of an isolated system does not decrease.
3. The minimal entropy of a system is achieved at the temperature of absolute zero.

Readers may recall encountering these famous laws (expressed in either this form or related equivalent forms) in a number of places – from chemistry to engineering courses – during their student careers. Yet it is our experience that knowledge of these laws does not necessarily translate into proficient understanding and application. The disconnect between knowledge and understanding may arise from the fact that while the laws are simple and straightforward, the quantities that they govern may be somewhat mysterious. For example, the so-called zeroth law makes a precise statement about the physical quantity temperature. Yet the quantity temperature is not defined, leaving us without an understanding of the physical significance of the statement.

Of the three quantities (temperature, energy, and entropy) that appear in the laws of thermodynamics, it seems on the surface that only energy has a clear definition, which arises from mechanics. In our study of thermodynamics a number of additional quantities will be introduced. Some of these quantities (for example, pressure, volume, and mass) may be defined from a non-statistical (non-thermodynamic) perspective. Others (for example Gibbs free energy and chemical potential) will require invoking a statistical view of matter, in terms of atoms and molecules, to define them. Our goals here are to see clearly how all of these quantities are defined thermodynamically and to make use of relationships between these quantities in understanding how biochemical systems behave.

To illustrate the potential disconnect between knowledge and understanding in the study of thermodynamics, consider the basic equation of macroscopic thermodynamics [88],

$$dE = TdS - PdV + \mu dN, \tag{1.1}$$

which relates infinitesimal changes in internal energy E, entropy S, volume V, and number of particles N in a system, to the temperature T, pressure P, and chemical potential μ. Armed with Equation (1.1) and the second law of thermodynamics, which we will encounter in several different forms throughout this book, one may develop all of the relevant formulas of macroscopic thermodynamics. For example, if the volume and number of particles in a system is held constant then the following relationship is apparent:

$$\left(\frac{\partial S}{\partial E}\right)_{N,V} = \frac{1}{T}. \tag{1.2}$$

In Equation (1.2) we have used the conventional notation $(\cdot)_{N,V}$ to specify that the expression in parentheses is computed holding N and V constant.

The definition of the Gibbs free energy $G = E - TS + PV$ may be combined with Equation (1.1) to show the following.

$$dG = dE - TdS - SdT + PdV + VdP$$
$$dG = \mu dN - SdT + VdP$$
$$\left(\frac{\partial G}{\partial N}\right)_{T,P} = \mu. \tag{1.3}$$

Yet mathematical manipulations such as above do not tell us why it is useful to introduce G as a thermodynamic variable. We shall see in Section 1.4.1 that systems held at constant temperature and pressure – i.e., typical laboratory conditions and a reasonable approximation for most biological systems – spontaneously move in the direction of lower G. Therefore gradients in Gibbs free energy represent the thermodynamic driving force for constant temperature and pressure (isothermal and isobaric) systems. To appreciate why this is the case, we need first to develop some ideas about how large numbers of microscopic particles interact and exchange energy under different macroscopically imposed conditions.

In the following sections we will see how temperature, entropy, and free energy are statistical properties that emerge in systems composed of large numbers of particles. In Chapter 12, the appendix to this book, we dig more deeply into statistical thermodynamics, derive a set of statistical laws that are used in this chapter, and show how Equation (1.1) – the fundamental equation of macroscopic thermodynamics – is in fact a statistical consequence of more fundamental principles operating at the microscopic level.

1.2 Isolated systems and the Boltzmann definition of entropy

An isolated system is defined to be a system that does not exchange material or energy with its environment. Thus the extensive thermodynamic variables N, V, and E are held fixed. Boltzmann's formula for the entropy of such a system is

$$S = k_B \ln \Omega(N, V, E), \tag{1.4}$$

where k_B is the Boltzmann constant, and $\Omega(N, V, E)$ is the total number of *microstates* that are available to the system at given values of N, V, and E. (The term *microstate* refers to the microscopic configuration of the system. For classical systems the microstate is defined by the positions and momenta of all particles making up the systems. For quantum mechanical systems the number $\Omega(N, V, E)$ can be obtained by counting the number of independent solutions to the Schrödinger equation that the system can adopt for a given eigenvalue E of the Hamiltonian [156].)

Similarly, if one is interested in a macroscopic thermodynamic state (i.e., a subset of microstates that corresponds to a macroscopically observable system with fixed mass, volume, and energy), then the corresponding entropy for the thermodynamic state is computed from the number of microstates compatible with the particular macrostate. All of the basic formulae of macroscopic thermodynamics can be obtained from Boltzmann's definition of entropy and a few basic postulates regarding the statistical behavior of ensembles of large numbers of particles. Most notably for our purposes, it is postulated that the probability of a thermodynamic state of a closed isolated system is proportional to Ω, the number of associated microstates. As a consequence, closed isolated systems move naturally from thermodynamic states of lower Ω to higher Ω. In fact for systems composed of many particles, the likelihood of Ω ever decreasing with time is vanishingly small and the second law of thermodynamics is immediately apparent.

Combining Equations (1.1) and (1.4), we can develop a statistical interpretation of the thermodynamic quantity temperature,

$$\left(\frac{\partial S}{\partial E}\right)_{N,V} = k_B \left(\frac{\partial \ln \Omega}{\partial E}\right)_{N,V} = \frac{1}{T}, \tag{1.5}$$

which states that temperature is a measure of the relationship between number of microstates and the internal energy of matter.

1.3 Closed isothermal systems

1.3.1 Helmholtz free energy

In biology and chemistry we are usually not interested in the study of isolated systems. Biological systems exchange energy and material with the environment and it is important to understand what are the thermodynamic driving forces in such systems. Since biochemical processes occur in an aqueous environment, we wish to treat the environment in a rigorous way without worrying about the details of the solvent molecules' conformations and interactions. The statistical thermodynamic approach to this problem was introduced by Josiah Gibbs [49].

In this section we study closed systems (closed to mass transport but not energy transfer) held at constant temperature. In statistical mechanics these systems are referred to as NVT systems (because the thermodynamic variables N, V, and T are held fixed). We shall see that the Helmholtz free energy represents the driving force for NVT systems. Just as an isolated system (an NVE system) evolves to increase its entropy, an NVT system evolves to decrease its Helmholtz free energy.

Since a system of constant volume and mass held at constant temperature exchanges energy with its surroundings, we can no longer define a fixed total internal

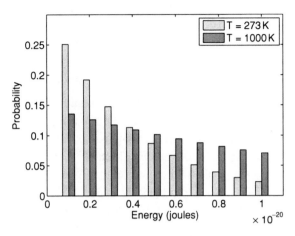

Figure 1.1 Illustration of the Boltzmann probability law of Equations (1.6) and (1.7). The state probability distribution is plotted at two different temperatures for a system with ten possible microstates with energy ranging from 10^{-21} to 10^{-20} joules. At the lower temperature ($T = 273$ K), the lower-energy states are significantly more probable than the higher-energy states. At the higher temperature ($T = 1000$ K), the energy distribution becomes more uniform than at the lower temperature.

energy E of the system. The probability of a microstate that has internal energy E is proportional to $e^{-E/k_B T}$, according to the Boltzmann probability law. (See Section 12.2.) Thus the probability of a state can be calculated

$$P(E) = \frac{e^{-\beta E}}{Q}, \qquad (1.6)$$

where

$$Q = \sum_i e^{-\beta E_i}, \qquad (1.7)$$

where the factor β is equal to $1/k_B T$. The summation in Equation (1.7) is over all possible states i. The Boltzmann probability law is illustrated in Figure 1.1. Using this probability law, we can determine an expression for the *average* internal energy U. In general, with the NVT probability distribution function defined according to Equation (1.6), we can calculate the expected value of a property of the system as

$$\langle f \rangle = \frac{\sum_i f_i e^{-\beta E_i}}{Q} \qquad (1.8)$$

where $\langle f \rangle$ is the expected value of some observable property f, and f_i is the value of f corresponding to the ith state. The average internal energy of a closed isothermal system is defined as the expected value of E:

$$U = \frac{\sum_i E_i e^{-\beta E_i}}{\sum_i e^{-\beta E_i}} = -\frac{\partial}{\partial \beta} \ln \left[\sum_i e^{-\beta E_i} \right] = -\frac{\partial}{\partial \beta} \ln Q. \qquad (1.9)$$

Similar to the case of an isolated system, which naturally moves to maximize Ω (the sum total number of accessible states), the NVT system naturally moves to maximize Q (the probability-weighted sum of states). Thus the second law takes a form in the NVT system analogous to that of the NVE system. The difference is that each state is not equally likely in the constant temperature case, as it is in the constant-energy case. Keeping in mind that we expect NVT systems to move from macrostates of lower Q to macrostates of higher Q, we shall now see why a certain *free energy* (specifically the Helmholtz free energy in the NVT system) is a useful quantity in determining the direction of spontaneous change in thermodynamic systems.

The Helmholtz free energy A is defined as $A = U - TS$, and incremental changes in A can be related to changes in average internal energy, temperature, and entropy by $dA = dU - T dS - S dT$. Substituting this into the basic thermodynamic formula accounting for the incremental changes in internal energy $dU = T dS - P dV + \mu dN$, results in:

$$dA = -S dT - P dV + \mu dN. \qquad (1.10)$$

Thus, the internal energy $U = A + TS$ can be expressed (through a clever manipulation [156]) as:

$$U = A - T \left(\frac{\partial A}{\partial T} \right)_{N,V} = -T^2 \left[\frac{\partial}{\partial T} \left(\frac{A}{T} \right) \right]_{N,V} = \left[\frac{\partial(A/T)}{\partial(1/T)} \right]_{N,V}. \qquad (1.11)$$

We can equate Equations (1.9) and (1.11) by recalling that $\beta = 1/k_B T$. Thus, the Helmholtz free energy can be calculated directly from the quantity Q in Equation (1.7):

$$A = -k_B T \ln Q. \qquad (1.12)$$

Equation (1.12) makes a powerful and useful statement:

NVT systems that minimize Helmholtz free energy maximize Q. Therefore A is a useful thermodynamic potential for NVT systems; NVT systems spontaneously move down gradients in Helmholtz free energy.

1.3.2 Entropy in an NVT system

We know that the probability of a given state i in an NVT system is given by:

$$P_i = e^{-\beta E_i}/Q. \tag{1.13}$$

Next we take the expected value of the logarithm of this quantity:

$$\langle \ln P_i \rangle = -\ln Q - \beta \langle E_i \rangle = -\ln Q - \beta U = \beta(A - U). \tag{1.14}$$

A useful relationship follows from Equation (1.14). Since $A - U = -TS$, $S = -k_B \langle \ln Pi \rangle$. The expected value of $\ln P_i$ is straightforward to evaluate, and

$$S = -k_B \sum_i P_i \ln P_i. \tag{1.15}$$

Interestingly, this equation applies equally to both the NVE and NVT cases. In the NVE system, each state is equally likely. Therefore $P_i = \Omega^{-1}$, and Equation (1.15) becomes

$$S = k_B \sum_i \Omega^{-1} \ln \Omega = k_B \ln \Omega, \tag{1.16}$$

which is the familiar definition of entropy for an isolated system. Equation (1.15) is called the Gibbs formula for entropy [104], while Equation (1.16) is the Boltzmann definition introduced in Section 1.2.[3]

1.3.3 Interpretation of temperature in the NVT system

In Sections 1.2 and 1.3 we discovered two ways to interpret the thermodynamic variable T. First we saw that temperature can be thought of as a measure of the amount of internal energy necessary to add to a system to allow it to obtain additional microstates. Second, given the definition $A = U - TS$, we can interpret T as a gauge that determines the relative contributions of U and S in driving spontaneous processes. At high temperature, entropy dominates the free energy and systems are driven to maximize internal disorder. In the low-temperature limit, free energy is dominated by internal energy and systems move in the direction of decreasing internal energy. A third interpretation of T arises from the equipartition of energy in constant temperature systems. The equipartition theorem states that the average kinetic energy of each particle is given by $3k_B T/2$. (See Section 12.2.4 for a detailed description of the equipartition theorem.) The higher the temperature, the higher the amount of energy stored in each internal degree of freedom.

---

[3] Incidentally, Equation (1.15) is also called the Shannon formula for entropy. Claude Shannon was an engineer who developed his definition of entropy, sometimes called "information entropy," as a measure of the level of uncertainty of a random variable. Shannon's formula is central in the discipline of information theory.

1.4 Isothermal isobaric systems

1.4.1 Gibbs free energy

We have seen that NVT systems spontaneously move to maximize the probability-weighted sum over all internal states Q and simultaneously to minimize Helmholtz free energy A. Yet the thermodynamics of biological systems is best understood in the context of systems held at constant pressure, rather than constant volume. What is the appropriate statistical law for constant-temperature (isothermal) constant-pressure (isobaric) systems? What free energy is the appropriate thermodynamic potential to apply in this situation? For this case, the Gibbs free energy represents the appropriate thermodynamic driving force. Gibbs free energy in NPT systems (constant N, P, and T) is the analog of the Helmholtz free energy in NVT systems.

The relative probability of a state in an NPT system is expressed as a function of thermodynamic quantity *enthalpy*, which is defined as $H = E + PV$. From Equation (1.1) we have

$$dH = TdS - VdP + \mu dN$$

and

$$\left(\frac{\partial S}{\partial H} \right)_{N,P} = \frac{1}{T}. \tag{1.17}$$

Thus in an NPT system enthalpy takes the place of internal energy in an NVT system. The probability law for an NPT system is

$$P = e^{-\beta H}/Z$$

$$Z = \sum_i e^{-\beta H_i}, \tag{1.18}$$

which is derived in Section 12.3. Thus the average enthalpy of an NPT system is computed

$$H = \frac{\sum_i H_i e^{-\beta H_i}}{\sum_i e^{-\beta H_i}} = -\frac{\partial}{\partial \beta} \ln \left[\sum_i e^{-\beta H_i} \right] = -\frac{\partial}{\partial \beta} \ln Z. \tag{1.19}$$

The Gibbs free energy is defined as $G = E - TS + PV$ or $G = H - TS$. Following logic analogous to that of Section 1.3.1, we have

$$H = G - T \left(\frac{\partial G}{\partial T} \right)_{N,V} = -T^2 \left[\frac{\partial}{\partial T} \left(\frac{G}{T} \right) \right]_{N,V} = \left[\frac{\partial (G/T)}{\partial (1/T)} \right]_{N,V}. \tag{1.20}$$

Equating (1.19) and (1.20), we obtain.

$$G = -k_B T \ln Z. \tag{1.21}$$

To complete the analogy to Helmholtz free energy in NVT systems:

NPT systems that minimize Gibbs free energy maximize Z. Therefore G is a useful thermodynamic potential for NPT systems; NPT systems spontaneously move down gradients in Gibbs free energy.

1.4.2 Entropy in an NPT system

In Section 1.3.2 we saw that the Gibbs formula for entropy applies equally well to an isolated system as to a closed isothermal system. To determine whether or not this formula applies to NPT systems as well, we can substitute the NPT probability law of Equation (1.18) into Equation (1.15):

$$S = -k_B \sum_i \frac{e^{-\beta H_i}}{Z} \ln \frac{e^{-\beta H_i}}{Z}$$

$$= k_B \sum_i \frac{e^{-\beta H_i}}{Z} (\beta H_i + \ln Z)$$

$$= k_B \beta H + k_B \ln Z. \tag{1.22}$$

Multiplying by T we have

$$ST = H + k_B T \ln Z = H - G, \tag{1.23}$$

which agrees with the definition $G = H - TS$.

1.5 Thermodynamic driving forces in different systems

In the previous three sections we have seen that three different thermodynamic driving forces are associated with three different systems (the NVE, NVT, and NPT systems). The driving forces for the three systems are summarized in Table 1.1. In the simplest case (the NVE system), in which the system does not exchange energy or material with its surroundings, the isolated system moves naturally to

Table 1.1 *Thermodynamic driving forces in three systems*

System	Thermodynamic driving force
NVE	Maximize entropy S
NVT	Minimize Helmholtz free energy $E - TS$
NPT	Minimize Gibbs free energy $E + PV - TS$

maximize its entropy by maximizing the number of microstates associated with its thermodynamic macrostate. The NVT and NPT systems require the introduction of the concept of free energy, with the Helmholtz and Gibbs free energies operating in these respective systems.

1.6 Applications and conventions in chemical thermodynamics

1.6.1 Systems of non-interacting molecules

To apply the preceding concepts of chemical thermodynamics to chemical reaction systems (and to understand how thermodynamic variables such as free energy vary with concentrations of species), we have to develop a formalism for the dependence of free energies and chemical potential on the number of particles in a system. We develop expressions for the change in Helmholtz and Gibbs free energies in chemical reactions based on the definition of A and G in terms of Q and Z. The quantities Q and Z are called the *partition functions* for the NVT and NPT systems, respectively.

Consider the case where we have an open system consisting of a single protein molecule in solution. This system could consist, for example, of a biological molecule in a bath of water held at constant temperature. If this molecule adopts a number of conformational states, its NVT and NPT partition functions are the familiar quantities:

$$Q_1 = \sum_i e^{-\beta E_i}$$

$$Z_1 = \sum_i e^{-\beta H_i}$$

where E_i and H_i are the energy and enthalpy associated with the ith state. The subscript "1" in the above formulas indicates that these are the single-molecule partition functions. (Note that it is possible that the solvation energy may change with molecular conformation; in this case these changes are assumed to be incorporated into the E_i and H_i.)

If our system consists of two identical and independent macromolecules, then the two-molecule canonical partition function is expressed

$$Q_2 = \frac{1}{2} \sum_i \sum_j e^{-\beta(E_i + E_j)} = \frac{1}{2} Q_1^2. \tag{1.24}$$

The factor of $1/2$ in the above equation for Q_2 is there so that we do not double-count identical states, assuming that the two molecules in the system are identical and indistinguishable. Generalizing from Equation (1.24) for the N-molecule case, we have[4]

$$Q_N = \frac{1}{N!} Q_1^N. \tag{1.25}$$

Equation (1.25) assumes that the system is *dilute*. Specifically this means that the molecules do not interact with one another in such a way that the energy associated with a given conformation of one molecule is not affected by the conformation of any other molecule in the system.

From here we can easily express the free energy

$$\begin{aligned} A &= -k_B T \ln(Q_N) = k_B T \left(\ln N! - N \ln Q_1 \right) \\ &= k_B T \left(N \ln N - N - N \ln Q_1 \right), \end{aligned} \tag{1.26}$$

where we have invoked the Stirling approximation that $\ln N! \approx N \ln N - N$ for large N. Using the formula $\mu = (\partial A / \partial N)_{V,T}$, we have

$$\mu = k_B T (\ln N - \ln Q_1). \tag{1.27}$$

The above equation can be written

$$\begin{aligned} \mu &= k_B T \ln(N/V) - k_B T \ln(Q_1/V) \\ \mu &= \mu^o + k_B T \ln c, \end{aligned} \tag{1.28}$$

where μ^o is a constant that depends on the molecular species and c is the concentration of molecules of that species.

One may be concerned about taking the log of a dimensional variable in Equation (1.28). Formally the values of μ^o are defined based on a specific reference concentration $c_o = 1$ Molar and the equation for chemical potential is $\mu = \mu^o + k_B T \ln(c/c_o)$.

1.6.2 Gibbs free energy of chemical reactions and chemical equilibrium

Because the Gibbs energy is the thermodynamic potential for the NPT system, the thermodynamic properties of chemical species and chemical reactions under laboratory conditions are formulated in terms of Gibbs free energy. Recall

$$dG = -SdT + VdP + \mu dN. \tag{1.29}$$

Or, at constant pressure and temperature

$$(dG)_{P,T} = \mu dN. \tag{1.30}$$

---

[4] See Exercise 1.4.

In dilute systems consisting of a number of different species, the Gibbs free energy is computed from the sum of independent contributions from N_s different species.

$$(dG)_{P,T} = \sum_{i=1}^{N_s} \mu_i dN_i \tag{1.31}$$

Next we consider a general chemical equation that be expressed as

$$-\sum_{i=1}^{N_s} \nu_i A_i \rightleftharpoons 0 \tag{1.32}$$

where $\{A_i\}$ represent chemical species and $\{\nu_i\}$ are the stoichiometric coefficients. For example, for the reaction

$$A_1 + 2A_2 \rightleftharpoons A_3 \tag{1.33}$$

the stoichiometric coefficients are $\nu_1 = -1$, $\nu_2 = -2$, and $\nu_3 = +1$.

If the general reaction of Equation (1.32) were to proceed an infinitesimal amount in the forward (left-to-right) direction, the change in free energy of the system is

$$(dG)_{P,T} = \sum_{i=1}^{N_s} \mu_i \nu_i d\phi, \tag{1.34}$$

where the variable ϕ represents the turnover or the number of times the reaction has progressed. From Equation (1.34) we make the following definition

$$\Delta_r G = \left(\frac{dG}{d\phi}\right)_{P,T} = \sum_{i=1}^{N_s} \mu_i \nu_i. \tag{1.35}$$

Using the expression for μ that was introduced in the previous section,

$$\Delta_r G = \sum_{i=1}^{N_s} \nu_i \left[\mu_i^o + k_B T \ln(c_i/c_o)\right]$$

$$= \Delta_r G^o + \sum_{i=1}^{N_s} \nu_i k_B T \ln(c_i/c_o), \tag{1.36}$$

where $\Delta_r G^o = \sum_{i=1}^{N_s} \nu_i \mu_i^o$. In these equations $\Delta_r G^o$ and $\Delta_r G$ are expressed as the change in free energy associated with a single turnover of the reaction. More conventionally, $\Delta_r G^o$ and $\Delta_r G$ are expressed in units of the change in free energy per mole of flux through a reaction. To express the above formulas in these units, we make use of the fact that the universal gas constant R is equal to $N_A k_B$, where

$N_A \approx 6.022 \times 10^{23}$ is the Avogadro constant:

$$\Delta_r G = \Delta_r G^o + \sum_{i=1}^{N_s} v_i RT \ln(C_i/C_o)$$

$$\Delta_r G^o = \sum_{i=1}^{N_s} v_i \Delta_f G_i^o, \tag{1.37}$$

where $\Delta_f G_i^o$ is the free energy of formation of species i. The value of $\Delta_f G_i^o$ for a given species depends on the environmental conditions, most notably temperature, pressure, and the ionic solution strength. The ionic strength affects the activity coefficients of charged species; this phenomenon is addressed in the next chapter.

Chemical equilibrium is achieved when the driving force for the reaction $\Delta_r G$ goes to zero. Equilibrium yields

$$-\frac{\Delta_r G^o}{RT} = \sum_{i=1}^{N_s} v_i \ln(C_i/C_o) = \ln \prod_{i=1}^{N_s} (C_i/C_o)^{v_i}$$

$$e^{-\Delta_r G^o/RT} = K_{eq} = \prod_{i=1}^{N_s} (C_i/C_o)^{v_i}, \tag{1.38}$$

where K_{eq} is the equilibrium constant for the reaction.

Equations (1.37) and (1.38) should be familiar to students of chemistry and biochemistry. Admittedly, the path that took us to Equations (1.37) and (1.38) was long; we may have avoided using considerable time and effort by simply writing these expressions down at the beginning of the chapter. However, in developing these ideas from first principles we have a deeper understanding of the meaning and assumptions behind them than we would otherwise have.

1.7 Applications of thermodynamics in biology

1.7.1 Enzyme reaction mechanisms

Thermodynamic concepts are useful to apply to the study of enzyme-mediated enzyme kinetics. Through a variety of reaction mechanisms, specific enzymes catalyze specific biochemical reactions to turn over faster than they would without the enzyme present. Making use of the fact that enzymes are not able to alter the overall thermodynamics (free energy, etc.) of a chemical reaction, we can develop sets of mathematical constraints that apply to the kinetic constants of enzyme reaction mechanism.

As an example we consider the overall chemical reaction, $S_1 + S_2 \rightleftharpoons P$, for which we denote the equilibrium constant K_{eq}. A possible enzyme-mediated

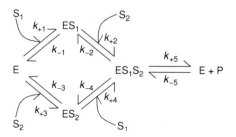

Figure 1.2 Illustration of an enzyme binding and reaction scheme for the reaction $S_1 + S_2 \rightleftharpoons P$.

reaction scheme for this reaction is illustrated in Figure 1.2, which assumes that substrates S_1 and S_2 bind to the enzyme E in arbitrary (random) order and the complex ES_1S_2 converts to free enzyme plus product, P.

The kinetic constants k_i in Figure 1.2 apply to mass-action kinetics for the state transition steps illustrated in the diagram. For example, the rate at which E binds to S_1 and generates the complex ES_1 is $k_{+1}[E][S_1]$, where [X] stands for the concentration of X. This process proceeds in the reverse direction at the rate $k_{-1}[ES_1]$.

It is straightforward to show that in chemical equilibrium

$$\left(\frac{[P]}{[S_1][S_2]} \right)_{eq} = K_{eq} = \frac{k_{+1}k_{+2}k_{+5}}{k_{-1}k_{-2}k_{-5}}. \tag{1.39}$$

This relationship is a consequence of the detailed balance condition that in equilibrium each of the reactions in this scheme is in equilibrium with its forward flux equal to reverse flux [127]. Similarly,

$$K_{eq} = \frac{k_{+3}k_{+4}k_{+5}}{k_{-3}k_{-4}k_{-5}}. \tag{1.40}$$

Equations 1.39 and 1.40 represent constraints on the kinetic constants for the enzyme mechanism that arise from the overall thermodynamics of the reaction that is catalyzed by the enzyme. Algebraic analysis of these two equations reveals that

$$\frac{k_{+1}k_{+2}k_{-3}k_{-4}}{k_{-1}k_{-2}k_{+3}k_{+4}} = 1, \tag{1.41}$$

which is a thermodynamic consequence of the closed reaction loop illustrated in Figure 1.2.

In addition to the thermodynamic constraints on the reaction kinetics, a number of assumptions (including quasi-equilibrium binding and quasi-steady state assumptions) are often invoked in computer modeling of enzyme kinetics. Analysis of enzyme kinetics is treated in greater depth in Chapter 4.

1.7.2 Electrostatic potential across a cell membrane

The electrostatic potential across cell membranes plays important roles in transport and in cell signaling. Muscle contraction is stimulated by depolarization of the cell membrane. Nerve cells communicate with other cells via propagated changes in membrane potential. Also the potential across membranes of intracellular organelles, such as mitochondria, can be central components of the function of the organelles.

In a state where the cell membrane is primarily permeable to a single ion, potassium for example, the membrane potential is primarily a function of the concentrations of that ion on either side of the membrane. If the potassium concentrations inside and outside of a cell are denoted $[K^+]_i$ and $[K^+]_o$, respectively, and if $[K^+]_i > [K^+]_o$, then there will exist a concentration-gradient driven thermodynamic driving force quantified by:

$$\frac{\mu_i - \mu_o}{k_B T} = \ln\left(\frac{[K^+]_i}{[K^+]_o}\right). \tag{1.42}$$

However, if ions move down the concentration gradient (from the inside to the outside of the cell) an electrostatic imbalance will be created, resulting in more positive charges outside of the cell than inside. The resulting electrostatic force will drive the positive potassium ions across the membrane from outside to inside. In thermodynamic equilibrium, the concentration driven potential is exactly balanced by the electrostatic potential, a situation illustrated in Figure 1.3.

The difference in electrostatic energy per ion between the outside and inside of the cell is given by $\Delta\Psi/k_B T$, where $\Delta\Psi$ is the electrostatic energy per charge. Expressing the electrostatic potential in conventional units of volts (joules per coulomb of charge), the concentration and electrostatic potential are equated in equilibrium:

$$\ln\left(\frac{[K^+]_i}{[K^+]_o}\right) = \frac{F\Delta\Psi}{N_A k_B T} = \frac{F\Delta\Psi}{RT}, \tag{1.43}$$

where $F \approx 9.65 \times 10^4$ coulomb per mole is the Faraday constant and $N_A \approx 6.022 \times 10^{23}$ is the Avogadro constant. The gas constant R is equal to $N_A k_B$. Thus considering the potassium ion as the only permeant ionic species, the equilibrium membrane potential is given by $\Delta\Psi = \frac{RT}{F} \ln([K^+]_i/[K^+]_o)$. Equation (1.43) is known as the Nernst equation and its predicted equilibrium potential is known as the Nernst potential.

If we note that the probability of finding a particle on one side of the membrane is proportional to the concentration of that side of the membrane, then we can arrive

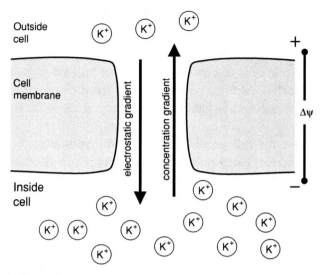

Figure 1.3 Illustration of a potassium ion concentration gradient across a cell membrane. The concentration gradient and the electrostatic potential oppose and balance one another in thermodynamic equilibrium.

at Equation (1.43) using the Boltzmann probability law:

$$\frac{[K^+]_i}{[K^+]_o} = \frac{P_i}{P_o} = e^{F\Delta\Psi/RT}. \tag{1.44}$$

In living cells the potential dynamically changes as different ion transporters and channels controlled by various mechanisms open and close. Approaches to and examples of modeling cellular electrophysiology are covered in greater depth in Chapter 7.

Concluding remarks

This chapter has introduced foundational concepts of statistical thermodynamics and physical chemistry for analysis of systems involving chemical reactions, molecular transitions, and material transport. A few simple examples of applications of thermodynamic concepts to biological systems were illustrated in Section 1.7. The remainder of this book focuses on applications to the analysis of biological systems.

Interested readers can find an introduction to statistical thermodynamics, with derivations of the statistical concepts used in this chapter, in Chapter 12. For deeper study of statistical mechanics and physical chemistry a number of excellent texts are available. See for example the texts on statistical mechanics of Hill [86] and Pathria [156] and the comprehensive physical chemistry text of McQuarrie and Simon

[138]. The elegant monograph by Hill [91] treats the subject of kinetic analysis of biochemical cycles in greater detail than has been provided in Section 1.7.1. In addition, specialized theories applicable to certain problems will be introduced at the appropriate places in the remainder of this text. That said, our introduction of the foundational theories is for the most part complete. It is expected that at this point readers are proficient with the quantitative tools for characterizing the thermodynamic processes that drive chemical processes and possess some degree of intuition for those processes. In the coming chapters we begin our study of the behavior of biochemical systems in earnest.

Exercises

1.1 In Planck's natural unit system, entropy is a unitless variable and Boltzmann's constant takes the value 1. In this system, what are the physical units of temperature?

1.2 The values of *extensive* thermodynamic variables, such as N, V, and E, are proportional to the size of the system. If we combine NVE subsystems into a larger system, then the total N, V, and E are computed as the sums of N, V, and E of the subsystems. Temperature, pressure, and chemical potential are *intensive* variables, for which values do not depend on the size of the system. Show that entropy is an extensive variable.

1.3 Verify that the equalities in Equations (1.11) and (1.12) are indeed correct.

1.4 Equations (1.24) and (1.25) assume that the number of possible single-molecule states is so high that the probability of two molecules having identical states is vanishingly small. Show how this assumption leads to these equations.

1.5 Consider a gel that carries a certain concentration $c_i(\mathbf{r})$ of immobile negative charges and is immersed in an aqueous solution. The bulk solution carries monovalent mobile ions of concentration $c_+(\mathbf{r})$ and $c_-(\mathbf{r})$. Away from the gel, the concentration of the salt ions achieves the bulk concentration, denoted c_o. What is the difference in electrical potential (known as the Donnan potential) between the bulk solution and the interior of the gel? [Hint: assume that inside the gel the overall concentration of positive salt ion balances immobile gel charge concentration.]

2

Conventions and calculations for biochemical systems

Overview

Biochemical species, from small-molecule metabolites such as inorganic phosphate to large proteins, reversibly bind hydrogen and metal ions, altering their thermo-chemical properties. Inorganic phosphate, for example, may exist in solution in several different states of protonation, including $H_2PO_4^-$ and HPO_4^{2-}. Yet in the biochemical literature it is standard practice to refer to the concentration of inorganic phosphate without explicitly considering the different species that contribute to its overall concentration.

In this chapter we describe how to derive expressions for *apparent* equilibrium constants and *apparent* Gibbs free energies for biochemical reactions expressed in terms of reactants that are made up of sums of rapidly interconverting species. We will see that the calculation of free energies, pH, and other variables that change as reactions evolve depends on the constraints that are imposed on a reaction system – is the system closed, or is the system maintained in a non-equilibrium steady state by transporting material into and out of the system? Is pH held constant by injecting or removing hydrogen ions or does the pH vary as the reactant concentrations change? We will see that consideration of these questions will be important in developing simulations of both in vivo and in vitro biochemical systems.

2.1 Conventional notation in biochemical thermodynamics

In this book, we follow the convention that the term *species* refers to a unique chemical compound, while a *reactant* is a biochemical compound that may be present as a number of related and rapidly inter-converting species. In a modern biochemistry textbook a reaction, such as the ATP hydrolysis reaction

$$ATP \rightleftharpoons ADP + PI, \tag{2.1}$$

is written in terms of biochemical reactants, such as ATP, ADP, and inorganic phosphate (PI).[1] Equation (2.1) is not balanced in terms of mass or charge, as detailed below in Section 2.2.2. The standard free energy of the ATP hydrolysis reaction is reported to be approximately $\Delta_r G'^o = -30.5$ kJ mol^{-1}; and the reaction free energy is calculated

$$\Delta_r G' = \Delta_r G'^o + RT \ln \frac{[\text{ADP}][\text{PI}]}{[\text{ATP}]}, \qquad (2.2)$$

where the concentrations [ATP], [ADP], and [PI] refer to summed concentrations of species that make up the reactants ATP, ADP, and inorganic phosphate, respectively. Because the thermochemical properties of species that make up [ATP], [ADP], [PI], and other biochemical reactants (and the relative concentrations of the species that make up the reactants) vary as functions of pH and concentration of other ionic species in the solution, biochemists define a *biochemical standard state* for which the solution is dilute ([H$_2$O] = 55.5 M), pH is constant at 7 ([H$^+$] = 10^{-7} M), and magnesium ion concentration [Mg^{2+}] is held constant (typically around 1 mM). Therefore calculations of reaction free energies do not depend explicitly on water, hydrogen ion, or magnesium ion concentrations – these concentrations in the biochemical standard state are implicitly incorporated into the definitions of the standard transformed reaction Gibbs free energies. The prime notation on free energies is used to denote variables and constants based on the biochemical standard state. Lehninger's *Principles of Biochemistry* explains, "Physical constants based on this biochemical standard state are called standard transformed constants and are written with a prime (such as $\Delta_r G'^o$ and K'_{eq}) to distinguish them from the untransformed constants used by chemists and physicists" ([147] p. 491).

As we will see later in this book, the so-called biochemical standard state is not necessarily maintained in vivo or in vitro. In vitro systems are often closed, meaning that as reactions progress pH may change as hydrogen ions are bound and unbound and incorporated and released from various species. Similarly the pH may vary in vivo. For example, certain subcellular compartments are maintained at pH values significantly different from neutral; the cytoplasmic pH may drop into the neighborhood of 6 during certain cardiac pathologies. Thus for certain applications we will need to perform calculations for systems that deviate from the biochemical standard state.

---

[1] The reactant inorganic phosphate is sometimes abbreviated P$_i$. Because in later chapters we use subscripts to indicate compartment, we stick to the convention here of abbreviating reactants using all capital letters.

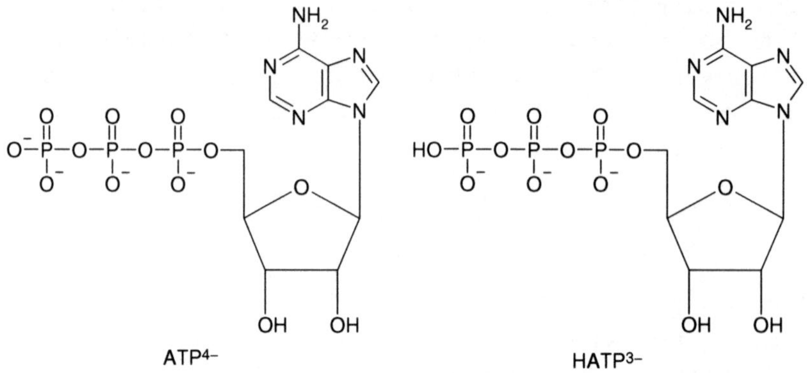

Figure 2.1 Illustration of two species of ATP: ATP^{4-} and $HATP^{3-}$.

2.2 Reactants and reactions in biochemistry

2.2.1 An example of a biochemical reactant

As an example of a biochemical reactant, consider the ubiquitous compound adenosine triphosphate (ATP). In its fully unprotonated state there are four negatively charged hydroxyl groups associated with the three phosphate groups on the ATP molecule. In solution near neutral pH, one of the phosphates may be protonated, as illustrated in Figure 2.1. Here we denote the two species illustrated in the figure ATP^{4-} and $HATP^{3-}$.[2]

In solutions of ionic strength of approximately 0.25 M, a hydrogen ion dissociates from $HATP^{3-}$ with an acid-base pK of approximately 6.47 [5], where pK is defined as $-\log_{10} K_{eq}$ and K_{eq} is the dissociation reaction equilibrium constant. The equilibrium expression is:

$$K_{eq} = 10^{-pK} = \frac{[ATP^{4-}][H^+]}{[HATP^{3-}]}. \tag{2.3}$$

If we denote total ATP concentration $[ATP] = [ATP^{4-}] + [HATP^{3-}]$ and assume that the system is in equilibrium, then

$$[ATP] = [ATP^{4-}] + [HATP^{3-}]$$
$$= [ATP^{4-}] + \frac{[ATP^{4-}][H^+]}{K_{eq}}$$
$$= [ATP^{4-}]\left(1 + \frac{[H^+]}{K_{eq}}\right). \tag{2.4}$$

---

[2] For biochemical species we use the convention of indicated charge with a superscript number and sign. This way, even an uncharged species such as $CoQH_2^0$ (reduced ubiquinone) is distinguished from the reactant $CoQH_2$, which could be made up of several rapidly interconverting species.

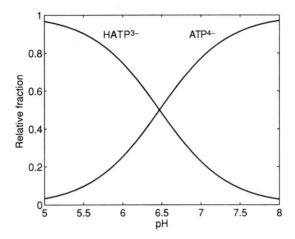

Figure 2.2 Mole fractions of ATP^{4-} and $HATP^{3-}$ as functions of pH, predicted by Equation (2.5), given pK = 6.47.

The molar fractions of ATP that are in the two forms are given by

$$\frac{[ATP^{4-}]}{[ATP]} = \frac{1}{1 + [H^+]/K_{eq}}; \quad \frac{[HATP^{3-}]}{[ATP]} = \frac{[H^+]/K_{eq}}{1 + [H^+]/K_{eq}}. \qquad (2.5)$$

Figure 2.2 illustrates the relative fractions of each of these species as a function of pH ranging from 5 to 8. At low pH, the fraction $[HATP^{3-}]/[ATP]$ approaches 1, as expected. At high pH, the majority of the ATP is in the unprotonated form.

The above analysis may be generalized to treat any biochemical reactant with two distinct species with different levels of proton binding. However, many reactants (including ATP) have multiple sites of proton association. Under acidic conditions (pH less than approximately 5), a second hydrogen ion may become associated with the ATP molecule, forming H_2ATP^{2-}. To treat the general case, where multiple hydrogen ions may bind and there exist multiple protonated forms, we denote the state with i protons bound to reactant L as $[LH_i]$. Denoting the total concentration of L as [L], we have:

$$[L] = [LH_0] + [LH_1] + [LH_2] + \cdots = \sum_{i=0}^{N}[LH_i], \qquad (2.6)$$

where N is the maximum number of protons bound. If K_1 is the dissociation constant for the first protonation, then

$$[LH_1] = [LH_0]\frac{[H^+]}{K_1}. \qquad (2.7)$$

If the second dissociation constant is denoted K_2, then

$$[\mathrm{LH}_2] = [\mathrm{LH}_1]\frac{[\mathrm{H}^+]}{K_2} = \left([\mathrm{LH}_0]\frac{[\mathrm{H}^+]}{K_1}\right)\frac{[\mathrm{H}^+]}{K_2}. \tag{2.8}$$

In general,

$$[\mathrm{LH}_i] = [\mathrm{LH}_0]\frac{[\mathrm{H}^+]^i}{\prod\limits_{j\leq i} K_j}. \tag{2.9}$$

Thus the total concentration may be expressed

$$[\mathrm{L}] = [\mathrm{LH}_0]\left(1 + \frac{[\mathrm{H}^+]}{K_1} + \frac{[\mathrm{H}^+]^2}{K_1 K_2} + \cdots\right)$$

$$= [\mathrm{LH}_0]\left(1 + \sum_{i=1}^{N} \frac{[\mathrm{H}^+]^i}{\prod_{j\leq i} K_j}\right), \tag{2.10}$$

and the molar fraction of LH_i is

$$\frac{[\mathrm{LH}_i]}{[\mathrm{L}]} = \frac{\frac{[\mathrm{H}^+]^i}{\prod_{j\leq i} K_j}}{\left(1 + \sum\limits_{i=1}^{N} \frac{[\mathrm{H}^+]^i}{\prod_{j\leq i} K_j}\right)} = \frac{\frac{[\mathrm{H}^+]^i}{\prod_{j\leq i} K_j}}{P_L([\mathrm{H}^+])}. \tag{2.11}$$

The denominator of Equation (2.11) is called the binding polynomial for reactant L and is denoted $P_L([\mathrm{H}^+])$.

The pK's for the first and second proton binding for ATP are 6.47 and 3.83, respectively [5]. Given these values, the predicted molar fractions of $\mathrm{H}_2\mathrm{ATP}^{2-}$, HATP^{3-}, and ATP^{4-} are plotted in Figure 2.3. The state $\mathrm{H}_2\mathrm{ATP}^{2-}$ becomes a significant fraction of total ATP at extremely low pH values.

2.2.2 An example of a biochemical reaction

Based on our exploration of a biochemical reactant in the previous section, we now recognize that the familiar biochemical reaction

$$\mathrm{ATP} \rightleftharpoons \mathrm{ADP} + \mathrm{PI} \tag{2.12}$$

is not stoichiometrically balanced in terms of mass or charge. We define the following reference reaction written in terms of species that is stoichiometrically balanced:

$$\mathrm{ATP}^{4-} + \mathrm{H}_2\mathrm{O} \rightleftharpoons \mathrm{ADP}^{3-} + \mathrm{HPO}_4{}^{2-} + \mathrm{H}^+. \tag{2.13}$$

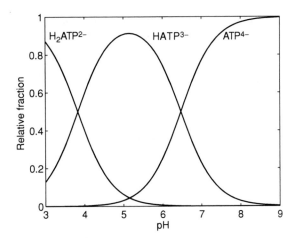

Figure 2.3 Mole fractions of ATP^{4-} and $HATP^{3-}$ as functions of pH, predicted by Equation (2.11), given $pK_1 = 6.47$, $pK_2 = 3.83$.

Equation (2.12) is the convenient form for biochemists who are interested in tracking reactant concentrations and apparent equilibrium constants and free energies; Equation (2.13) is the physicochemically rigorous equation from which we can build relationships for the apparent thermodynamics of Equation (2.12).

Equation (2.13) has associated with it a K_{eq} and a $\Delta_r G^o$ that do not depend on pH:

$$K_{eq} = e^{-\Delta_r G^o/RT} = \left(\frac{[\text{ADP}^{3-}][\text{HPO}_4{}^{2-}][\text{H}^+]}{[\text{ATP}^{4-}]} \right)_{eq}, \qquad (2.14)$$

where the above expression makes the assumption of a dilute solution, and each concentration is expressed in molar units. Note that, as discussed in Sections 2.4 and 2.5, these thermodynamic variables vary with the temperature and ionic concentration of the solution. At an ionic strength of 0.25 M and T = 298.15 K,[3] the equilibrium constant of Equation (2.14) is approximately $K_{eq} = 0.1$. The concentrations $[\text{ATP}^{4-}]$, $[\text{HADP}^{3-}]$, and $[\text{HPO}_4{}^{2-}]$, may be computed

$$[\text{ATP}^{4-}] = [\text{ATP}] \left(1 + \frac{[\text{H}^+]}{K_1^{ATP}} + \frac{[\text{H}^+]^2}{K_1^{ATP} K_2^{ATP}} \right)^{-1} = \frac{[\text{ATP}]}{P_{ATP}([\text{H}^+])}$$

$$[\text{HADP}^{3-}] = [\text{ADP}] \left(1 + \frac{[\text{H}^+]}{K_1^{ADP}} + \frac{[\text{H}^+]^2}{K_1^{ADP} K_2^{ADP}} \right)^{-1} = \frac{[\text{ADP}]}{P_{ADP}([\text{H}^+])}$$

$$[\text{HPO}_4{}^{2-}] = [\text{PI}] \left(1 + \frac{[\text{H}^+]}{K_1^{PI}} \right)^{-1} = \frac{[\text{PI}]}{P_{PI}([\text{H}^+])}. \qquad (2.15)$$

---

[3] The ionic strength is a function of the total concentrations of all ionic species in a solution. It is defined $I = (1/2) \sum_i z_i^2 c_i$, where z_i and c_i are the charge number and concentration of each dissolved species.

Table 2.1 *Approximate pKs for hydrogen binding to ATP, ADP, and PI*

	pK$_1$	pK$_2$
ATP	6.47	3.83
ADP	6.33	3.79
PI	6.65	–

Values are obtained from [4] and correspond to $T = 298.15$ K and ionic strength $I = 0.25$ M.

Again [ATP], [ADP], and [PI] denote the total (sums of species) concentrations of ATP, ADP, and PI. The hydrogen binding polynomials for ATP, ADP, and PI are denoted $P_{ATP}([\text{H}^+])$, $P_{ADP}([\text{H}^+])$, and $P_{PI}([\text{H}^+])$. In Equation (2.15), K_i^{ATP}, K_i^{ADP}, and K_i^{PI} denote the hydrogen ion binding constants for ATP, ADP, and PI, respectively. Table 2.1 lists the approximate pKs at ionic concentration of 0.25 M. Binding of a second hydrogen ion to phosphate is highly unfavorable for pH greater than 5. Therefore there is only one pK for phosphate.

Combining Equations (2.14) and (2.15), we have,

$$K_{eq} = e^{-\Delta_r G^0/RT} = \left(\frac{[\text{ADP}][\text{PI}][\text{H}^+]}{[\text{ATP}]} \right)_{eq} \cdot \frac{P_{ATP}([\text{H}^+])}{P_{ADP}([\text{H}^+]) \, P_{PI}([\text{H}^+])}. \quad (2.16)$$

The transformed $\Delta_r G'^o$ and K'_{eq} may be computed[4]

$$K'_{eq} = e^{-\Delta_r G'^o/RT} = \left(\frac{[\text{ADP}][\text{PI}]}{[\text{ATP}]} \right)_{eq} = K_{eq} \frac{P_{ADP}([\text{H}^+]) \, P_{PI}([\text{H}^+])}{[\text{H}^+] \, P_{ATP}([\text{H}^+])}. \quad (2.17)$$

Figure 2.4 illustrates how the transformed equilibrium Gibbs free energy $\Delta_r G'^o$ varies with pH of the solution. Because of the hydrogen ion generated in the reference reaction of Equation (2.13), the reaction becomes more favorable as pH increases. Near pH of 7, the $\Delta_r G'^o$ is approximately -36 kJ mol^{-1}.

Next we explore the question of how the apparent thermodynamic variables change as a reaction progresses from a non-equilibrium initial condition toward equilibrium in a closed system. Imagine a closed system initially contains ATP, ADP, and PI, at concentrations of 10 mM, 1 mM, and 1 mM, respectively, at neutral pH of 7. To see how pH and related thermodynamic properties change as the reaction progresses toward equilibrium, imagine that an ATP hydrolysis enzyme is present in our system, and the reference reaction of Equation (2.13) moves in

---

[4] Equations (2.16) and (2.17) describe equilibria in terms of *concentrations* rather than *activities*. The concept of activity versus concentration of a species is discussed in Section 2.5.

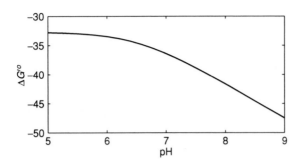

Figure 2.4 Transformed equilibrium Gibbs free energy as a function of pH, predicted from Equation (2.17), using the reference $K_{eq} = 0.1$, and the pK values from Table 2.1.

the forward (left-to-right) direction until equilibrium is achieved. At any time, the total concentration of hydrogen ion bound to reactant L may be computed from Equation (2.11)

$$\left[H_{bound}^{L}\right] = \sum_{i=1}^{N} i[LH_i] = [L] \left(\frac{\sum_{i=1}^{N} i \dfrac{[H^+]^i}{\prod_{j \leq i} K_j}}{1 + \sum_{i=1}^{N} \dfrac{[H^+]^i}{\prod_{j \leq i} K_j}} \right) = \frac{[L] \sum_{i=1}^{N} i \dfrac{[H^+]^i}{\prod_{j \leq i} K_j}}{P_L([H^+])}. \quad (2.18)$$

The total concentration of bound hydrogen ion is computed by summing over all reactants: $[H_{bound}^{total}] = \sum_L [H_{bound}^L]$.

If the system is closed then the total number of protons in the system is conserved. This conservation statement may be expressed

$$\text{(Bound H}^+) + \text{(Free H}^+) = \text{(Initial Bound H}^+) + \text{(Initial Free H}^+)$$
$$+ \text{(Generated H}^+) \quad (2.19)$$

or

$$\left[H_{bound}^{total}\right] + 10^{-pH} = \left[H_{bound}^{total}\right]_o + 10^{-7} + ([ATP]_o - [ATP]), \quad (2.20)$$

where concentrations are expressed in Molar. In Equation (2.20), $\left[H_{bound}^{total}\right]_o$ and $[ATP]_o$ denote the initial concentration of bound hydrogen ions and ATP, respectively. The third term on the right-hand side of the mass balance equation computes the number of times the reference reaction has turned over, generating hydrogen ions. Given a set of initial concentrations, Equation (2.20) can be numerically solved for pH at any ATP concentration.

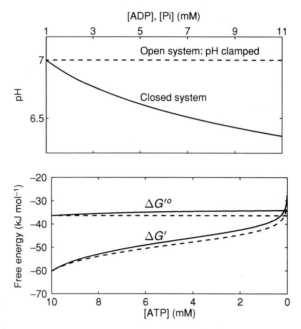

Figure 2.5 Upper panel shows change in pH as a function of ATP, ADP, and PI concentrations predicted by Equation (2.20). The lower panel illustrates $\Delta_r G'^o$ and $\Delta_r G'$ as function of reactant concentration. The solid line represents predictions for the closed system, where pH varies with reactant concentrations; dashed line represents a system with pH clamped.

Figure 2.5 illustrates how the pH varies in the closed system as ATP concentration is reduced from the initial value of 10 mM. Note that as ATP concentration is reduced, ADP and PI concentrations increase in proportion. As the ATP concentration approaches zero, the pH drops below 6.5. Figure 2.5 also shows how the $\Delta_r G'$ and $\Delta_r G'^o$ vary as the reaction progresses and pH is reduced. The dashed lines in Figure 2.5 illustrate how the system would behave if it were assumed that pH is held at the constant value of 7.

When pH is held constant $\Delta_r G'^o$ remains constant at all concentrations of ATP, ADP, and PI. On the other hand, if the system is closed and pH drops as the reaction progresses, then the impact of the pH variation is apparent on the computed $\Delta_r G'^o$ and $\Delta_r G'$.

2.3 Effects of pH and ion binding on biochemical reaction thermodynamics

The previous section illustrated how to calculate apparent equilibrium properties for biochemical reactions expressed in terms of biochemical reactants that are sums of rapidly inter-converting species. The species considered consisted of different proton-binding states. As a result the apparent thermodynamic properties vary as functions of pH. Yet in addition to binding hydrogen ions, biochemical reactants may bind to other ionic species in solution. Important examples are magnesium and potassium ions, Mg^{2+} and K^+. Because of its $+2$ charge and its relative abundance in cells, Mg^{2+} interacts significantly with ATP^{4-}, ADP^{3-}, and HPO_4^{2-}. Potassium ion tends to bind negatively charged biochemical species with a lower affinity (lower pKs) than either hydrogen or magnesium ion. However K^+ tends to be present in many cells in concentrations of the order of 0.1 M, while Mg^{2+} and H^+ are present in concentrations of the order of 10^{-3} M and 10^{-7} M, respectively. The overall result is that potassium ion binding can be as significant as magnesium ion binding in cells [122]. Thus in biological systems binding to K^+ and Mg^{2+} must be accounted for in the calculation of apparent thermodynamic properties.

The impact of K^+ and Mg^{2+} binding on the apparent equilibrium constant may be calculated by incorporating binding of these species into the binding polynomials for the reactants in a given reaction. For example, for the ATP hydrolysis reaction, the binding polynomials for the three reactants become:

$$P_{ATP}([H^+], [K^+], [Mg^{2+}]) = 1 + \frac{[H^+]}{K_1^{ATP}} + \frac{[K^+]}{K_{K\text{-}ATP}} + \frac{[Mg^{2+}]}{K_{Mg\text{-}ATP}}$$

$$P_{ADP}([H^+], [K^+], [Mg^{2+}]) = 1 + \frac{[H^+]}{K_1^{ADP}} + \frac{[K^+]}{K_{K-ADP}} + \frac{[Mg^{2+}]}{K_{Mg\text{-}ADP}}$$

$$P_{PI}([H^+], [K^+], [Mg^{2+}]) = 1 + \frac{[H^+]}{K_1^{PI}} + \frac{[Mg^{2+}]}{K_{Mg-PI}}, \tag{2.21}$$

where here we have ignored states with more than one exchangeable H^+ bound to ADP and ATP. (This is a reasonable approximation near pH $= 7$.) The above expressions assume that Mg^{2+} associates with the unprotonated species ATP^{4-}, ADP^{3-}, and HPO_4^{2-}, and that K^+ associates with the unprotonated species ATP^{4-} and ADP^{3-}. Dissociation constants are given in Table 2.2.

When K^+ and Mg^{2+} binding are considered, the apparent equilibrium free energy and equilibrium constant are computed as functions of the K^+- and Mg^{2+}-dependent binding polynomials.

$$K'_{eq} = e^{-\Delta_r G'^o/RT} = K_{eq} \frac{P_{ADP}([H^+], [K^+], [Mg^{2+}]) \, P_{PI}([H^+], [K^+], [Mg^{2+}])}{[H^+] \, P_{ATP}([H^+], [K^+], [Mg^{2+}])}.$$
$$\tag{2.22}$$

Using the values listed in Table 2.2, Equation (2.22) yields $\Delta_r G'^o = -36.4 \, kJ$ mol^{-1} at $[K^+] = [Mg^{2+}] = 0$, and $\Delta_r G'^o = -31.2 \, kJ \, mol^{-1}$ at $[K^+] = 150 \, mM$ and $[Mg^{2+}] = 1 \, mM$ (at pH $= 7$). Because K^+ and Mg^{2+} bind to unprotonated ATP

Table 2.2 *Approximate dissociation constants for*
K^+ *and* Mg^{2+} *binding to ATP, ADP, and PI*

Reactant (L)	K_{K-L}	K_{Mg-L}
ATP	1.78×10^{-2} M	2.63×10^{-4} M
ADP	3.63×10^{-2} M	2.46×10^{-3} M
PI	–	2.63×10^{-2} M

Values are obtained from [134] and correspond to $T = 298.15$ K and ionic strength $I = 0.25$ M.

(with its -4 charge) with affinities higher than those of other species, potassium and magnesium binding tends to move the equilibrium of the reference reaction of Equation (2.13) from right to left. Higher free magnesium concentrations result in a lower magnitude of the apparent equilibrium Gibbs free energy for the ATP hydrolysis reaction.

2.4 Effects of temperature on biochemical reaction thermodynamics

As we saw in Chapter 1, the standard free energy for a chemical reaction of the form

$$-\sum_{i=1}^{N_s} \nu_i A_i \rightleftharpoons 0$$

is computed based on the standard free energy of formation of the species

$$\Delta_r G^o = \sum_{i=1}^{N_s} \nu_i \Delta_f G_i^o. \tag{2.23}$$

For example, for the chemical reaction

$$ATP^{4-} + H_2O \rightleftharpoons ADP^{3-} + HPO_4^{2-} + H^+,$$

Equation (2.23) yields

$$\Delta_r G^o = \Delta_f G_{ADP}^o + \Delta_f G_{PI}^o - \Delta_f G_{ATP}^o - \Delta_f G_{H_2O}^o - \Delta_f G_H^o. \tag{2.24}$$

Substituting tabulated values [4] for the $\Delta_f G^o$'s yields

$$\Delta_r G^o = -1906.13 - 1096.10 + 2768.10 + 237.19 + 0 \text{ kJ mol}^{-1}$$
$$= +3.06 \text{ kJ mol}^{-1}, \tag{2.25}$$

where the above values correspond to zero ionic strength and $T = 298.15$ K.

Defining enthalpy and entropy of formation according to the definition of Gibbs free energy, we have

$$\Delta_f G_i^o = \Delta_f H_i^o - T \Delta_f S_i^o. \tag{2.26}$$

If the standard enthalpies of formation for the species are known and assumed to remain approximately constant (over some range of temperature), then Equation (2.26) yields the following expression for calculating $\Delta_f G_i^o$ as a function of temperature:

$$\Delta_f G_i^o(T_2) = \left(1 - \frac{T_2}{T_1}\right) \Delta_f H_i^o + \left(\frac{T_2}{T_1}\right) \Delta_f G_i^o(T_1). \tag{2.27}$$

Given information on the standard enthalpies of formation for the species in a reaction, this expression may be used to estimate the standard free energies of formation at one temperature T_2 as functions of the values of $\Delta_f G_i^o$ at another temperature T_1.[5] The temperature-corrected estimates of $\Delta_f G_i^o$ yield temperature-corrected estimates of $\Delta_r G^o$ through Equation (2.23). This ability to correct for temperature is important because the majority of biochemical reaction data are tabulated at 25°C, while human beings and many other warm-blooded creatures operate at temperatures near 37°C.

2.5 Effects of ionic strength on biochemical reaction thermodynamics

Because the energy associated with solvation depends on the ionic properties of the solution, the values of $\Delta_f G_i^o$ depend on the ionic properties of the solution. This phenomenon is formally accounted for by defining the *activity* of species i as

$$a_i = \gamma_i c_i, \tag{2.28}$$

where γ_i is called the *activity coefficient*, which is a function of the ionic solution; the activity coefficient is 1 in the limit of zero ionic strength.

We construct a formulation of the values of $\Delta_f G_i^o$ for species and $\Delta_r G^o$ that absorbs the concept of activity by defining the $\Delta_f G_i^o$ as a function of ionic strength, I:

$$\Delta_f G_i^o(I) = \Delta_f G_i^o(I = 0) + RT \ln \gamma_i(I), \tag{2.29}$$

so that

$$\Delta_r G^o(I) = \sum_{i=1}^{N_s} \nu_i \Delta_f G_i^o(I) \tag{2.30}$$

---

[5] Values of Gibbs free energy and enthalpy of formation for a number of species (at T = 298.15 K and zero ionic strength) are tabulated by Alberty [4].

and the usual equilibrium expression holds:

$$\left(\prod_{i=1}^{N_s} c_i^{\nu_i}\right)_{eq} = e^{-\Delta_r G^o/RT}. \tag{2.31}$$

The extended Debye–Hückel equation [138] provides a useful approximation for $\gamma_i(I)$ for non-dilute solutions:

$$\ln \gamma_i = -\frac{A z_i^2 I^{1/2}}{1 + B I^{1/2}}, \tag{2.32}$$

where z_i is the charge number of species i and the parameter B is equal to 1.6 $M^{-1/2}$ [3].

The quantity A varies with temperature; an empirical function that reproduces apparent thermodynamic properties over a temperature range of approximately 273 to 313 K is [33, 5]:

$$A = 1.10708 - (1.54508 \times 10^{-3})T + (5.95584 \times 10^{-6})T^2. \tag{2.33}$$

Appropriate values of $\Delta_f G_i^o$ can be computed at given temperature and ionic strength by applying Equations (2.27) and (2.29) in concert. First, Equation (2.27) is used to compute $\Delta_f G_i^o(T, I = 0)$; then Equation (2.27) is used to compute $\Delta_f G_i^o(T, I)$.

2.6 Treatment of CO_2 in biochemical reactions

Carbon dioxide in solution does not significantly bind to hydrogen ions or other metal ions. Thus it is possible to treat CO_2 as a reactant composed of a single species in biochemical reactions, without introducing a binding polynomial or multiple ion-bound states for CO_2. However, CO_2 can be hydrolyzed to H_2CO_3 via the reaction

$$CO_2 + H_2O \rightleftharpoons H_2CO_3 \tag{2.34}$$

and protons can dissociate from the species H_2CO_3, forming HCO_3^- and CO_3^{2-}. The reactions and dissociation constants associated with CO_2 acid–base buffering are listed in Table 2.3.

For certain applications it is convenient to express apparent thermodynamic properties for reactions involving CO_2 in terms of total carbon dioxide, which is defined

$$[\Sigma CO_2] = [CO_2] + [H_2CO_3] + [HCO_3^-] + [CO_3^{2-}]. \tag{2.35}$$

As an illustrative example, let us consider a reaction of the form

$$A \rightleftharpoons B + CO_2, \tag{2.36}$$

Table 2.3 *Dissociation constants for CO_2 buffering reactions*

Reaction	Dissociation constants ($I = 0$ M)	($I = 0.25$ M)
$CO_2 + H_2O \rightleftharpoons H_2CO_3$	$K_h = 2.714 \times 10^{-3}$	2.714×10^{-3}
$HCO_3^- \rightleftharpoons CO_3^{2-} + H^+$	$K_1 = 5.891 \times 10^{-11}$ M	1.602×10^{-10} M
$H_2CO_3 \rightleftharpoons HCO_3^{2-} + H^+$	$K_2 = 1.748 \times 10^{-4}$ M	2.439×10^{-4} M

Values computed for $T = 310$ K and $I = 0.25$ M based on data in [3].

which is a chemical reference reaction with A, B, and CO_2 denoting chemical species. The apparent thermodynamic properties of this reaction are computed

$$K'_{eq} = e^{-\Delta_r G'^o / RT} = \left(\frac{[\Sigma B][CO_2]}{[\Sigma A]} \right)_{eq} = K_{eq} \cdot \frac{P_B}{P_A}, \qquad (2.37)$$

where $[\Sigma A]$ and $[\Sigma B]$ denote the sum of concentrations of all species of reactants A and B, P_A and P_B are the binding polynomials for reactants A and B, and K_{eq} is the equilibrium constant for the reference reaction. Thus Equation (2.37) determines the apparent equilibrium properties for the biochemical reaction $\Sigma A \rightleftharpoons \Sigma B + CO_2$.

Alternatively, one may be interested in the apparent equilibrium properties for the biochemical reaction

$$\Sigma A \rightleftharpoons \Sigma B + \Sigma CO_2, \qquad (2.38)$$

in which the total carbon dioxide concentration appears as a reactant. To analyze this case, we express $[\Sigma CO_2]$ in terms of the concentration of the reference species CO_3^{2-}:

$$[\Sigma CO_2] = \left[CO_3^{2-} \right] \left(1 + \frac{[H^+]}{K_1} + \frac{[H^+]^2}{K_1 K_2} + \frac{[H^+]^2}{K_h K_1 K_2} \right)$$
$$= \left[CO_3^{2-} \right] \cdot P_{\Sigma CO_2}([H^+]), \qquad (2.39)$$

where the dissociation constants K_h, K_1, and K_2 are defined in Table 2.3.

Since we have expressed $[\Sigma CO_2]$ in terms of $[CO_3^{2-}]$, we must consider a reference reaction in terms of CO_3^{2-} rather than CO_2. For the biochemical reaction of Equation (2.38), we have:

$$A + H_2O \rightleftharpoons B + CO_3^{2-} + 2 H^+, \qquad (2.40)$$

where the H_2O and H^+ appear in order to balance the reaction in terms of mass and charge. (Equation (2.40) is the sum of chemical reference reactions $A \rightleftharpoons B + CO_2$ and $CO_2 + H_2O \rightleftharpoons CO_3^{2-} + 2 H^+$.) This reference reaction has well defined $\Delta_r G^o$

and K_{eq} that are not functions of pH. Thus the apparent equilibrium properties for the biochemical reaction are computed in terms of the K_{eq} for the reference reaction:

$$K'_{eq} = e^{-\Delta_r G'^o / RT} = \left(\frac{[\Sigma B][\Sigma CO_2]}{[\Sigma A]} \right)_{eq} = K_{eq} \cdot \frac{P_B P_{\Sigma CO_2}}{[H^+]^2 P_A}. \qquad (2.41)$$

2.7 pH variation in vivo

2.7.1 *In vivo deviation from the standard state*

Cells contain numerous species that buffer pH, including the reactants studied in this chapter. In addition to the small molecule metabolites such as PI, ATP, and ADP, large proteins contain sites of hydrogen ion association/dissociation and serve as important intracellular buffers. Although pH in many living systems is tightly regulated under normal conditions, the pH within cells is rarely (if ever) maintained at exactly the so-called biochemical standard state. In the human body, pH and concentrations of ionic species vary throughout the different body fluids and cell types and vary with physiological state. As we have seen, the thermodynamic properties of biochemical reactants and reactions change as systems move away from the standard state conditions.

There are numerous important examples where physiological pHs may vary greatly from the standard state. In exercising skeletal muscle the pH may drop below 6.5 as net ATP hydrolysis generates excess hydrogen ions via Equation (2.13). The intracellular pH in cardiac muscle may drop as low as 6.0 when blood supply to tissue is compromised, such as during a heart attack or during surgery. In addition, certain intracellular compartments are maintained at pH values different from the rest of the cell. For example, the pH of the mitochondrial matrix is thought to be maintained at a value between 7 and 8 [149]. Thus it is important to understand how the thermodynamic properties of biochemical reactions in these systems vary with pH.

2.7.2 *The bicarbonate system in vivo*

An important physiological buffering system active in the blood is the bicarbonate system. Bicarbonate (H_2CO_3) is a weak acid that is involved in maintaining the pH of human blood in the neighborhood of 7.4. The acid–base equilibration for bicarbonate is expressed

$$[H^+] = K_2 \frac{[H_2CO_3]}{[HCO_3^-]}, \qquad (2.42)$$

where K_2 is the dissociation constant given in Table 2.3. The two sources of bicarbonate in the blood are bicarbonate salts such as $NaHCO_3$ (which dissociates into Na^+ and HCO_3^-), and carbon dioxide. Carbon dioxide is converted into bicarbonate through the hydrolysis reaction of Equation (2.34), which is catalyzed by the enzyme carbonic anhydrase. If carbonic anhydrase maintains Equation (2.34) in equilibrium with an equilibrium constant K_h, then the H^+ concentration is

$$[H^+] = K_h K_2 \frac{[CO_2]}{[HCO_3^-]}. \tag{2.43}$$

The lumped pK for the bicarbonate buffer system, $pK = -\log(K_h K_2)$, is approximately 6.18. Thus the pH may be expressed:

$$pH = 6.18 + \log \frac{[HCO_3^-]}{[CO_2]}. \tag{2.44}$$

Carbon dioxide levels in the blood vary with the rate of respiration, while bicarbonate concentration is regulated primarily by the kidney [75]. Thus the pH of the blood is maintained by the coordinated function of the lungs and the kidneys in maintaining the ratio $[HCO_3^-]/[CO_2]$ at a value of approximately 18 [75].

Concluding remarks

This chapter has presented the basics of how thermodynamics are treated for biochemical systems, with an emphasis on the impact of pH and ion binding on apparent equilibria and Gibbs free energy functions. This field owes much to the work of Robert Alberty; an extensive study of the field is presented in Alberty's text, *Thermodynamics of Biochemical Systems* [4]. In our study of the theory and simulation of biochemical systems, we will usually be concerned with biochemical reactants such as ATP and ADP, although the detailed breakdown of these reactants into individual species will be important for many applications.

Exercises

2.1 Make a plot of the molar fractions of the species H_2ADP^-, $HADP^{2-}$, and ADP^{3-} as a function of pH, given the pK values listed in Table 2.1.

2.2 Write a computer program that solves Equation (2.20) for pH as a function of the concentrations of reactants ATP, ADP, and PI. Verify your results by comparison to Figure 2.5.

2.3 There are conservation statements for total magnesium and potassium concentrations that are analogous to that for hydrogen ion of Equation (2.20). Write the potassium and magnesium conservation statements in terms of the binding polynomials.

2.4 Ethylene glycol tetraacetic acid (EGTA) is commonly used as a chelator of calcium ions in biochemical solutions. Given ion-binding pKs for EGTA show in the table below, corresponding to 298.15 K and $I = 0.1$ M, plot the free $[Ca^{2+}]$ as a function of total concentration for solutions containing 1 mM EGTA, 10 mM NaCl, 130 mM KCl, and 1 mM MgCl at pH $= 6.5, 7.0,$ and 7.5.

Ion-dissociation data for EGTA:

Ion	pK
H^+ (first)	9.32
H^+ (second)	8.71
K^+	1.38
Na^+	1.38
Mg^{2+}	5.28
Ca^{2+}	10.86

3

Chemical kinetics and transport processes

Overview

Differential equations are the backbone of computer modeling and simulation in fields ranging from astrophysics to ecology. Essentially, when differential equations are used to model the behavior of any system, including chemical reaction systems, a set of model equations is developed that mimics as faithfully as possible the essential behavior of the system. When a physical system, such as a living cell or tissue, is to be simulated, it is important not only that the behavior of the system is mimicked by the model equations, but also that the model equations are physically reasonable. Thus the principles of physical chemistry and thermodynamics presented in the previous chapters provide both laws that biochemical models must obey and a framework for building simulations that make physical sense. Our emphasis on physically realistic simulations is not necessarily appreciated by all practitioners in the field. Yet we believe that this emphasis is crucial for building simulations that not only mimic observed behaviors in biological systems, but also *predict* behaviors that are not easily observable or have not yet been observed.

Here we focus on the issue of how to build computational models of biochemical reaction systems. The two foci of the chapter are on modeling chemical kinetics in well mixed systems using ordinary differential equations and on introducing the basic mathematics of the processes that transport material into and out of (and within) cells and tissues. The tools of chemical kinetics and mass transport are essential components in the toolbox for simulation and analysis of living biochemical systems.

Instead of digging into the details of differential equations and numerical analysis, which we leave to specialized books on those topics, we show examples of how mathematical models may be simulated using tools such as Matlab, a high-level and easy-to-use programming environment. Thus our focus here is on building

41

differential equation-based models – setting up the governing equations and related conditions – and not on mathematical analyses of model systems.

3.1 Well mixed systems

The first step in developing general methodologies for the theory and simulation of chemical reaction systems is the study of kinetics in *well mixed* systems. A system is considered to be well mixed when spatial gradients in concentration are considered negligible. (Imagine that the system is stirred rapidly and the concentrations of all species in the system remain homogenous.) When the progress of reaction and transport processes are deterministic (not random) in a well mixed system, and the concentrations of species and reactants are treated as continuous (not discrete), its behavior is governed by a set of ordinary differential equations. Ordinary differential equations have a single independent variable – in this case, time.

3.1.1 Differential equations from mass conservation

Differential equations governing the kinetics of chemical reaction systems may be thought of as arising from statements of mass conservation. For example, consider the well mixed system illustrated in Figure 3.1, containing reactants A and B in a dilute system of constant volume, V.

The statement of mass conservation of any reactant has the form:

$$\text{rate of change} = \text{rate of appearance} - \text{rate of disappearance.} \qquad (3.1)$$

The mass of reactant A is calculated from the volume times the concentration: $V \cdot [A]$; and the rate of change, assuming that the volume is constant, is $V d[A]/dt$. The rates of appearance and disappearance are computed from the mass fluxes (the "Γ's") illustrated in Figure 3.1. If each flux is expressed in units of mass per unit time, then the differential equation governing [A] is

$$V\frac{d[A]}{dt} = \left(\Gamma_A^{in} + \Gamma_r^-\right) - \Gamma_r^+, \qquad (3.2)$$

where Γ_A^{in} is the flux of A into the system, and Γ_r^+ and Γ_r^- are the forward and reverse fluxes through the reaction $A \rightleftharpoons B$, respectively. Similarly, the differential

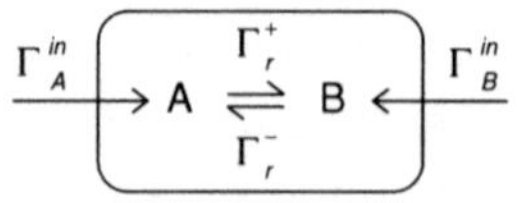

Figure 3.1 Illustration of a system containing molecules A and B, undergoing reaction $A \rightleftharpoons B$.

equation for [B] is

$$V\frac{d[B]}{dt} = \left(\Gamma_B^{in} + \Gamma_r^+\right) - \Gamma_r^-. \tag{3.3}$$

Summing Equations (3.2) and (3.3), we arrive at

$$V\left(\frac{d[A]}{dt} + \frac{d[B]}{dt}\right) = \Gamma_A^{in} + \Gamma_B^{in}. \tag{3.4}$$

If the system is closed, then the transport fluxes Γ_A^{in} and Γ_B^{in} are zero and $\frac{d([A]+[B])}{dt} = 0$, and the total mass of reactants in state A plus reactants in state B in the system remains constant.

Here the symbol Γ has been used to denote mass flux in units of mass per unit time. Throughout this book we use the symbol J to denote a chemical concentration flux for constant volume systems. Chemical flux J has units of concentration per unit time. Writing Equations (3.2) through (3.4) in terms of chemical fluxes, we have

$$\frac{d[A]}{dt} = \left(J_A^{in} + J_r^-\right) - J_r^+, \tag{3.5}$$

$$\frac{d[B]}{dt} = \left(J_B^{in} + J_r^+\right) - J_r^-, \tag{3.6}$$

and

$$\left(\frac{d[A]}{dt} + \frac{d[B]}{dt}\right) = J_A^{in} + J_B^{in}, \tag{3.7}$$

where $J_A^{in} = \Gamma_A^{in}/V$, $J_B^{in} = \Gamma_B^{in}/V$, $J_r^+ = \Gamma_r^+/V$, and $J_r^- = \Gamma_r^-/V$.

3.1.2 Reaction thermodynamics revisited

Simulation of the kinetic behavior of systems such as the one illustrated in Figure 3.1 involves determining mathematical expressions for the fluxes in Equations (3.2) and (3.3), which are typically expressed as functions of the concentrations of reactants A and B. However, before studying mechanisms that govern reaction fluxes, and in turn control reaction kinetics, it is valuable to dig a bit more deeply into the general thermodynamic properties of a chemical reaction such as A $\rightleftharpoons$ B. We have seen in Chapter 1 that the Gibbs free energy change per mole of molecules that transform from state A to state B is expressed

$$\Delta G = \Delta G^o + RT \ln(N_B/N_A), \tag{3.8}$$

where N_A and N_B are the number of molecules in states A and B, respectively. In equilibrium, the ratio N_B/N_A is equal to $e^{-\Delta G^o/RT} = K_{eq}$ and the net reaction flux is $J = J^+ - J^- = 0$, where J^+ and J^- are the forward and reverse reaction fluxes, respectively. When $\Delta G < 0$, the net flux J is positive.

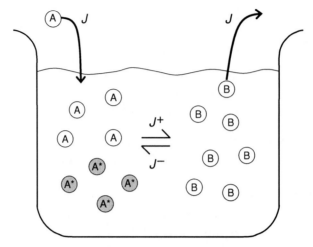

Figure 3.2 Illustration of a system containing molecules A and B, undergoing reaction A $\rightleftharpoons$ B. Molecules in state A are pumped into the system and molecules in state B are pumped out of the system at steady state flux J. Molecules converted from B to A via the reverse reaction are labeled A*.

To determine how flux and free energy are related for systems not in equilibrium we consider, without loss of generality, the case where $N_B/N_A < K_{eq}$ and $J > 0$. In a non-equilibrium steady state N_A and N_B are held constant by pumping A molecules into the system, and pumping B molecules out of the system, at the steady state flux rate J.

Next imagine that we are able to place a label on each molecule that converts from state B to state A. These particles we denote by A*. Apart from the label, A* molecules are identical in every way to unlabeled A molecules in this thought experiment. In addition, imagine that A* molecules lose their label when they convert to B molecules. Thus if we continue to pump A and B molecules into and out of the system at the constant flux J, then a steady state will be reached for which N_{A*}, the number of labeled molecules in state A*, is less than or equal to N_A, the total number of labeled plus unlabeled molecules in state A.

This system is illustrated in Figure 3.2. The steady state is reached when the rate of conversion of labeled A* molecules into state B is equal to the rate of conversion from B to A*. Since there is no transport of A* into or out of the system, then in the steady state the N_{A*} molecules in state A* will be in equilibrium with the N_B molecules in state B. Next, we define

$$j^+ = J^+/N_A, \ j^- = J^-/N_B,$$

the *per-particle* forward and reverse fluxes. The flux j^+ is the rate at which a single molecule in state A transforms to state B; j^- is the rate at which a single molecule in state B transforms to state A. (The per-particle forward rate is also the inverse of the average time that a particle in state A waits to transform to state B.)

At the steady state conservation of mass requires that $j^+ \cdot N_{A*} = j^- \cdot N_B$. Because the ensemble of N_{A*} molecules in state A^* is in equilibrium with the ensemble of N_B molecules in state B,

$$\frac{j^+}{j^-} = \frac{N_B}{N_{A*}} = K_{eq}. \tag{3.9}$$

Equation (3.9) holds for a reaction operating in any steady state, including thermodynamic equilibrium. In equilibrium, $J^+ = J^-$, and

$$\frac{J^+}{J^-} = 1 = \frac{j^+}{j^-} \left(\frac{N_A}{N_B} \right)_{eq}. \tag{3.10}$$

Thus it is trivial that Equation (3.9) holds in equilibrium. The more interesting case is a non-equilibrium steady state for which

$$\frac{j^+}{j^-} = \frac{J^+}{J^-} \frac{N_B}{N_A} = K_{eq}. \tag{3.11}$$

From Equation (3.11), we have $J^+/J^- = \frac{N_A}{N_B} K_{eq}$, from which follows the relationship

$$\Delta G = -RT \ln(J^+/J^-). \tag{3.12}$$

Equation (3.12) is an identity that does not depend on the details of the kinetic reaction mechanism that is operating in a particular system [19]. We [19] have shown that Equation (3.12) is intimately related to the Crooks fluctuation theorem [41] – an important result in non-equilibrium statistical thermodynamics – as well as to theories developed by Hill [87, 90], Ussing [201], and Hodgkin and Huxley [95].

3.1.3 Reaction kinetics

The last step to building a simulation of a well mixed system such as the one illustrated in Figure 3.1 is to define the mathematical form of the reaction fluxes in Equations (3.2) and (3.3). The simplest possible rate law for reaction fluxes is the well-known *law of mass action*.

3.1.3.1 Mass action kinetics

The law of mass action is built on the assumption that the per-particle forward and reverse fluxes j^+ and j^- are independent of the number of particles in a system.

In other words, for the forward reaction the rate at which an individual A molecule transforms into a B molecule in the reaction $A \rightleftharpoons B$ does not depend on N_A or [A]. This assumption is valid if each molecule of A does not interact with other A molecules (even indirectly) in transforming from A to B. Similarly for the reverse reaction, if j^- does not depend on [B] the reverse kinetics are governed by the law of mass action. (The mass-action assumption is not valid for the overall reaction, for example, if the reaction is catalyzed by an enzyme and the number of sites available for interaction with the enzyme depends on the total number of A and B molecules in the system competing for the enzyme. We shall see that when an enzyme catalyzes a reaction $A \rightleftharpoons B$, the overall reaction is typically modeled by a number of subreactions, each of which is governed by mass action.)

The law of mass action implies

$$J^+ = k_+[A]$$
$$J^- = k_-[B] \tag{3.13}$$

for the reaction $A \rightleftharpoons B$, where k_+ and k_- are constants. These expressions yield

$$\frac{J^+}{J^-} = \frac{k_+}{k_-}\frac{[A]}{[B]}. \tag{3.14}$$

Combining Equations (3.14) and (3.12) yields $k_+/k_- = K_{eq}$. It is straightforward to verify that a closed system governed by mass action kinetics will approach the proper equilibrium when the kinetic constants satisfy this relationship.

Using Equation (3.13) we rewrite Equations (3.5) and (3.6) as

$$\frac{d[A]}{dt} = \left(J_A^{in} + k_-[B] - k_+[A]\right)$$
$$\frac{d[B]}{dt} = \left(J_B^{in} + k_+[A] - k_-[B]\right). \tag{3.15}$$

As shown in Section 3.1.1, if the system is closed and the transport fluxes are zero, $\frac{d([A]+[B])}{dt} = 0$. Thus $[A] + [B] = X_o$, where X_o is some constant and the closed system may be reduced to a one-dimensional system:

$$\frac{d[A]}{dt} = -(k_- + k_+)[A] + k_- X_o. \tag{3.16}$$

The steady state solution to this equation is obtained by setting $\frac{d[A]}{dt} = 0$, which yields

$$\left(\frac{[B]}{[A]}\right)_{eq} = \frac{k_+}{k_-}. \tag{3.17}$$

Thus the relationship $k_+/k_- = K_{eq}$ is verified.

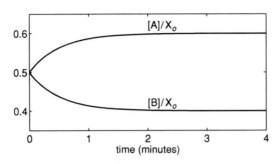

Figure 3.3 Simulation of open system kinetics governed by Equations (3.18) for the reaction A $\rightleftharpoons$ B, with constant flux J. Solution is plotted from Equation (3.19) for parameter values $X_o = 1$ mM, $k_+ = 1$ min^{-1}, $k_- = 1$ min^{-1}, $V = 1$ ml, and $J = 0.2 \times 10^{-6}$ M min^{-1}. The initial condition is $A_o = X_o/2$.

Thermodynamic equilibrium is a special-case steady state that is obtained by closed systems. Open systems with steady (constant) transport fluxes may approach stable steady states that are not equilibrium states. For example, a *non-equilibrium steady state* is achieved by the system of Equations (3.15) when $J_A^{in} = -J_B^{in} = J =$ constant. As in the case of the closed system, $\frac{d([A]+[B])}{dt} = 0$ and $[A] + [B] = X_o$ under this assumption and

$$\frac{d[A]}{dt} = -(k_- + k_+)[A] + J + k_- X_o. \tag{3.18}$$

Equation (3.18) has solution

$$A(t) = \frac{J + k_- X_o - [J + k_- X_o - A_o(k_- + k_+)]\, e^{-(k_-+k_+)t}}{(k_- + k_+)} \tag{3.19}$$

where the variable $A(t)$ represents the concentration $[A]$ as a function of time, t, and A_o is the initial concentration of A.

The steady state concentrations can be obtained by setting Equation (3.18) to zero, which yields

$$[A]_{ss} = A(t \to \infty) = \frac{(J + k_- X_o)}{(k_- + k_+)}. \tag{3.20}$$

Figure 3.3 illustrates the predicted time courses of $[A]$ and $[B]$, for parameter values indicated in the figure legend. The initial condition used corresponds to thermodynamic equilibrium ($[B]/[A] = k_+/k_-$); since the flux is positive (left-to-right in Figure 3.1), the ratio $[A]/[B]$ increases to a value greater than $1/K_{eq}$ in the steady state.

(Note that it is possible to set the constant flux J high enough that Equation (3.19) predicts that $[A]$ will become greater than X_o, and $[B]$ will simultaneously become negative. Of course a negative number of molecules (or a negative concentration) does not make physical sense. Nonsense predictions from Equations (3.15) arise when $[B] \to 0^-$, at which point it is not possible to remove B molecules from the system at a constant flux.)

The mass-action assumption that j^+ and j^- do not depend on reactant concentrations can be applied to reactions other than the uni-unimolecular reaction $A \rightleftharpoons B$. For example, the mass-action rate laws for the reaction

$$A + E \underset{k_-}{\overset{k_+}{\rightleftharpoons}} C \tag{3.21}$$

are

$$J^+ = k_+[A][E]; \quad J^- = k_-[C], \tag{3.22}$$

which we indicate in the chemical equation using the symbol $\underset{k_-}{\overset{k_+}{\rightleftharpoons}}$.

It is straightforward to write down the governing differential equations for this system. For example, if the system is closed,

$$\frac{d[A]}{dt} = (-J^+ + J^-)$$
$$\frac{d[E]}{dt} = \frac{d[A]}{dt}$$
$$\frac{d[C]}{dt} = -\frac{d[A]}{dt}. \tag{3.23}$$

The above set of equations is reduced to one independent differential equation by defining two conservation relationships

$$X_o = [A] + [C] = \text{constant, and, } Y_o = [E] + [C] = \text{constant.}$$

Given these definitions, the governing equation for [A] is

$$\frac{d[A]}{dt} = -k_+[A]^2 - (k_+(Y_o - X_o) + k_-)[A] + k_- X_o. \tag{3.24}$$

This equation is non-linear, which is a general property of kinetics of mass-action reaction systems other than uni-unimolecular reaction systems.

3.1.3.2 *Complex rate laws*

Rate laws that are different from simple mass action often arise in chemical and biochemical applications. Important examples in biochemistry are enzyme and transporter mediated reactions where it is often assumed that a number of discrete steps are involved in converting substrates to products. The individual steps may be governed by mass action, but the overall steady state flux through an enzyme can take a more complex form.

As an example in this section, we consider the well-known Michaelis–Menten enzyme mechanism:

$$A + E \underset{k_{-1}}{\overset{k_{+1}}{\rightleftharpoons}} C \underset{k_{-2}}{\overset{k_{+2}}{\rightleftharpoons}} B + E, \tag{3.25}$$

in which E is an enzyme involved in converting substrate A into product B. The enzyme-mediated mechanism is composed of two steps: (i) binding of A to the enzyme forming a substrate–enzyme complex denoted C, and (ii) conversion of C to product B and unbound enzyme E. For convenience in analyzing this system, we denote the concentrations $a = [A]$, $b = [B]$, $e = [E]$, and $c = [C]$. In addition, we make the assumption that the total concentration of free plus substrate-bound enzyme is conserved: $e + c = E_o = $ constant. The kinetic equations, if the system is closed, are

$$da/dt = -k_{+1}ae + k_{-1}c$$
$$dc/dt = -(k_{-1} + k_{+2})c + (k_{+1}a + k_{-2}b)e$$
$$db/dt = +k_{+2}c - k_{-2}be$$
$$de/dt = da/dt + db/dt. \tag{3.26}$$

(Again, chemical fluxes are defined as mass per unit volume per unit time. For example $J_1^+ = k_{+1}ae$, where a and e have units of M and k_{+1} has units $M^{-1} \sec^{-1}$.)

Using the conservation statement $e = E_o - c$, we eliminate one of the redundant dependent equations and arrive at

$$da/dt = -k_{+1}E_o a + (k_{-1} + k_{+1}a)c$$
$$dc/dt = -(k_{-1} + k_{+2})c + (k_{+1}a + k_{-2}b)(E_o - c)$$
$$db/dt = -k_{-2}E_o b + (k_{+2} + k_{-2}b)c. \tag{3.27}$$

The equilibrium constant for the overall reaction $A \rightleftharpoons B$ can be found by setting da/dt, dc/dt, and db/dt to zero. Alternatively, one may use the fact that, in equilibrium, each of the steps in Equation (3.25) is in thermodynamic equilibrium, and

$$\left(\frac{c}{ae}\right)_{eq} \left(\frac{be}{c}\right)_{eq} = \left(\frac{b}{a}\right)_{eq} = \frac{k_{+1}k_{+2}}{k_{-1}k_{-2}} = K_{eq}. \tag{3.28}$$

The behavior of the Michaelis–Menten enzyme system is illustrated in Figure 3.4; parameter values and initial conditions are listed in the figure legend. For the parameter values indicated $K_{eq} = 10$, which corresponds to the final ratio of b/a for a closed system. From the figure it is apparent that for the given set of parameters, the enzyme complex concentration c changes at a rate much smaller in magnitude than da/dt and db/dt. Based on this observation we can introduce the

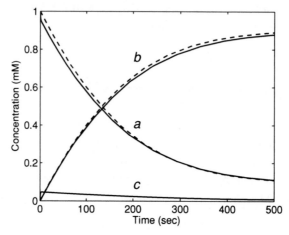

Figure 3.4 Simulation of Michaelis–Menten enzyme mechanism kinetics in closed system. Solid lines correspond to solution of Equations (3.27) with parameter values $k_{+1} = 1000 \, \mathrm{M}^{-1} \sec^{-1}$, $k_{-1} = 1.0 \sec^{-1}$, $k_{+2} = 0.1 \sec^{-1}$, $k_{-2} = 10 \, \mathrm{M}^{-1} \sec^{-1}$, and $E_o = 0.1$ mM. The initial conditions are $a(0) = 1$ mM, $b(0) = 0$, and $c(0) = 0$. Dashed lines correspond to the solution obtained by Equations (3.32).

approximation that $c(t)$ remains in a quasi-steady state, while a and b change more rapidly. This approximation leads to a simplified approximation for the enzyme mechanism. Setting $dc/dt = 0$, we obtain

$$c(a, b) = \frac{E_o \, (k_{+1}a + k_{-2}b)}{k_{-1} + k_{+2} + k_{+1}a + k_{-2}b}. \tag{3.29}$$

With c expressed as a function of a and b, we express the forward and reverse fluxes through the enzyme as:

$$J^+ = k_{+1}a(E_o - c) = k_{+2}c$$

and

$$J^- = k_{-2}b(E_o - c) = k_{-1}c. \tag{3.30}$$

From these equations, we have

$$J_{MM}(a, b) = J^+ - J^- = \frac{E_o \, (k_{+1}k_{+2}a - k_{-1}k_{-2}b)}{k_{-1} + k_{+2} + k_{+1}a + k_{-2}b} = \frac{k_f a - k_r b}{1 + a/K_a + b/K_b}, \tag{3.31}$$

where $k_f = E_o k_{+1}k_{+2}/(k_{-1} + k_{+2})$, $k_r = E_o k_{-1}k_{-2}/(k_{-1} + k_{+2})$, $K_a = (k_{-1} + k_{+2})/k_{+1}$, and $K_b = (k_{-1} + k_{+2})/k_{-2}$. For the parameters listed in the legend of

Figure 3.4, $k_f = 9.09 \times 10^{-3}\,\text{sec}^{-1}$, $k_r = 9.09 \times 10^{-4}\,\text{sec}^{-1}$, $K_a = 1.10\,\text{mM}$, and $K_b = 110\,\text{mM}$.

Equation (3.31) is the standard form for the steady state flux though a simple reversible Michaelis–Menten enzyme. This expression obeys the equilibrium ratio arrived at above: $(b/a)_{eq} = K_{eq} = k_{+1}k_{+2}/(k_{-1}k_{-2})$, when $J_{MM}(a, b) = 0$. In addition, from the positive and negative one-way fluxes in Equation (3.30), we note that the relationship $J^+/J^- = K_{eq}(a/b) = e^{-\Delta G/RT}$ is maintained whether or not the system is in equilibrium. Thus, as expected, the general law of Equation (3.12) is obeyed by this reaction mechanism.

Based on Equation (3.31) we can model the kinetics of the system (ignoring the dc/dt equation) as

$$da/dt = -J_{MM}(a, b)$$
$$db/dt = +J_{MM}(a, b). \tag{3.32}$$

The solution obtained using this system of equations is plotted as dashed lines in Figure 3.4. The solution based on this quasi-steady state approximation closely matches the solution obtained by solving the full kinetic system of Equations (3.27). The major difference between the two solutions is that the quasi-steady approximation does not account explicitly for enzyme binding. Therefore $a + b$ remains constant in this case, while in the full kinetic system $a + b + c$ remains constant. Since the fraction of reactant A that is bound to the enzyme is small ($c/a \ll 1$), the quasi-steady approximation is relatively accurate.

Obtaining quasi-steady approximations for fluxes through reaction mechanisms, including mechanisms more complex than the simple Michaelis–Menten system studied in this section, is a major component of the study of enzyme kinetics. This topic will be treated in some detail in Chapter 4.

3.1.3.3 Net flux for nearly irreversible reactions is proportional to reverse flux

In computational modeling of biochemical systems, the approximation that certain reactions are irreversible is often invoked. In this section, we explore the consequences of such an approximation, and show that the flux through nearly irreversible enzyme-mediated reactions is proportional to the reverse reaction flux.

Our analysis of nearly irreversible systems begins with Equation (3.12), from which we have

$$J^+/J^- = e^{-\Delta G/RT} \tag{3.33}$$

for a given reaction. The net flux through the system is $J = J^+ - J^-$; thus, $e^{-\Delta G/RT} = J/J^- + 1$, which leads to the approximation

$$J = J^- e^{-\Delta G/RT} = J^- K_{eq}\frac{a}{b} \tag{3.34}$$

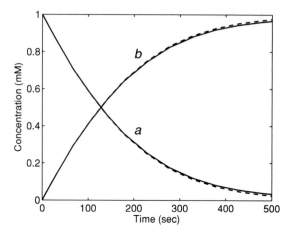

Figure 3.5 Simulation of a nearly irreversible Michaelis–Menten enzyme system. Solid lines correspond to solution of Equations (3.32) with parameter values $k_f = 9.09 \times 10^{-3}\,\mathrm{sec}^{-1}$, $k_r = 9.09 \times 10^{-5}\,\mathrm{sec}^{-1}$, $K_a = 1.1\,\mathrm{mM}$, and $K_b = 1.10\,\mathrm{M}$. The initial conditions are $a(0) = 1\,\mathrm{mM}$ and $b(0) = 0$. Dashed lines correspond to the simulation of the system governed by irreversible kinetics of Equations (3.35) and (3.36).

for nearly irreversible reactions ($J \gg J^-$). Thus the flux through the enzyme is proportional to the reverse flux.

For the quasi-steady approximation of Equation (3.31) the reverse flux is $J^- = k_r b/(1 + a/K_a + b/K_b)$; thus for a nearly irreversible reaction

$$J = \frac{k_r K_{eq} a}{1 + a/K_a + b/K_b} = \frac{k_f a}{1 + a/K_a + b/K_b},$$

which is the expression that we would arrive at by setting $k_r = 0$ in Equation (3.31). When the reaction is nearly irreversible and the reverse mechanism is not near saturation, we can approximate the flux as

$$J_{IR} = \frac{k_f a}{1 + a/K_a}; \tag{3.35}$$

and the kinetics of a and b can be modeled by

$$da/dt = -J_{IR}(a)$$
$$db/dt = +J_{IR}(a). \tag{3.36}$$

Figure 3.5 illustrates the comparison between a system governed by the reversible Michaelis–Menten kinetics of Equations (3.31) and (3.32), the irreversible kinetics of Equations (3.35) and (3.36). The parameter values are indicated in the legend. The values used correspond to the same set of values as used in Figure 3.4 with the exception that k_{-2} is changed from $10\,\mathrm{M}^{-1}\,\mathrm{sec}^{-1}$ in Figure 3.4 to $1.0\,\mathrm{M}^{-1}\,\mathrm{sec}^{-1}$ in

Figure 3.5. These values ensure that the reaction simulated in Figure 3.5 is nearly irreversible over most of the concentration range of a and b, with an equilibrium constant of $K_{eq} = 100$. It is apparent that the solution obtained for this system using the irreversible approximation is nearly identical to the simulation using reversible kinetics over most of the concentration range. As a and b approach equilibrium the assumption that $J \gg J^-$ becomes less reliable, and the irreversible approximation becomes less accurate.[1]

3.1.3.4 Net flux for highly reversible reactions is proportional to reverse flux

Near equilibrium (for $|\Delta G| \ll RT$) the flux can be approximated as linearly proportional to the thermodynamic driving force: $J = -X\Delta G$, where X is called the *Onsager coefficient* [154, 155]. When the near-equilibrium approximation $|\Delta G| \ll RT$ holds, the flux ratio J^+/J^- is approximately equal to 1. In this case Equation (3.12) is approximated:

$$\Delta G = -RT(J^+/J^- - 1) = \frac{RT}{J^-}(J^+ - J^-). \qquad (3.37)$$

From this expression, we have

$$J = -\frac{J^-}{RT}\Delta G. \qquad (3.38)$$

Therefore for highly reversible systems, the net flux is proportional to the reverse flux times the thermodynamic driving force; the Onsager coefficient is equal to J^-/RT.

3.1.4 Using computers to simulate chemical kinetics

3.1.4.1 Example: simulating Michaelis–Menten enzyme kinetics

As we have pointed out in the introduction, our focus in this chapter is on how to build models of biochemical systems, and not on mathematical analysis of models. As an example, consider the system of Equations (3.27), which represents a model for the reactions of Equation (3.25). It is possible to analyze these equations using a number of mathematical techniques. For example Murray [146] presents an elegant asymptotic analysis of a model of an irreversible (with $k_{-2} = 0$) Michaelis–Menten enzyme. Such analyses invariably yield mathematical insights into the behavior of

---

[1] Although the irreversible approximation successfully simulates the enzymatic flux in the range in which the reverse flux is small compared to the forward flux, the impact of approximating nearly irreversible reactions as entirely irreversible in simulations of reaction *systems* can be significant. It has been shown that feedback of product concentration in nearly irreversible reactions, either through reverse flux or product inhibition, is necessary for models of certain reaction networks to reach realistic steady states [36].

model systems, as illustrated in Section 4.2.2 of the following chapter on enzyme kinetics. However, the goal of the present chapter is to develop techniques for building models of a scale and complexity that are inaccessible to traditional mathematical analysis.

For large-scale problems, the most widely useful mathematical tool available is computational/numerical simulation. A great number of computer tools are available for simulation of ordinary differential equation (ODE) based models, such as Equations (3.27). Here we demonstrate how this system may be simulated using the ubiquitous Matlab software package.

The major requirement to simulate an ODE-based model using Matlab (or most other packages) is to supply a code that computes the time derivatives of the state variables – the right-hand side of Equations (3.27). An example of a function that computes the time derivatives for this model using Matlab syntax is given below.[2]

```
function [f] = dCdT_mm1(t,x,kp1,km1,kp2,km2,Eo);
% FUNCTION dCdT_mm1
%   Inputs: t - time (seconds)
%           x - vector of concentrations {a,b,c} (M)
%           kp1 - forward rate constant (M^{-1} sec^{-1})
%           km1 - reverse rate constant (sec^{-1})
%           kp2 - forward rate constant (sec^{-1})
%           km2 - reverse rate constant (M^{-1} sec^{-1})
%           Eo - total enzyme concentration (M)
%
%   Outputs: f - vector of time derivatives
%               {da/dt,dc/dt,db/dt}
a = x(1);
c = x(2);
b = x(3);
f(1) = -kp1*Eo*a + (km1 + kp1*a)*c;
f(2) = -(km1 + kp2)*c + (kp1*a + km2*b)*(Eo - c);
f(3) = -km2*Eo*b +(kp2 + km2*b)*c;
f = f';
```

---

[2] We make use of Matlab to present the applications in this section because of its widespread use and because its syntax represents a convenient pseudo-code format for representing algorithms [29]. For this same reason we will continue to use Matlab syntax for certain examples throughout this book. However, Matlab is certainly not the only available software package for simulating ODE systems. There exist a great many freely available packages based in Fortran, C, or other computing languages, for integrating systems of ODEs. The web repository www.netlib.org maintained by Oak Ridge National Laboratory is a useful resource to search for codes for integrating ODEs and other applications in computing. Two packages that are widely used by the metabolic modeling community are Jarnac [179] and Copasi [99].

The function "dCdTmm1" accepts as inputs: a variable representing time, a vector that stores the three state variables, and variables that represent the model parameters, as indicated in the comments in the code. (Comments follow the symbol "%" at the beginning of a given line of code.) The output is the vector "f", which lists the time derivatives da/dt, dc/dt, and db/dt.

The above code represents "the model" for Equations (3.27); with this code we can use a package such as Matlab to simulate the behavior of the model. The simulation illustrated in Figure 3.4 is completed using the following commands in Matlab:

```
% Set parameter values:
kp1 = 1000; km1 = 1.0; kp2 = 0.1; km2 = 10.0; Eo = 1e-4;
% Set initial Condition:
xo = [0.001 0 0];
% Integrate ODE:
[t,x] = ode15s(@dCdT_mm1,[0 500],xo,[],kp1,km1,kp2,km2,Eo);
% Plot results:
plot(t,x(:,1)*1e3,t,x(:,2)*1e3,t,x(:,3)*1e3);
```

The commands above set the parameters to the values indicated in the figure, then introduce a variable "xo", in which are entered the initial conditions for the simulation. Next, the ODE system is integrated over the interval $t \in [0, 500 \text{ sec}]$ using the built-in ODE solver "ode15s". The ODE-solver package accepts as inputs the name of the function used to compute the derivatives, the time range over which to simulate the system, and the values of the initial conditions and parameters to use in the simulation. The final command plots the three state variables, in units of mM, as functions of time.

3.1.4.2 Example: non-linear oscillations in glycolysis

As a next example, we consider the model of Termonia and Ross [193] for the non-linear kinetics of glycolysis. This model simulates the kinetics of the following reactions

$$\text{rxn 1} : \text{GLU} + \text{ATP} \rightleftharpoons \text{F6P} + \text{ADP}$$
$$\text{rxn 2} : \text{F6P} + \text{ATP} \rightleftharpoons \text{FBP} + \text{ADP}$$
$$\text{rxn 3} : \text{FBP} + 2\,\text{NAD} + 2\,\text{ADP} + 2\,\text{PI} \rightleftharpoons 2\,\text{PEP} + 2\,\text{ATP} + 2\,\text{NADH}$$
$$\text{rxn 4} : \text{PEP} + \text{ADP} \rightleftharpoons \text{PYR} + \text{ATP} \tag{3.39}$$

where the first reaction combines the reactions of hexokinase and phosphoglucose isomerase; the second reaction is phosphofructokinase; the third reaction combines the reactions of aldolase, triose phosphate isomerase, glyceraldehyde-3-phosphate dehydrogenase, phosphoglycerate kinase, phosphoglycerate mutase, and enolase; the fourth reaction is the reaction of pyruvate kinase. The reactants in this system are: glucose (GLU), fructose-6-phosphate (F6P), fructose-1,6-bisphosphate (FBP), phosphoenolpyruvate (PEP), pyruvate (PYR), NAD, NADH, ATP, ADP, and inorganic phosphate (PI).

In the model of Termonia and Ross, glucose, inorganic phosphate, NAD, and NADH concentrations are not modeled explicitly and assumed fixed. The kinetic equations for the remaining reactants are

$$d[\text{F6P}]/dt = J_1 - J_2$$
$$d[\text{FBP}]/dt = J_2 - J_3$$
$$d[\text{PEP}]/dt = 2J_3 - J_4$$
$$d[\text{PYR}]/dt = J_4 - J_5$$
$$d[\text{ATP}]/dt = -J_1 - J_2 + 2*J_3 + J_4 - J_6$$
$$\frac{[\text{ATP}] \cdot [\text{AMP}]}{[\text{ADP}]^2} = K$$
$$A_o = [\text{ATP}] + [\text{ADP}] + [\text{AMP}], \tag{3.40}$$

where [AMP] is the adenosine monophosphate concentration, K is the apparent equilibrium constant for the adenylate kinase reaction ($2\,\text{ADP} \rightleftharpoons \text{AMP} + \text{ATP}$), and A_o is total conserved concentration of adenine nucleotide in the system. The J's in Equation (3.40) represent the fluxes through the corresponding reaction; J_5 and J_6 represent transport of pyruvate out of the system and the rate of ATP consumption, respectively.

The above set of model equations is a combined set of differential and algebraic expressions. The algebraic equation $\frac{[\text{ATP}] \cdot [\text{AMP}]}{[\text{ADP}]^2} = K$ arises because the assumption is made that the adenylate kinase reaction rapidly achieves equilibrium and is maintained in equilibrium at all times. We will see below how this algebraic expression is treated in simulating the kinetics of Equation (3.40).

The expressions for the fluxes in this model are

$$J_1 = \text{constant} = 2.0\,\text{mM min}^{-1}$$
$$J_2 = \frac{V_2([\text{F6P}]/(1\,\text{mM}))^n}{(K_2 + K_2 R_2([\text{ATP}]/[\text{AMP}])^n + ([\text{F6P}]/(1\,\text{mM}))^n)}$$
$$J_3 = k_3([\text{FBP}]/(1\,\text{mM}))^\alpha - \bar{k}_3([\text{PEP}]/(1\,\text{mM}))^\beta$$
$$J_4 = \frac{V_4([\text{PEP}]/(1\,\text{mM}))^\gamma}{(K_4 + K_4 R_4([\text{ATP}]/[\text{FBP}])^m + ([\text{PEP}]/(1\,\text{mM}))^\gamma)}$$
$$J_5 = k_5 \cdot [\text{PYR}]$$
$$J_6 = k_6 \cdot [\text{ATP}], \tag{3.41}$$

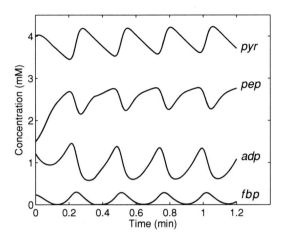

Figure 3.6 Simulation of the model of Termonia and Ross [193] based on the Equation (3.40). $K = 1$, $A_o = 40.95$ mM, $n = 2$, $K_2 = 0.0016$, $R_2 = 0.1736$, $V_2 = 12.5$ mM min^{-1}, $k_3 = 5.8$ mM min^{-1}, $\bar{k}_3 = 0.01$ mM min^{-1}, $\alpha = 0.05$, $\beta = 6$, $\gamma = 1$, $m = 4$, $K_4 = 0.1296$, $R_4 = 2.1389 \times 10^{-7}$, $V_4 = 100$ mM min^{-1}, $k_5 = 1.0$ min^{-1}, and $k_6 = 0.1$ min^{-1}. The assumed initial conditions are [F6P] = 13 mM, [FBP] = 0.24 mM, [PEP] = 1.5 mM, [PYR] = 4.0 mM, and [ATP] = 39.7 mM.

where the assumed parameter values are listed in the legend of Figure 3.6.

The rapid-equilibrium algebraic expression may be handled in this model by solving the system of equations

$$[ATP] \cdot [AMP]/[ADP]^2 = K$$

$$A_o = [ATP] + [ADP] + [AMP]$$

for [AMP] and [ADP] as a function of [ATP] and the parameter values. Assuming $K = 1$ yields

$$[AMP] = (2A_o - [ATP] - [(2A_o - [ATP])^2 - 4([ATP] - A_o)^2]^{1/2})/2$$

$$[ADP] = A_o - [ATP] - [AMP]. \tag{3.42}$$

Equation (3.42) provides [ADP] and [AMP] in terms of the state variable [ATP]. Therefore these expressions can be used to compute the fluxes and the derivatives of the state variables as explicit functions of the values of the state variables.

An example simulation of this model is illustrated in Figure 3.6, using the parameter values and initial conditions indicated in the legend. Note that the model predicts sustained non-linear oscillations, which have been observed in yeast cells and in extracts from yeast and also mammalian cells [82].

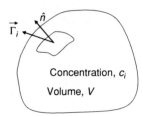

Figure 3.7 Enclosed volume V with the continuum vector field mass flux density $\vec{\Gamma}_i$ and scalar concentration field c_i.

3.2 Transport processes

When biochemical systems are studied in vitro, it is typically under well mixed conditions. Yet the contents of living cells are not necessarily well mixed and the biochemical workings within cells are inseparably coupled to the processes that transport material into, out of, and within cells. The three processes responsible for mass transport in living systems are advection, diffusion, and drift. Characterizing which, if any, of these processes is active in a given system is an important component of building differential equation-based models of living biochemical systems.

In general, the equations governing various transport processes, like equations for chemical kinetics in well mixed systems, are built upon the foundation of mass conservation [23].

To derive equations for mass transport we introduce the quantity $\vec{\Gamma}_i$, which we define as the mass flux density of species i, expressed in units of mass per unit area per unit time. Given $\vec{\Gamma}_i$, we express the rate of change of total mass inside volume V as equal to the rate of mass flux into the volume:

$$\frac{\partial}{\partial t} \int c_i \, dV = - \oint \hat{n} \cdot \vec{\Gamma}_i \, dS, \tag{3.43}$$

where the concentration of species i, c_i, is a continuous spatial concentration field, the symbol $\oint$ indicates a surface integral over the surface S enclosing the volume V, and $\hat{n}$ is the outward-point unit normal vector on the surface of the volume V. The concept is illustrated in Figure 3.7.

Using Gauss' theorem, Equation (3.43) can be rewritten

$$\frac{\partial}{\partial t} \int c_i \, dV = - \int \nabla \cdot \vec{\Gamma}_i \, dV. \tag{3.44}$$

Since within a continuum system the volume V is arbitrary in Equation (3.44), the statement of mass conservation may be expressed in differential form:

$$\frac{\partial c_i}{\partial t} = -\nabla \cdot \vec{\Gamma}_i. \tag{3.45}$$

Equation (3.45) is the general continuum statement of mass conservation of species i written in terms of the mass flux density $\vec{\Gamma}_i$ and the concentration field c_i.

3.2.1 Advection

Advection is the process by which material contained in a flowing fluid is transported by bulk motion of the fluid. An important example is blood flow, which delivers oxygen and nutrients to the tissues of the body. Maintaining blood flow is essential to maintaining life in higher organisms.

In advection the mass flux is driven by a continuous velocity field. Given a velocity field $\vec{v}$ and concentration field c_i, the mass flux density is $\vec{\Gamma}_i = c_i \vec{v}$. Substitution of this expression into Equation (3.45) yields

$$\frac{\partial c_i}{\partial t} = -\nabla \cdot (c_i \vec{v}) = -c_i \nabla \cdot \vec{v} - \vec{v} \cdot \nabla c_i. \tag{3.46}$$

For incompressible fluids $\nabla \cdot \vec{v} = 0$, and this equation simplifies to

$$\frac{\partial c_i}{\partial t} = -\vec{v} \cdot \nabla c_i. \tag{3.47}$$

Equation (3.47) is known as the advection equation. For one-dimensional fluid flow the advection equation reduces to

$$\frac{\partial c_i}{\partial t} = -v \frac{\partial c_i}{\partial x}, \tag{3.48}$$

which has the general solution: $c_i(x, t) = c_i(x - vt, 0)$. One-dimensional advection is illustrated in Figure 3.8, which demonstrates that concentration profiles governed by the advection equation translate in space with velocity v.

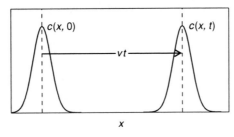

Figure 3.8 Illustration of advection governed by Equation (3.48).

One-dimensional advection will be used in Chapter 8 as a component in models of cellular biochemical systems that are coupled to blood–tissue solute exchange.

3.2.2 Diffusion

According to Fick's first law of diffusion [56], the mass flux density due to molecular diffusion is proportional to the gradient in concentration. Formally,

$$\vec{\Gamma}_i = -D_i \nabla c_i, \tag{3.49}$$

where D_i is called the molecular diffusion coefficient. Substitution into Equation (3.45) yields

$$\frac{\partial c_i}{\partial t} = \nabla \cdot (D_i \nabla c_i), \tag{3.50}$$

which is called the diffusion equation. If the system studied is homogenous and D_i is not a function of position, then the diffusion equation becomes $\partial c_i / \partial t = D_i \nabla^2 c_i$.

Equation (3.50) is used below to develop an expression for the passive flux of solute across a thin homogeneous membrane. In addition, diffusion-driven processes will appear in our study of spatially distributed systems in Chapter 8.

3.2.3 Drift

The term drift is used to describe mass transport of charged species driven by an electric field. The drift mass flux density is

$$\vec{\Gamma}_i = \frac{z_i}{|z_i|} u_i c_i \vec{E}, \tag{3.51}$$

where z_i is the valence of the ion, u_i is the electrokinetic mobility, and $\vec{E}$ is the electric field vector. The factor $z_i / |z_i| = \pm 1$ is the sign of the charge of species i. Positive ions are driven in the direction of $\vec{E}$, negatively charged ions are driven in the opposite direction. For uncharged species, $z_i = 0$ and the drift flux is zero.

In biology, we are typically concerned with the study of electrostatic systems in which electromagnetic interactions are ignored and the electric field is the gradient of the electrostatic potential: $\vec{E} = -\nabla \phi$. The differential equation for electrokinetic drift follows from Equation (3.45).

$$\frac{\partial c_i}{\partial t} = \frac{z_i}{|z_i|} u_i \nabla \cdot (c_i \nabla \phi) = \frac{z_i}{|z_i|} u_i c_i \nabla^2 \phi + \frac{z_i}{|z_i|} u_i (\nabla \phi) \cdot (\nabla c_i) \tag{3.52}$$

3.2.3.1 Einstein relation relates molecular diffusivity and electrokinetic mobility

We have seen in Chapter 1 that, in equilibrium, the probability of a state is expressed as a function of the energy of the state according to the Boltzmann law: $P_i = e^{-E_i/k_B T}$. In a dilute system of non-interacting charged particles, the energy associated with a particle of species i located at position $\vec{x}_i$ is $E_i = z_i F \phi(\vec{x}_i)/N_A$, where $\phi(\vec{x}_i)$ is the electrostatic potential expressed in units of energy per unit of charge and $F/N_A \approx 1.602 \times 10^{-19}$ coulomb is the proton charge. The equilibrium particle concentration field may be thought of as a measure of the relative probability of finding a particle in a given position in space. According to Boltzmann statistics, the equilibrium concentration is $c_i(\vec{x}) = c_o e^{-z_i F \phi(\vec{x})/RT}$.

In equilibrium the net particle flux is zero; when drift and diffusion are the two active transport processes, $\vec{\Gamma}_i^{drift} + \vec{\Gamma}_i^{diffusion} = 0$, or

$$-D_i \nabla c_i - \frac{z_i}{|z_i|} u_i c_i \nabla \phi = 0. \tag{3.53}$$

This equation can be re-expressed:

$$\frac{\nabla c_i}{c_i} = -\frac{z_i u_i}{|z_i| D_i} \nabla \phi, \tag{3.54}$$

which may be integrated:

$$\ln c_i = -\frac{z_i u_i}{|z_i| D_i} \phi + a_o, \tag{3.55}$$

where a_o is a constant. This equation is rewritten

$$c_i = c_o e^{-\frac{z_i u_i}{|z_i| D_i} \phi}, \tag{3.56}$$

where c_o is a constant. Equating Equation (3.56) with the Boltzmann equilibrium distribution yields

$$\frac{u_i}{D_i} = \frac{|z_i| F}{RT}, \tag{3.57}$$

which is called the Einstein relation. Equation (3.57) is a form of the fluctuation-dissipation theorem, which relates the magnitude of thermal fluctuations (in this case molecular diffusivity) to the system response to macroscopic driving forces (in this case current flow in response to an electric field).

3.2.4 Example: passive permeation across a membrane

If a biological membrane is treated as a homogeneous isotropic medium, then transport of a solute in the membrane is governed by the diffusion equation

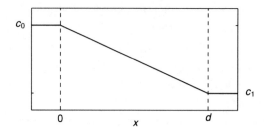

Figure 3.9 Steady state concentration profile due to passive diffusion in a homogeneous membrane of width d. Linear concentration profile is given by Equation (3.59).

$\partial c/\partial t = D_m \nabla^2 c$, where D_m is the molecular diffusion coefficient of the solute in the membrane. If the membrane is thin compared to other space scales in the system, then the transport in the membrane is effectively represented by a one-dimensional equation $\partial c/\partial t = D_m \partial^2 c/\partial x^2$. In the steady state, we have $D_m d^2 c/dx^2 = 0$, or $dc/dx = a_1$, where a_1 is a constant. Therefore the concentration profile in the membrane is linear:

$$c(x) = a_1 x + a_o. \tag{3.58}$$

If the membrane has width d and the concentrations on either side of the membrane are given by $c(0) = c_0$, $c(d) = c_1$, then the concentration profile in the membrane is given by

$$c(x) = c_1 + (c_1 - c_0)x/d, \tag{3.59}$$

which is illustrated in Figure 3.9.

The flux through the membrane is calculated

$$\Gamma = -D_m \frac{dc}{dx} = \frac{D_m}{d}(c_0 - c_1). \tag{3.60}$$

Therefore the steady state flux through the membrane is linearly proportional to the concentration difference across the membrane. Defining the membrane permeability $p = D_m/d$, we have $\Gamma = p\Delta c$.

3.2.5 Example: coupled diffusion and drift in a membrane

To treat the coupled diffusion and drift of a charged ion in a homogeneous membrane, we include both molecular diffusion and electrokinetic drift in the governing

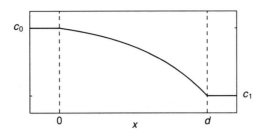

Figure 3.10 Steady state concentration profile due to molecular diffusion and electrokinetic drift in a homogeneous membrane of width d. Concentration profile is given by Equation (3.63) with $zF\Delta\Phi/RT = 2$.

transport equation:

$$\frac{\partial c}{\partial t} = -\nabla \cdot (\vec{\Gamma}^{drift} + \vec{\Gamma}^{diffusion})$$

$$= -\nabla \cdot \left(-D_m \nabla c - \frac{z}{|z|} uc\nabla\phi\right). \qquad (3.61)$$

For a thin membrane in the steady state, and assuming a uniform electric field directed across the membrane,[3] we have:

$$D_m \frac{d^2c}{dx^2} + \frac{zu}{|z|}\left(\frac{d\phi}{dx}\right)\left(\frac{dc}{dx}\right) = 0$$

$$D_m \frac{d^2c}{dx^2} - \frac{zu}{|z|}\left(\frac{\Delta\Phi}{d}\right)\left(\frac{dc}{dx}\right) = 0, \qquad (3.62)$$

where $\Delta\Phi = \phi(0) - \phi(d)$. Equation (3.62), with boundary conditions $c(0) = c_0$ and $c(d) = c_1$, has the solution:

$$c(x) = \frac{c_o e^{\frac{zF\Delta\Phi}{RT}} - c_1 + (c_1 - c_0)e^{\frac{zF\Delta\Phi x}{RTd}}}{e^{\frac{zF\Delta\Phi}{RT}} - 1}, \qquad (3.63)$$

where we have used the Einstein relation. Equation (3.63) is illustrated in Figure 3.10 for the case $zF\Delta\Phi/RT = 2$.

The flux through the membrane is calculated

$$\Gamma = -D_m\frac{dc}{dx} + \frac{z}{|z|}u\frac{\Delta\Phi}{d}c,$$

---

[3] The uniform-field approximation is valid when the width of the membrane (or length of the channel through which the ion passes) is small compared with the *Debye length*, which is the length scale over which mobile ions screen the electric field in the membrane. The Debye length is inversely proportional to the square root of the ionic strength [138]. At high ionic concentrations, a linear-conductance model (introduced in Chapter 7) is more appropriate than the uniform-field approximation.

or

$$\Gamma = \frac{zDF\Delta\Phi}{dRT}\left(\frac{c_0 e^{\frac{zF\Delta\Phi}{RT}} - c_1}{e^{\frac{zF\Delta\Phi}{RT}} - 1}\right). \tag{3.64}$$

Equation (3.63) is known as the the Goldman–Hodgkin–Katz equation for passive flux of an ion through a membrane [108, 123].

Concluding remarks

This chapter has developed the theory of how to build differential equation-based models of biological systems based on simulating chemical kinetics and transport phenomena. The coming chapters build on these techniques in analyzing successively larger and more complex biochemical reaction systems in Chapters 4 through 7. The transport processes active in living systems will be represented at varying levels of detail depending on the application. In Chapters 4 through 7, cells are represented as one or more well mixed compartments with transport into, out of, and between compartments governed by fluxes of solutes across thin membranes. Chapter 8 explicitly considers spatially distributed transport in the cells, tissues, and organs of living organisms. For a study of the field of mass transport that delves deeply into how mass and energy are transported in physical systems of all sorts, we highly recommend Bird, Stewart, and Lightfoot's canon *Transport Phenomena* [23].

Exercises

3.1 Verify that the relationship $k_+/k_- = K_{eq}$ follows from Equations (3.14) and (3.12).

3.2 Verify that Equation (3.19) is the solution of Equation (3.18), and that $\frac{d([A]+[B])}{dt} = 0$, $[A] + [B] = X_o$, and Equation (3.18) is valid when $J_B^{in} = -J_A^{in} = J = $ constant.

3.3 Derive Equation (3.24) from the stated assumptions.

3.4 Although Equation (3.24) is non-linear, a convenient closed form solution is available. Find the solution and make a plot of [A], [E], and [C], given an assumed set of parameters and initial conditions.

3.5 What is the equilibrium constant for the association of reactant A to the enzyme for the kinetic parameters used in Figure 3.4? How close is the reaction $A + E \rightleftharpoons C$ to equilibrium during the simulation that is illustrated? How does the quasi-steady approximation depend on the equilibrium constant for enzyme binding?

3.6 Use computational simulation to investigate the validity of the quasi-steady assumption for different sets of parameters and initial conditions for the Michaelis–Menten system. Under what circumstance(s) does the approximation fail?

3.7 Write a computer program to generate the simulation shown in Figure 3.6 using the computing language and environment of your choice.

3.8 The flux expressions for this model of Section 3.1.4.2 violate certain physical constraints on reaction fluxes. What constraints are violated? Under what circumstances will these violations be important?

3.9 Show that $\nabla \cdot \vec{v} = 0$ for an incompressible fluid. [Hint: start with the continuity equation for fluid density ρ: $\partial \rho / \partial t = -\nabla \cdot (\rho \vec{v})$.]

3.10 Show that in the limit $\Delta \Phi \to 0$, Equation (3.64) reduces to Equation (3.60).

3.11 For a given concentration ratio c_0/c_1, at what value of $\Delta \Phi$ does Equation (3.64) predict that the flux through the membrane will be zero? Provide a physical explanation of why flux is zero at this value of membrane potential.

Part II

Analysis and modeling of biochemical systems

4

Enzyme-catalyzed reactions: cycles, transients, and non-equilibrium steady states

Overview

There is almost no biochemical reaction in a cell that is not catalyzed by an *enzyme*. (An enzyme is a specialized protein that increases the flux of a biochemical reaction by facilitating a mechanism [or mechanisms] for the reaction to proceed more rapidly than it would without the enzyme.) While the concept of an enzyme-mediated kinetic mechanism for a biochemical reaction was introduced in the previous chapter, this chapter explores the action of enzymes in greater detail than we have seen so far. Specifically, catalytic cycles associated with enzyme mechanisms are examined; non-equilibrium steady state and transient kinetics of enzyme-mediated reactions are studied; an asymptotic analysis of the fast and slow timescales of the Michaelis–Menten mechanism is presented; and the concepts of cooperativity and hysteresis in enzyme kinetics are introduced.

While the majority of these concepts are introduced and illustrated based on single-substrate single-product Michaelis–Menten-like reaction mechanisms, the final section details examples of mechanisms for multi-substrate multi-product reactions. Such mechanisms are the backbone for the simulation and analysis of biochemical systems, from small-scale systems of Chapter 5 to the large-scale simulations considered in Chapter 6. Hence we are about to embark on an entire chapter devoted to the theory of enzyme kinetics. Yet before delving into the subject, it is worthwhile to point out that the entire theory of enzymes is based on the simplification that proteins acting as enzymes may be effectively represented as existing in a finite number of discrete states (substrate-bound states and/or distinct conformational states). These states are assumed to inter-convert based on the law of mass action. The set of states for an enzyme and associated biochemical reaction is known as an *enzyme mechanism*. In this chapter we will explore how the kinetics of a given enzyme mechanism depend on the concentrations of reactants and enzyme states and the values of the mass action rate constants associated with the mechanism.

4.1 Simple Michaelis–Menten reactions revisited

4.1.1 Steady state enzyme turnover kinetics

As was demonstrated computationally in Section 3.1.3.2, when the rate of change in the concentration of an enzyme–substrate complex (dc/dt in the example) is much smaller in magnitude than the rate of change in the concentrations of substrates and products, (da/dt and db/dt in the example) then kinetics of the enzyme mechanism may be approximated using the quasi-steady approximation. Specifically, the approximation that $dc/dt = 0$ is invoked. For the example in Section 3.1.3.2, the initial concentration of the substrate is ten times greater than that of the enzyme: $a(0) = 1$ mM and $E_o = 0.1$ mM. In many biological settings, the ratio of reactant to enzyme concentration is much greater than ten and the quasi-steady approximation tends to be valid.

In fact, certain enzymatic reactions in cells may involve only a handful of enzyme molecules, with as few as several hundred or several thousand enzyme molecules present compared to many more substrate and product molecules. This condition implies that the substrate molecules compete for binding to the relatively few enzyme molecules, while the enzymes function more or less independently of one another. The assumption that the molecules of a particular enzyme act independently significantly simplifies the modeling of biochemical kinetics. In addition, if reactants are present in quantities much greater than enzymes then one can reasonably treat enzyme turnover kinetics by assuming that the substrate concentration remains effectively constant over the timescale of enzyme turnover.

While precise information on the numbers of copies of all the enzymes present in specific cells is not yet available, a recent study in yeast has shown that the number of copies of all proteins present ranges from about 100 to 250 000. (These numbers correspond to total proteins of all types. Enzymes are specialized proteins that catalyze specific reactions. Not all cellular proteins serve as enzymes.) Figure 4.1 shows the frequency distribution of number of protein copies per cell from Ghaemmaghami *et al.* [64]. The peak in the frequency distribution is around 3000 molecules per cell. In other words, for a typical protein approximately 3000 copies are present in an individual cell. Some proteins are present in greater quantities; some in smaller quantities.

The quasi-steady approximation is generally valid when the amount that enzyme complex concentrations change is less than the amount that reactant concentrations change over the timescale of interest. This is true, for example, in Section 3.1.3.2 as long as $dc/dt \ll J_{MM}$. Thus the stricter condition that reactant concentrations are large compared to enzyme concentrations (a condition that is by no means universally true in vivo) is not necessarily required to apply the approximation.

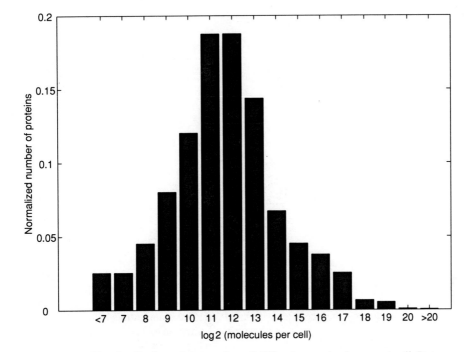

Figure 4.1 The distribution of the number of different proteins in a yeast cell. Data are from Ghaemmaghami *et al.* [64].

Consider again the simple irreversible Michaelis–Menten scheme:

$$\text{E} + \text{S} \underset{k_{-1}}{\overset{k_{+1}}{\rightleftharpoons}} \text{ES} \overset{k_{+2}}{\rightarrow} \text{E} + \text{P}. \tag{4.1}$$

This mechanism can be represented as the catalytic cycle illustrated in Figure 4.2A, in which an individual enzyme molecule converts between two states: free (unbound) enzyme E and the enzyme–substrate complex ES. The catalytic cycle is represented by the following kinetic mechanism.

$$\text{E} \underset{k_{-1}}{\overset{k_{+1}[\text{S}]}{\rightleftharpoons}} \text{ES} \overset{k_{+2}}{\rightarrow} \text{E} \tag{4.2}$$

and the kinetic equations for E and ES are

$$\frac{d[\text{E}]}{dt} = -k_{+1}[\text{S}][\text{E}] + (k_{-1} + k_{+2})[\text{ES}] \tag{4.3a}$$

$$\frac{d[\text{ES}]}{dt} = k_{+1}[\text{S}][\text{E}] - (k_{-1} + k_{+2})[\text{SE}]. \tag{4.3b}$$

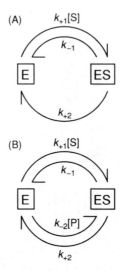

Figure 4.2 Kinetic mechanism of a Michaelis–Menten enzyme. (A) The reaction mechanism for the irreversible case – Equation (4.1) – is based on a single intermediate-state enzyme complex (ES) and an irreversible conversion from the complex to free enzyme E and product P. (B) The reaction mechanism for the reversible case – Equation (4.7) – includes the formation of ES complex from free enzyme and product P. For both the irreversible and reversible cases, the reaction scheme is illustrated as a catalytic cycle.

Here, $k_{+1}[S]$ serves as the apparent mass-action rate constant for the conversion E $\rightarrow$ ES. Each time an enzyme cycles from state E to ES and back to E again, one molecule of S is converted to P. If the rate of turnover of the catalytic cycle is significantly greater than the rate of change of reactant (S and P) concentrations, then the apparent mass-action constant $k_{+1}[S]$ in Equation (4.2) remains effectively constant over the timescale of the catalytic cycle. This is true, for example, when the enzyme concentration is small compared to reactant concentrations, many catalytic cycles are required to produce a significant change in reactant concentrations.

Since the catalytic cycle operates with relatively rapid kinetics, E and ES will obtain a steady state governed by Equations (4.2) and (4.3) and the quasi-steady state concentrations of enzyme and complex will change rapidly in response to relatively slow changes in [S]. Thus the quasi-steady approximation is justified based on a difference in timescales between the catalytic cycle kinetics and the overall rate of change of biochemical reactions.

The steady state populations for [E] and [ES] are readily obtained by setting $\frac{d[E]}{dt} = \frac{d[ES]}{dt} = 0$:

$$[E] = \frac{k_{-1} + k_{+2}}{k_{-1} + k_{+2} + k_{+1}[S]} E_o, \quad [ES] = \frac{k_{+1}[S]}{k_{-1} + k_{+2} + k_{+1}[S]} E_o, \quad (4.4)$$

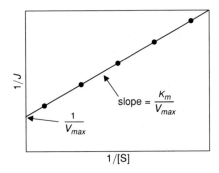

Figure 4.3 Lineweaver–Burk, or double-reciprocal, plot for single-substrate irreversible Michaelis–Menten enzyme. A plot of $1/J$ versus $1/[S]$ yields estimates of V_{max} and K_M, as illustrated in the figure.

where $E_o = E + ES$: the total enzyme concentration. The steady state flux, which is the flux for the cycle $E \rightarrow ES \rightarrow E$, is the rate of product formation:

$$J = k_{+2}[ES] = k_{+1}[S][E] - k_{-1}[ES] = \frac{k_{+1}k_{+2}[S]E_o}{k_{-1} + k_{+2} + k_{+1}[S]}. \qquad (4.5)$$

Equation (4.5), which can be rewritten

$$J = \frac{V_{max}[S]}{K_M + [S]}, \qquad (4.6)$$

which is the by-now familiar Michaelis–Menten equation, exhibiting a hyperbolic dependence of J as a function of $[S]$. The Michaelis–Menten parameters are $V_{max} = k_{+2}E_o$ and $K_M = (k_{-1} + k_{+2})/k_{+1}$, which may be estimated from the so-called Lineweaver–Burk (or double-reciprocal) plot of $1/J$ versus $1/[S]$, as illustrated in Figure 4.3 [132]. The quantity $1/J$ varies with $1/[S]$ according to $1/J = 1/V_{max} + (K_m/V_{max}) \cdot (1/[S])$. Thus the slope and intercept of the Lineweaver–Burk plot provide estimates of V_{max} and K_M as illustrated in the figure.[1]

4.1.2 Reversible Michaelis–Menten kinetics

The derivation of the Michaelis–Menten equation in the previous section differs from the standard treatment of the subject found in most textbooks in that the quasi-steady approximation is justified based on the argument that the catalytic cycle kinetics is rapid compared to the overall biochemical reactant kinetics. In

---

[1] Note that V_{max} and K_M may be estimated from data on steady state flux and substrate concentration based on a number of different ways of plotting J and $[S]$. Cornish-Bowden illustrates that the Lineweaver–Burk plot (or *double-reciprocal* plot) is not recommended when one would like to minimize the effect of experimental error on parameter estimates. For a detailed discussion see Section 2.6 of [35].

Section 4.2 we explore the quasi-steady approximation with somewhat more mathematical rigor. However, before undertaking that analysis, let us analyze the reversible enzyme mechanism studied in Chapter 3 from the perspective of cycle kinetics.

In Chapter 3 we determined the conditions under which it is and is not appropriate to treat a reaction as irreversible. Using the notation of cycle kinetics and apparent mass-action constants, the reversible mechanism of Equation (3.25) is represented

$$\text{E} \underset{k_{-1}}{\overset{k_{+1}[\text{S}]}{\rightleftharpoons}} \text{ES} \underset{k_{-2}[\text{P}]}{\overset{k_{+2}}{\rightleftharpoons}} \text{E}. \tag{4.7}$$

Assuming again that the cycle kinetics are rapid and maintain enzyme and complex in a rapid quasi-steady state, we can obtain the steady state velocity for the reversible Michaelis–Menten enzyme kinetics:

$$J = \frac{(k_{+1}k_{+2}[\text{S}] - k_{-1}k_{-2}[\text{P}])\text{E}_o}{k_{-1} + k_{+2} + k_{+1}[\text{S}] + k_{-2}[\text{P}]}. \tag{4.8}$$

With the Michaelis–Menten parameters given by

$$V_{max}^f = k_{+2}\text{E}_o, \quad V_{max}^r = k_{-1}\text{E}_o, \quad K_{M.S} = \frac{k_{-1} + k_{+2}}{k_{+1}}, \quad K_{M.P} = \frac{k_{-1} + k_{+2}}{k_{-2}}, \tag{4.9}$$

we have

$$J = \frac{\frac{V_{max}^f[\text{S}]}{K_{M.S}} - \frac{V_{max}^r[\text{P}]}{K_{M.P}}}{1 + \frac{[\text{S}]}{K_{M.S}} + \frac{[\text{P}]}{K_{M.P}}}. \tag{4.10}$$

These expressions are equivalent to those derived in Section 3.1.3.2.

One of the important predictions from Equation (4.10) is the Haldane relation, which states that

$$\frac{V_{max}^f/K_{M.S}}{V_{max}^r/K_{M.P}} = \frac{k_{+1}k_{+2}}{k_{-1}k_{-2}} = K_{eq}, \tag{4.11}$$

where K_{eq} is the equilibrium constant between P and S. This relation follows from the fact that $J = 0$ when P and S are in chemical equilibrium. Since the apparent equilibrium constant for S $\rightleftharpoons$ P is independent of the enzyme, the feasible values of the kinetic parameters for the enzyme mechanism are constrained by the equilibrium thermodynamics of the reaction.

4.1.3 Non-equilibrium steady states and cycle kinetics

As discussed in Chapter 3, living cells exist away from thermodynamic equilibrium. When a reaction such as S $\rightleftharpoons$ P is maintained in a steady state, we refer to this state as a non-equilibrium steady state. Such a state is characterized by non-zero net flux, which necessarily is maintained by input of substrate and removal of product from the system. When the reaction is part of a network of coupled reactions, the whole system is sustained by input and output of material across its boundary. In other words, biological systems are *open systems* – open to the exchange of material and energy with their environments.

For the example of the reversible Michaelis–Menten enzyme catalyzing $S \rightleftharpoons P$, in the steady state S and P are transported into and out of the system at the constant rate J. The positive and negative fluxes of the catalytic cycle are given by

$$J^+ = \frac{\frac{V^f_{max}[S]}{K_{M,S}}}{1 + \frac{[S]}{K_{M,S}} + \frac{[P]}{K_{M,P}}} \tag{4.12a}$$

$$J^- = \frac{\frac{V^r_{max}[P]}{K_{M,P}}}{1 + \frac{[S]}{K_{M,S}} + \frac{[P]}{K_{M,P}}}, \tag{4.12b}$$

where the net flux is given by $J = J^+ - J^-$. These positive and negative fluxes correspond to the positive and negative cycle fluxes illustrated in Figure 4.2B. The forward cycle flux J^+ is the rate at which the cycle $E \rightarrow ES \rightarrow E$ proceeds in the clockwise direction; the reverse cycle flux is the rate at which the cycle proceeds in the counterclockwise direction.

From Equations (4.12), the identity of Equation (3.12) is obeyed: $\Delta G = -RT \ln(K_{eq}[S]/[P]) = -RT \ln(J^+/J^-)$. From this identity follows the corollary

$$J \cdot \Delta G = -(J^+ - J^-) \cdot RT \ln \left(\frac{J^+}{J^-} \right) \le 0. \tag{4.13}$$

The equality in Equation (4.13) holds if and only if $J = 0$ and $\Delta G = 0$, i.e., when the reaction is in equilibrium.

When an enzyme-catalyzed biochemical reaction operating in an isothermal system is in a non-equilibrium steady state, energy is continuously dissipated in the form of heat. The quantity $J \cdot \Delta G$ is the rate of heat dissipation per unit time. The inequality of Equation (4.13) means that the enzyme can extract energy from the system and dissipate heat and that an enzyme cannot convert heat into chemical energy. This fact is a statement of the second law of thermodynamics, articulated by William Thompson (who was later given the honorific title Lord Kelvin), which states that with only a single temperature bath T, one may convert chemical work to heat, but not vice versa.

In Chapter 9 we will see that the second-law inequality of Equation (4.13) will form a cornerstone of the constraint-based approach to analyzing biochemical networks.

4.2 Transient enzyme kinetics

The quasi-steady approximation, which was introduced in Section 3.1.3.2 and justified on the basis of rapid cycle kinetics in Section 4.1.1, forms the basis of the study of enzyme mechanisms, a field with deep historical roots in the subject of biochemistry. In later chapters of this book, our studies make use of this approximation in building models of biochemical systems. Yet there remains something unsatisfying about the approximation. We have seen in Section 3.1.3.2 that the approximation is not perfect. Particularly during short-time transients, the quasi-steady approximation deviates significantly from the full kinetics of the Michaelis–Menten system described by Equations (3.25)–(3.27). Here we mathematically analyze the short timescale kinetics of the Michaelis–Menten system and reveal that a different quasi-steady approximation can be used to simplify the short-time kinetics.

4.2.1 Rapid pre-equilibrium

If in the mechanism of Equation (4.1) the first-order rate constant k_{+2} is much smaller than the first-order constant k_{-1}, then the association reaction $E + S \rightleftharpoons ES$ occurs on a much faster timescale than the transformation $ES \rightarrow E + P$. Hence, the first reaction may be assumed to achieve a rapid equilibrium

$$[ES] = \frac{k_{+1}}{k_{-1}}[E][S] = \frac{k_{+1}}{k_{-1}}(E_o - [ES])[S]. \tag{4.14}$$

Solving for [E] and [ES] yields:

$$[E] = \frac{k_{-1}E_o}{k_{+1}[S] + k_{-1}}, \quad [ES] = \frac{k_{+1}[S]E_o}{k_{+1}[S] + k_{-1}}. \tag{4.15}$$

The timescale for the association step $E + S \rightleftharpoons ES$ is $(k_{+1}[S] + k_{-1})^{-1}$ (which is even smaller than $1/k_{-1}$), while the timescale for the transformation $ES \rightarrow E + P$ is $1/k_{+2}$. Thus, if $k_{-1} \gg k_{+2}$, the association step equilibrates rapidly compared to the rate of product formation.

Under the assumption of rapid pre-equilibrium, the rate of product formation is then

$$J = k_{+2}[ES] = \frac{k_{+1}k_{+2}[S]E_o}{k_{+1}[S] + k_{-1}}. \tag{4.16}$$

In this case the Michaelis–Menten parameters are $K_M = k_{-1}/k_{+1}$, which is the dissociation constant, and $V_{max} = k_{+2}E_o$. Here we have arrived at the familiar Michaelis–Menten result without making the assumption that the substrate concentration [S] is in excess of the enzyme concentration. Recall that the quasi-steady analysis yielded $K_M = (k_{-1} + k_{+2})/k_{+1}$. Thus when $k_{-1} \gg k_{+2}$, the two approaches yield essentially the same result.

Kinetics involving rapid pre-equilibrium steps finds numerous applications both within and beyond the study of enzyme kinetics. Other important examples are the theory of proton–deuterium exchange kinetics of a protein [169] and gene activation involving DNA looping [186]. Because of its central importance in biological kinetics, let us provide a more complete mathematical treatment of the problem in a short digression.

We consider the simple kinetics of

$$A \underset{\beta}{\overset{\alpha}{\rightleftharpoons}} B \overset{\gamma}{\rightarrow} C. \tag{4.17}$$

The kinetic equations are

$$\frac{d}{dt} \begin{pmatrix} [A] \\ [B] \\ [C] \end{pmatrix} = \begin{pmatrix} -\alpha & \beta & 0 \\ \alpha & -(\beta + \gamma) & 0 \\ 0 & \gamma & 0 \end{pmatrix} \begin{pmatrix} [A] \\ [B] \\ [C] \end{pmatrix}. \tag{4.18}$$

This system has two eigenvalues:

$$\lambda_{1,2} = \frac{1}{2}\left(\alpha + \beta + \gamma \pm \sqrt{(\alpha + \beta + \gamma)^2 - 4\alpha\gamma}\right). \tag{4.19}$$

A third eigenvalue is zero, due to the sum $[A] + [B] + [C]$ being constant in the kinetics. Now if $\beta \gg \gamma$, then

$$
\begin{aligned}
\lambda_{1,2} &= \frac{1}{2}\left(-\alpha - \beta - \gamma \pm \sqrt{(\alpha + \beta + \gamma)^2 - 4\alpha\gamma}\right) \\
&\approx -\frac{\alpha + \beta + \gamma}{2} \pm \left(\frac{\alpha + \beta + \gamma}{2} - \frac{\alpha\gamma}{(\alpha + \beta)}\right) \\
\lambda_1 &\approx -(\alpha + \beta), \quad \lambda_2 \approx -\frac{\alpha\gamma}{\alpha + \beta}.
\end{aligned} \tag{4.20}
$$

Kinetics on the timescale of rapid equilibrium (determined by α and β) is governed by the larger eigenvalue λ_1. The slower timescale is governed by λ_2, which is equal to γ multiplied by the rapid equilibrium fraction of β.

This simple example shows us that when rapid pre-equilibrium steps exist, one simply solves the equilibrium among the rapid steps and calculates equilibrium fractions of each state. When doing this, one neglects all the slow steps (i.e., γ), and the fast eigenvalue(s) λ_1 determine(s) the relaxation rate for the rapid $A \rightleftharpoons B$ when $\gamma = 0$. In the above example, the equilibrium fractions are $\frac{\alpha}{\alpha+\beta}$ and $\frac{\beta}{\alpha+\beta}$. Then for the slower kinetics one can simply lump all the states in pre-equilibrium together as a single state, with rate constant(s) being the average(s) according to the pre-equilibrium. That is, λ_2 is the rate constant for $(A \cup B) \rightarrow C$:

$$\frac{d[C]}{dt} = \gamma[B] = \frac{\gamma[B]}{[A] + [B]}[A \cup B] = \lambda_2[A \cup B]. \tag{4.21}$$

Approaching problems of this sort based on separation of timescales can be powerful. The following analysis is based on a formalization of this kind of analysis.

4.2.2 A singular perturbation approach to Michaelis–Menten kinetics

The flux expression in Equation (4.16) displays the canonical Michaelis–Menten hyperbolic dependence on substrate concentration [S]. We have shown that this dependence can be obtained from either rapid pre-equilibration or the assumption that $[S] \gg [E]$. The rapid pre-equilibrium approximation was the basis of Michaelis and Menten's original 1913 work on the subject [140]. In 1925 Briggs and Haldane [24] introduced the quasi-steady approximation, which follows from $[S] \gg [E]$. (In his text on enzyme kinetics [35], Cornish-Bowden provides a brief historical account of the development of this famous equation, including outlines of the contributions of Henri [80, 81], Van Slyke and Cullen [203], and others, as well as those of Michaelis and Menten, and Briggs and Haldane.)

A more cogent mathematical treatment of this problem was given in the 1970s by several mathematical biologists. For details see books by Lin and Segel [130] and Murray [146]. Here we provide a brief account of this approach. The approach uses the somewhat advanced mathematical method of singular perturbation analysis, but does provides a deep appreciation of the Michaelis–Menten enzyme kinetics.

Let us return to the kinetics of Equation (4.3). From this mechanism we have the equations

$$\frac{d[S]}{dt} = -k_{+1}[S][E] + k_{-1}[ES], \tag{4.22a}$$

$$\frac{d[ES]}{dt} = k_{+1}[S][E] - (k_{-1} + k_{+2})[ES]. \tag{4.22b}$$

The total enzyme concentration is $[E] + [ES] = E_o$, and the initial concentrations are assumed to be $[E] = E_o$ and $[S] = S_o$ at $t = 0$. Thus at $t = 0$, substrate is present in concentration S_o and the enzyme complex concentration [ES] is zero.

For a mathematically convenient analysis of these equations, we express the concentrations of E and S as *unitless* quantities. Specifically, we express their concentrations relative to the initial contractions: $u = [S]/S_o$ and $v = [ES]/E_o$.

From these definitions, Equations (4.22) are recast

$$\frac{du}{d\tau} = -u(1 - v) + \left(\frac{k_{-1}}{k_{+1}S_o}\right)v, \tag{4.23a}$$

$$\epsilon\frac{dv}{d\tau} = u(1 - v) - \frac{K_M}{S_o}v, \tag{4.23b}$$

where $\tau = k_{+1}E_o t$ and $\epsilon = E_o/S_o$. The variables u and v are unitless state variables representing substrate and enzyme complex levels and τ is the unitless time variable.

If $\epsilon \ll 1$ then the second equation in Equation (4.23) reaches steady state much faster than the first equation. Hence, $u(1 - v) - (K_M/S_o)v = 0$, which yields

$$v = \frac{uS_o}{K_M + uS_o}. \tag{4.24}$$

This implies that when u changes with time, v follows Equation (4.24) nearly instantaneously. Substituting Equation (4.24) into the first equation in (4.23) we have

$$\frac{du}{d\tau} = -\left(\frac{k_{+2}}{k_{+1}}\right)\frac{u}{K_M + uS_o}. \tag{4.25}$$

In terms of the original variables with physical units, we have

$$\frac{d[S]}{dt} = -\frac{k_{+2}E_o[S]}{K_M + [S]} = -J([S]) \tag{4.26}$$

where J, as a function of $[S]$, is the familiar function of Equation (4.6). Equation (4.25) can be solved to obtain $u(\tau)$, which can then be substituted into Equation (4.24) to obtain $v(\tau)$.

However, the above approach remains incomplete: at $\tau = 0$ (and $t = 0$), the initial conditions that we have imposed are $u(0) = 1$ and $v(0) = 0$ (or $[S] = S_o$ and $[ES] = 0$). However these initial conditions are inconsistent with Equation (4.24), $v(0) = S_o/(K_M + S_o) \neq 0$. This conflict arises from setting the *small* (but finite) ϵ equal to zero in Equation (4.23). In doing so, we have neglected the early kinetics in which $dv/d\tau$ may be large compared to $du/d\tau$. Hence at $\tau = 0$, $\epsilon(dv/d\tau)$ is not small!

To remedy this inconsistency, applied mathematicians have developed an elegant approach. The idea is to analyze Equations (4.23) in a short timescale on the order of ϵ. If we let $\hat{\tau} = \tau/\epsilon$, then we have

$$\frac{du}{d\hat{\tau}} = \epsilon\left(-u(1 - v) + \left(\frac{k_{-1}}{k_{+1}S_o}\right)v\right), \tag{4.27a}$$

$$\frac{dv}{d\hat{\tau}} = u(1 - v) - \frac{K_M}{S_o}v. \tag{4.27b}$$

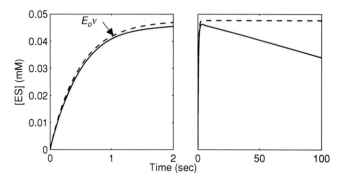

Figure 4.4 Plot of enzyme complex concentration as a function of time for the Michaelis–Menten mechanism of Equations (4.22). The concentration of ES predicted from a kinetic simulation of Equations (4.22) is plotted as a solid line. The parameter values used are $k_{+1} = 1000 \, \mathrm{M}^{-1} \, \mathrm{sec}^{-1}$, $k_{-1} = 1.0 \, \mathrm{sec}^{-1}$, $k_{+2} = 0.1 \, \mathrm{sec}^{-1}$, and $E_o = 0.1 \, \mathrm{mM}$. The left plot illustrates the fast-time kinetics. The fast-time variable $v(\hat{\tau})$ predicted by Equation (4.29) is plotted as a dashed line.

Now if we now let $\epsilon \to 0$, we are saying that $du/d\hat{\tau} = 0$. In other words, on the short timescale, $u(\hat{\tau}) = 1$ remains essentially constant, while $v(\hat{\tau})$ changes with time $\hat{\tau}$. Therefore, we have

$$\frac{dv}{d\hat{\tau}} = (1 - v) - \frac{K_M}{S_o} v, \qquad (4.28)$$

which has the solution (with initial value $v(0) = 0$) of

$$v(\hat{\tau}) = \frac{S_o}{K_M + S_o} \left(1 - e^{-\frac{K_M + S_o}{S_o} \hat{\tau}} \right). \qquad (4.29)$$

Notice that when $\hat{\tau} \to \infty$, $v \to S_o/(K_M + S_o)$. This is exactly the value of v that we arrived at for $\tau = 0$. Thus as $\hat{\tau} \to \infty$ (on the fast timescale), v approaches the derived initial condition for the slow timescale ($\tau = 0$). Hence, the entire transient for Michaelis–Menten kinetics can be represented by combining the short timescale result, Equation (4.29), with the long timescale result, the solution to Equation (4.25). The two results match seamlessly at $\hat{\tau} = \infty$ and $\tau = 0$. This is known as asymptotic matching in singular perturbation analysis [110].

Figure 4.4 illustrates how the fast time kinetics of this system are represented by the approximation of Equation (4.29). In the figure the enzyme complex concentration as a function of time is plotted based on a numerical simulation of the reactions of Equation (4.22). The parameter values are identical to those used in the simulation of the irreversible system in Section 3.1.3.3 of Chapter 3. (See Figure 3.5 for the corresponding plot of substrate and product concentrations.) The ES concentration predicted by Equation (4.29) approaches the value $E_o S_o/(S_o + K_m) \approx 0.048 \, \mathrm{mM}$ with the exponential time constant of $1/(k_{+1}(K_m + S_o))$, which is approximately equal to 0.48 seconds for the parameter values listed in the figure legend.

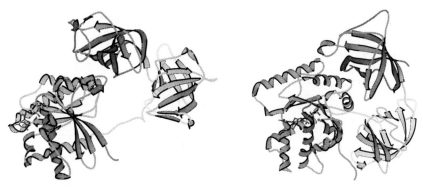

Figure 4.5 Conformational change of an enzyme that is responsible for elongation in biosynthesis in bacteria: EF-Tu (elongation factor-Tu). The protein can bind either GDP (left) or GTP (right). There are two conformations of the protein associated with the two different nucleotides. In biochemical modeling, different conformational states of the enzyme are usually modeled as two distinct states or species. Image on the left is from the GDP bound structure published by Heffron *et al.* [79]. Image on the right is from a structure bound to the GTP analog GPPNHP [84]. Both structure images rendered with the KiNG program.

4.3 Enzyme with multiple binding sites: cooperativity

Cooperativity is an important concept that links enzyme kinetics at the molecular level with functional biological processes at the cellular level. An essential idea in cooperative enzyme kinetics is allosteric regulation, which involves binding of ligands to enzymes resulting in changes in the enzymes' molecular conformations, as illustrated in Figure 4.5. The different conformational states of an enzyme, which affect its function, can be observed using techniques such as x-ray crystallography [47]. Historically, the idea of allosteric regulation originated with Jacques Monod, a pioneer in the field of molecular genetics [143].

The tendency of proteins in biological systems to exist in a number of well-defined discrete conformational states sets the study of protein kinetics apart from the general theory of polymer physics, such as presented by Grosberg and Khokhlov [73]. A rich theory of biological polymers, including of proteins and their conformational transitions, is presented in Cantor and Schimmel's *Biophysical Chemistry* [27].

4.3.1 Sigmoidal equilibrium binding

To illustrate the phenomenon of cooperative binding, let us consider an enzyme E, which has two binding sites for its substrate S:

$$E + S \overset{K_1}{\rightleftharpoons} ES_1; \quad ES_1 + S \overset{K_2}{\rightleftharpoons} ES_2. \tag{4.30}$$

The upper-case K_1 and K_2 are equilibrium association constants for the reactions. The fraction of sites occupied as a function of the concentration of the free S is determined based on the equilibrium expressions:

$$\frac{[ES_1]}{[E][S]} = K_1,$$

$$\frac{[ES_2]}{[ES_1][S]} = K_2,$$

$$([E] + [ES_1] + [ES_2]) = E_o.$$

The fraction of enzyme binding sites that are occupied is calculated

$$f_b = \frac{[ES_1] + 2[ES_2]}{2E_o} = \frac{K_1[S] + 2K_1K_2[S]^2}{2(1 + K_1[S] + K_1K_2[S]^2)}. \tag{4.31}$$

Two special cases of Equation (4.31) are particularly worth mentioning: (i) If the two sites are identical and independent, then $K_1 = 2K$ and $K_2 = K/2$, where K is the association constant for a single binding site. (K_1 is equal to $2K$ because there are two sites for binding the first substrate; $K_2 = K/2$ because both sites can release a substrate.) In this case we have

$$f_b = \frac{K_1[S] + 2K_1K_2[S]^2}{2(1 + K_1[S] + K_1K_2[S]^2)} = \frac{K[S]}{1 + K[S]} \tag{4.32}$$

(ii) The second situation is $K_1 \approx 0$ while $K_2 \approx \infty$ such that $K_1K_2 = K^2$ is finite. In this case the two sites are highly cooperative: either none binds or both bind to substrate. Here we have

$$f_b = \frac{K_1[S] + 2K_1K_2[S]^2}{2(1 + K_1[S] + K_1K_2[S]^2)} = \frac{K^2[S]^2}{1 + K^2[S]^2} \tag{4.33}$$

We see that, as a function of [S], Equation (4.33) has sigmoidal shape. This is the hallmark of the cooperativity: the fraction of site occupied has a more sharp response to changes in [S] compared to the case of independent identical binding.

4.3.2 Cooperativity in enzyme kinetics

Equation (4.31) gives us the equilibrium binding in the case of dual cooperativity, yet it does not tell us what cooperative binding has to do with enzyme kinetics. To illustrate the role of cooperativity in enzyme kinetics, consider the following

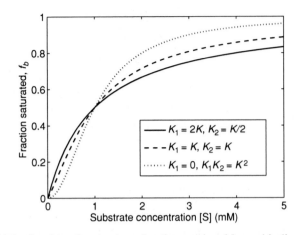

Figure 4.6 The fraction of saturation, f_b, of a protein with two binding sites binds its substrate as a function of the substrate concentration [S], according to Equation (4.31). Plots are for $K = 1$ mM. The solid line represents two independent sites and the dotted line represents two highly cooperative sites.

kinetic scheme:

$$E \underset{}{\overset{S}{\rightleftharpoons}} ES_1 \underset{}{\overset{S}{\rightleftharpoons}} ES_2 \xrightarrow{k_{f2}} ES_1 + P$$

$$k_{f1} \downarrow$$

$$E + P$$

where the enzyme E is assumed to have two catalytic sites. Product is generated from ES_1 with rate constant k_{f1} and from ES_2 with rate constant k_{f2}. If $k_{f1} = k_{f2}/2 = k_f/2$ and the two binding steps are modeled using rapid pre-equilibrium, then the flux expression will have the form $J = k_f E_o f_b([S])$, where $f_b([S])$ is given by Equation (4.33). If $k_{f1} \neq k_{f2}$, then the general form

$$J = \frac{a[S] + b[S]^2}{1 + c[S] + d[S]^2}$$

is still obeyed, where a, b, c, and d are constants.

4.3.3 The Hill coefficient

Figure 4.6 illustrates the effects of cooperativity on binding saturation. The level of cooperativity can be quantified in terms of the steepness of the binding curve at its 50 percent-saturation level. A parameter widely used to characterize the steepness of this curve is the Hill coefficient, which is defined as

$$h_n = 2 \left(\frac{d \ln f_b}{d \ln[S]} \right)_{f_b=0.5} \tag{4.34}$$

Thus the Hill coefficient is essentially a measure of the sensitivity of the binding saturation to the substrate concentration. For the saturation curve of Equation (4.33), $h_n = 2$, corresponding to theoretical maximum for the case of two binding sites.

4.3.4 Delays and hysteresis in transient kinetics

The previous section illustrated how allosteric cooperativity can result in a sigmoidal relationship between binding saturation and substrate concentration. In this section, we demonstrate how a sigmoidal relationship between product concentration and time can arise from enzyme kinetics with time lags.

Recall that in the standard Michaelis–Menten enzyme kinetics we approximate the kinetics of substrate and product using Equation (3.32) or (4.26) for the essentially irreversible case:

$$\frac{d[P]}{dt} = -\frac{d[S]}{dt} = \frac{k_{+2}E_o[S]}{K_M + [S]}. \tag{4.35}$$

Taking the second derivative of [P], we obtain

$$\frac{d^2[P]}{dt^2} = \frac{k_{+2}E_o K_M}{(K_M + [S])^2} \left(\frac{d[S]}{dt} \right) < 0.$$

Thus the initial phase of the [P] as a function of t is an increasing function with negative curvature. However, for certain enzyme kinetics, [P] as a function of t has a positive curvature, a lag, in its initial phase. This phenomenon is known as hysteresis, first discovered by Carl Frieden [60].

Strictly speaking, any multi-step kinetic scheme will involve a lag. However, realistically observing hysteresis in enzyme kinetics is always associated with the existence of one of several slow step(s) prior to the final step. This is because if all the steps prior to the final step were fast, then there would be a rapid pre-equilibriation and the rapid steps could be lumped into a single kinetic species (see Section 4.2.1).

To illustrate quantitatively the above discussion, we can again use the simple example given in Equation (4.17):

$$A \underset{\beta}{\overset{\alpha}{\rightleftharpoons}} B \overset{\gamma}{\to} C.$$

The kinetics of the appearance of C is given by

$$c(t) = 1 - \frac{\lambda_2 e^{-\lambda_1 t} - \lambda_2 e^{-\lambda_2 t}}{\lambda_2 - \lambda_1},$$ (4.36)

in which $\lambda_{1,2}$ are given in Equation (4.19) and $c = [C]$. The initial phase of $c(t)$ in Equation (4.36) has slope zero and curvature $\lambda_1 \lambda_2 = \alpha \gamma$.[2] Since both α and γ are positive, the initial curvature is positive. Hence, theoretically $c(t)$ always has a lag.

Realistically, however, this lag phase can be too small to be observed. To see how significant the lag phase is, we compute the inflection point at which the curvature turns from positive to negative. Setting $d^2 c(t^*)/dt^2 = 0$, we solve the inflection point $t^* = \frac{1}{\lambda_1 - \lambda_2} \ln(\frac{\lambda_1}{\lambda_2})$. (Whether one can observe a lag phase depends on whether the time resolution of the measurement is greater than t^*.)

Here we can see a pattern in the mathematics. Note that the initial slope of $c(t)$ is $\frac{dc(0)}{dt} = \gamma b(0)$. Hence, if $b(0) = 0$, then $\frac{dc(0)}{dt} = 0$, and $\frac{d^2 c(0)}{dt^2} = \gamma \frac{db(0)}{dt} = \gamma \alpha a(0)$. By the same argument, if a kinetic process starts with a species that is more than two steps away from the species C, then even the initial curvature in $c(t)$ will be zero.

An example of a kinetic scheme that displays hysteresis occurs when a slow conformation change between two enzyme–substrate complexes is required before the product can be released:

$$E + S \underset{k_{-1}}{\overset{k_{+1}}{\rightleftharpoons}} ES \underset{k_{-2}}{\overset{k_{+2}}{\rightleftharpoons}} (ES)^* \overset{k_{+3}}{\rightarrow} E + P.$$ (4.37)

Invoking the quasi-steady approximation, the steady state flux expression for this system can be shown to be:[3]

$$J_{SS} = \frac{k_{+3} E_o [S]}{[S] + K_S + K_o [S]}$$ (4.38)

where

$$E_o = [E] + [ES] + [(ES)^*]$$
$$K_S = \frac{k_{-1} k_{-2} + k_{+2} k_3 + k_{-1} k_3}{k_{+1} k_{+2}}$$
$$K_o = \frac{k_{+3} + k_{-2}}{k_{+2}}.$$ (4.39)

A simulation of this system is illustrated in Figure 4.7. The solid lines show the predicted time courses for [S] and [P] given initial conditions of [S] = 1 mM, and [P] = [ES] = [(ES)*] = 0 mM and parameter values listed in the figure legend. The

---

[2] See exercise 5.
[3] See exercise 6.

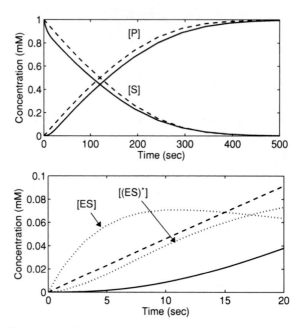

Figure 4.7 Simulation of the kinetics of the system of Equation (4.37). The solid lines show the predicted time courses for [S] and [P] given initial conditions of [S] = 1 mM, and [P] = [ES] = [(ES)*] = 0 mM. The dashed lines correspond to a simulation using the quasi-steady approximation, Equation (4.39). The parameter values used are $k_{+1} = 100\,\mathrm{M^{-1}\,sec^{-1}}$, $k_{-1} = 0.01\,\mathrm{sec^{-1}}$, $k_{+2} = 0.1\,\mathrm{sec^{-1}}$, $k_{-2} = 0.01\,\mathrm{sec^{-1}}$, $k_{+3} = 0.05\,\mathrm{sec^{-1}}$, and $E_o = 0.2\,\mathrm{mM}$.

dashed lines in the upper panel correspond to a simulation based on the steady state flux expression of Equation (4.38):

$$\frac{d[\mathrm{S}]}{dt} = -J_{SS}$$
$$\frac{d[\mathrm{P}]}{dt} = +J_{SS}. \tag{4.40}$$

Enzyme binding and conformational change cause a delay of approximately ten seconds in the appearance of product in the simulations based on the full kinetics of Equation (4.37) compared to the simplified kinetics of Equation (4.40). This delay results in a positive second derivative for [P] at early times.

4.4 Enzymatic fluxes with more complex kinetics

Detailed kinetic analyses of mechanisms such as in Equation (4.37) and the determination of flux expression such as Equation (4.38) is a central theme of several treatises on enzyme kinetics. Important texts include Segel's encyclopedic *Enzyme Kinetics* [183] and Cornish-Bowden's cogent *Fundamentals of Enzyme Kinetics* [35]. The essence of the field is to determine what mechanism is consistent (and which mechanisms are inconsistent) with the available data on a given enzyme. In the era before scientific computing became widely available this process was daunting. With the power of computing and numerical solutions to differential equations, we are able to compare quickly the dynamics predicted by a particular kinetic scheme with available experimental data. However, likely due to increased attention to reductionist molecular biology and qualitative methods in recent decades, the attention paid to enzyme kinetic studies has diminished since the 1970s. It is expected that in the coming years attention will be returned to this subject due to its central role in understanding the behavior of biological systems.

Often the key entity one is interested in obtaining in modeling enzyme kinetics is the analytical expression for the turnover flux in quasi-steady state. Equations (4.12) and (4.38) are examples. These expressions are sometimes called Michaelis–Menten "rate laws." Such expressions can be used in simulation of cellular biochemical systems, as is the subject of Chapters 5, 6, and 7 of this book. However, one must keep in mind that, as we have seen, these rates represent approximations that result from simplifications of the kinetic mechanisms. We typically use the approximate Michaelis–Menten-type flux expressions rather than the full system of equations in simulations for several reasons. First, often the quasi-steady rate constants (such as K_S and K_o in Equation (4.38)) are available from experimental data while the mass-action rate constants (k_{+1}, k_{-1}, etc.) are not. In fact, it is possible for different enzymes with different detailed mechanisms to yield the same Michaelis–Menten rate expression, as we shall see below. Second, in metabolic reaction networks (for example), reactions operate near steady state in vivo. Kinetic transitions from one in vivo steady state to another may not involve the sort of extreme shifts in enzyme binding that have been illustrated in Figure 4.7. Therefore the quasi-steady approximation (or equivalently the approximation of rapid enzyme turnover) tends to be reasonable for the simulation of in vivo systems.

4.4.1 Reciprocal of flux: the mean time of turnover

As we have seen, the catalytic cycle flux provides a useful metric for analyzing enzyme kinetics. In this section, we analyze the turnover time for catalytic cycles and show that the quasi-steady rate law arises from the mean cycle time [151]. In addition, we show that for arbitrary mechanisms for a single-substrate reaction, the steady state rate law can always be expressed using the Michaelis–Menten form of Equation (4.5) when the product unbinding step(s) is (are) approximated as irreversible.

Consider the irreversible catalytic cycle illustrated in Figure 4.2A. We can compute the mean time it takes to transition from state E to ES and back to E again as the sum of the mean dwell time in state E and the mean dwell time in state ES:

$$T_{\text{E}\to\text{ES}\to\text{E}} = \frac{1}{k_{+1}[\text{S}]} + \frac{1}{k_{-1} + k_{+2}}. \tag{4.41}$$

However, this expression does not provide the mean cycle time. Since a fraction of the transitions ES $\to$ E are through the reverse binding step ES $\to$ E + S, the mean cycle time is longer than $T_{\text{E}\to\text{ES}\to\text{E}}$ computed by the above equation.

The probability that a transition ES $\to$ E is through the reaction ES $\to$ E + P (rather than through ES $\to$ E + S) is given by:

$$P_{\text{forward}} = \frac{k_{+2}}{k_{+1} + k_{+2}} \tag{4.42}$$

and the mean cycle time is:

$$T_{\text{cycle}} = \frac{T_{\text{E}\to\text{ES}\to\text{E}}}{P_{\text{forward}}}. \tag{4.43}$$

Combining these equations, we get

$$T_{\text{cycle}} = \frac{k_{-1} + k_{+2}}{k_{+1}k_{+2}[\text{S}]} + \frac{1}{k_{+2}}, \tag{4.44}$$

which is the mean time of turnover of a single enzyme; $1/T_{\text{cycle}}$ is the rate of turnover, and E_o/T_{cycle} gives the flux of Equation (4.5).

For a more complex mechanism the expressions for K_M and V_{max} might differ, but $T_{\text{cycle}} = \frac{1}{V_{max}}(\frac{k_M}{[\text{S}]} + 1)$ still holds true for any irreversible reaction with a single substrate–enzyme binding step. In this case there may be several inter-conversion steps between different enzyme–substrate complexes. Yet since there is only a single substrate binding step, the only term that involves [S] in the T_{cycle} has to be proportional to $1/(k_{+1}[\text{S}])$. Hence the general form of Michaelis–Menten-type kinetics.

Next, we can use the mean turnover time to analyze the reversible catalytic cycle illustrated in Figure 4.2B. For this case the mean dwell time in state E plus the mean dwell time in state ES is

$$T_{E \to ES \to E} = \frac{1}{k_{+1}[S] + k_{-2}[P]} + \frac{1}{k_{-1} + k_{+2}}. \qquad (4.45)$$

In this case there is a total of four pathways for transition $E \to ES \to E$. The transition time $T_{E \to ES \to E}$ is the forward cycle time multiplied by the probabilities that the transitions $E \to ES$ and $ES \to E$ are in the forward (clockwise in Figure 4.2B) direction:

$$T_{cycle}^{+} \left(\frac{k_{+2}}{k_{-1} + k_{+2}} \right) \left(\frac{k_{+1}[S]}{k_{+1}[S] + k_{-2}[P]} \right) = \frac{1}{k_{+1}[S] + k_{-2}[P]} + \frac{1}{k_{-1} + k_{+2}}, \qquad (4.46)$$

where T_{cycle}^{+} is the forward cycle time.

Solving for T_{cycle}^{+} we obtain

$$T_{cycle}^{+} = \frac{1}{k_{+2}} + \frac{k_{-1} + k_{+2}}{k_{+1}k_{+2}[S]} + \frac{k_{-2}[P]}{k_{+1}k_{+2}[S]}, \qquad (4.47)$$

which is E_o / J^{+} in Equation (4.12).

4.4.2 The method of King and Altman

In 1956, King and Altman introduced a shortcut method for determining expressions for steady state concentrations and fluxes from diagrams of catalytic mechanisms [112]. This procedure is useful in determining flux expressions for complex mechanisms and is used in the remaining sections of this chapter. Rather than providing a general proof, which would be too cumbersome for our purposes, here we illustrate how the method is applied to the specific catalytic mechanism illustrated in Figure 4.8. This is an example of an enzyme catalyzing the reaction $A \rightleftharpoons B$, with four distinct enzyme states, E_1, EA, EB, and E_2.

To effectively express concentrations and fluxes for this example, we introduce the shorthand notations ⊔, ⊔, ⊓, ⊓, ⊐, ⊐, ⊏, ⊏, ⊔, ⊔, ⊓, ⊓, ⊐, ⊐, ⊏, and ⊏, where each of these symbols represents a product of three first-order or pseudo-first-order rate constants. These symbols (called *directional diagrams*) represent the product of rate constants along the path defined by the diagram. The symbol ⊔ represents the product $k_{+2}k_{+3}k_{+4}$. Where substrate or product concentrations participate in one of the state transitions for a diagram, the pseudo-first-order rate constant (which incorporates the steady state reactant concentration) is used. Therefore the diagram ⊔ represents the product $k_{-4}k_{-3}k_{-2}[B]$,

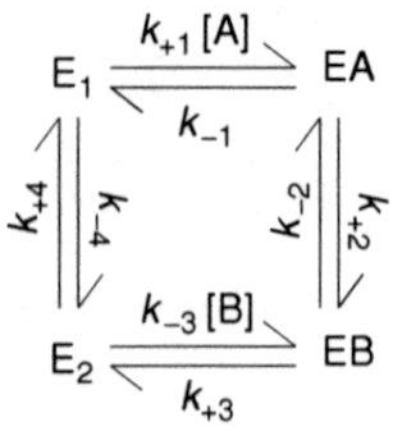

Figure 4.8 Four-state catalytic mechanism for the reaction $A \rightleftharpoons B$.

where $k_{-3}[B]$ is the pseudo-first-order rate constant for the state transition from E_2 to EB.

For the example illustrated in Figure 4.8 the full set of directional diagrams is defined:

⊔↑	$= k_{-4}k_{-3}k_{-2}[B]$	⊔↓	$= k_{+2}k_{-3}k_{-4}[B]$
⊔	$= k_{+2}k_{+3}k_{+4}$	⊔	$= k_{+2}k_{+3}k_{-4}$
⊓↓	$= k_{+4}k_{+1}k_{+2}[A]$	⊓↑	$= k_{+1}k_{-2}k_{+4}[A]$
⊓	$= k_{-2}k_{-1}k_{-4}$	⊓	$= k_{-1}k_{-2}k_{+4}$
⊐	$= k_{-3}k_{-2}k_{-1}[B]$	⊐	$= k_{+1}k_{-2}k_{-3}[A][B]$
⊐	$= k_{+1}k_{+2}k_{+3}[A]$	⊐	$= k_{+1}k_{+2}k_{-3}[A][B]$
⊏	$= k_{+3}k_{+4}k_{+1}[A]$	⊏	$= k_{-1}k_{+3}k_{+4}$
⊏	$= k_{-1}k_{-4}k_{-3}[B]$	⊏	$= k_{-1}k_{+3}k_{-4}$

for this mechanism.

The steady state concentrations of any of the enzyme states E_1, EA, EB, or E_2 can be expressed as the sum of all the directional diagrams that feed into a given state divided by the sum of all diagrams. For example, the directional diagrams that feed into state EA are ⊏, ⊔↑, ⊓↑, and ⊐. The concentration [EA] is given by

$$\frac{[EA]}{E_o} =$$

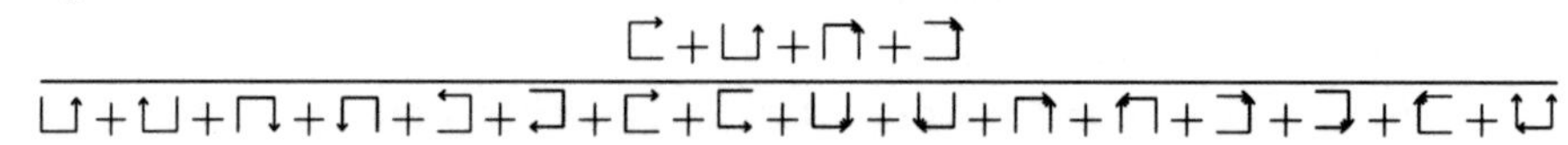

$$\frac{⊏ + ⊔↑ + ⊓↑ + ⊐}{⊔↑ + ⊔ + ⊓↓ + ⊓ + ⊐ + ⊐ + ⊏ + ⊏ + ⊔↓ + ⊔ + ⊓↑ + ⊓ + ⊐ + ⊐ + ⊏ + ⊔}$$

$$= \frac{(k_{+3}k_{+4}k_{+1} + k_{+1}k_{-2}k_{+4})[A] + k_{-4}k_{-3}k_{-2}[B] + k_{+1}k_{-2}k_{-3}[A][B]}{\Sigma}$$

(4.48)

where $E_o = [E_1] + [EA] + [EB] + [E_2]$ is the total enzyme concentration and Σ represents the sum of all diagrams.

Similarly the steady state concentration of E_2 is given by

$$\frac{[E_2]}{E_o} =$$

$$\frac{\sqcap + \sqsubset + \sqcup + \sqsupset}{\sqcup + \sqcup + \sqcap + \sqcap + \sqsupset + \sqsupset + \sqsubset + \sqsubset + \sqcup + \sqcup + \sqcap + \sqcap + \sqsupset + \sqsupset + \sqsubset + \sqcup}$$

$$= \frac{k_{+4}k_{+1}k_{+2}[A] + (k_{-1}k_{-4}k_{-3} + k_{+2}k_{-3}k_{-4})[B] + k_{+1}k_{+2}k_{-3}[A][B]}{\Sigma}$$

(4.49)

and the steady state flux can be expressed $J = k_{+2}[EA] - k_{-2}[E_2]$.

For this and other mechanisms, the set of directional diagrams used in this analysis is the set of all directed graphs that include the maximal number of edges (lines in the graph) while forming no closed loops and no diverging edges. The requirement of no diverging edges means that for any node in a diagram (corresponding to an enzyme state) with more than one edge connected, the directions associated with the edges do not diverge. For example, the diagram $\sqcup$ is not allowed.

To illustrate this point, we consider another four-state mechanism for $A \rightleftharpoons B$:

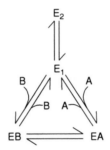

which involves two unbound enzyme states E_1 and E_2. The complete set of directional diagrams for this mechanism is:

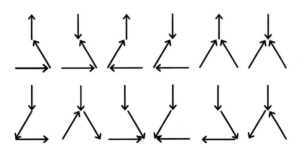

4.4.3 Enzyme-catalyzed bimolecular reactions

Previous sections of this chapter have focused on developing general principles for enzyme-catalyzed reactions based on analysis of single-substrate enzyme systems. Yet the majority of biochemical reactions involve multiple substrates and products. With multiple binding steps, competitive and uncompetitive binding interactions, and allosteric and covalent activations and inhibitions possible, the complete set of possible kinetic mechanisms is vast. For extensive treatments on a great number of mechanisms, we point readers to Segel's book [183]. Here we review a handful of two-substrate reaction mechanisms, with detailed analysis of the compulsory-order ternary mechanism and a cursory overview of several other mechanisms.

4.4.3.1 Compulsory-order ternary mechanism

Consider first the case where two substrates (A and B) bind to an enzyme in an ordered manner and two products (P and Q) dissociate in an ordered manner as well:

$$E + A \underset{k_{-1}}{\overset{k_{+1}}{\rightleftharpoons}} EA, \quad E \cdot A + B \underset{k_{-2}}{\overset{k_{+2}}{\rightleftharpoons}} E \cdot AB,$$

$$E \cdot AB \underset{k_{-3}}{\overset{k_{+3}}{\rightleftharpoons}} E \cdot Q + P, \quad E \cdot Q \underset{k_{-4}}{\overset{k_{+4}}{\rightleftharpoons}} E + Q.$$

This mechanism is known as the "ordered bi-bi" mechanism ("bi-bi" denotes a bi-substrate bi-product reaction), or the "compulsory-order ternary mechanism", where the term "ternary" refers to the three-species complex formed by the binding of two substrates to the enzyme.

This catalytic mechanism is illustrated in Figure 4.9. Here E represents free (unbound) enzyme; E·A represents the complex formed between enzyme and the

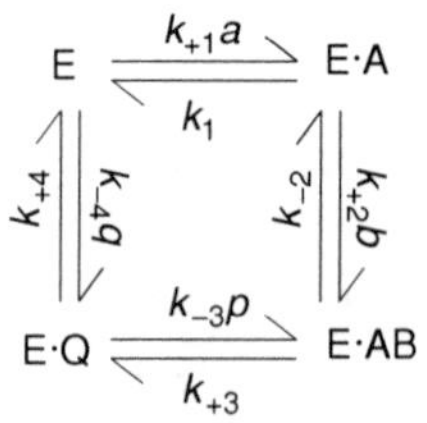

Figure 4.9 Basic compulsory-order ternary-complex mechanism. The basic ordered mechanism for the general reaction $A + B \rightleftharpoons P + Q$, with $a = [A]$, $b = [B]$, $p = [P]$, and $q = [Q]$ is illustrated. The four states are unbound enzyme (E), enzyme–substrate A complex (E·A), enzyme–substrate A-substrate B complex (E·AB), and enzyme–product Q complex (E·Q).

species A, which binds first; E·AB is the ternary complex that represents enzyme bound to both substrates or both products; and E·Q represents the complex formed between enzyme and the species Q. In Figure 4.9 the substrate and product concentrations are denoted $a = [A]$, $b = [B]$, $p = [P]$, and $q = [Q]$ and the reactant concentrations are incorporated into apparent mass-action rate constants for the state transitions between the four enzyme states.

From the four-state diagram of Figure 4.9, the expression for the steady state flux through the reaction can be obtained from the diagrammatic method of King and Altman [112]. The flux J may be expressed

$$J = \frac{N}{D} \qquad (4.50)$$

where

$$N = E_o k_{+1} k_{+2} k_{+3} k_{+4} (ab - pq/K_{eq}) \qquad (4.51)$$

and

$$\begin{aligned} D = &(k_{+2}k_{+3}k_{+4}b + k_{-1}k_{+3}k_{+4} + k_{-2}k_{-1}k_{+4} + k_{-3}k_{-2}k_{-1}p) \, I_1 \\ &+ (k_{-4}k_{-3}k_{-2}pq + k_{+3}k_{+4}k_{+1}a + k_{+4}k_{+1}k_{-2}a + k_{+1}k_{-2}k_{-3}ap) \, I_2 \\ &+ (k_{-4}k_{-3}k_{+2}bpq + k_{-1}k_{-4}k_{-3}pq + k_4 1 k_{+1}k_{+2}ab + k_{+1}k_{+2}k_{-3}abp) \, I_3 \\ &+ (k_{-4}k_{+3}k_{+2}pq + k_{-1}k_{-4}k_{+3}q + k_{-1}k_{-2}k_{-4}q + k_{+1}k_{+2}k_{+3}ab) \, I_4. \end{aligned}$$

$$(4.52)$$

The constant K_{eq} is the equilibrium constant for the reaction; $E_o = [E] + [E \cdot A] + [E \cdot AB] + [E \cdot Q]$ is the total enzyme concentration; and the I_i factors in Equation (4.52) account for non-productive binding (inhibition) of inhibitors to each of the enzyme states. These inhibition factors are computed

$$I_i = 1 + \sum_j c_j K_{ij}, \qquad (4.53)$$

where K_{ij} is the binding constant for non-productive binding of species j to enzyme state i and c_j is the concentration of species j.

Defining

$$n = \frac{N}{k_{-1}k_{+4}(k_{+3} + k_{-2})}$$

and

$$d = \frac{D}{k_{-1}k_{+4}(k_{+3} + k_{-2})}$$

The flux equation is $J = n/d$, where the numerator and denominator written in terms of kinetic constants are

$$n = \frac{V_m}{K_{eA} K_{mB}} (ab - pq/K_{eq}),$$ (4.54)

and

$$d = \left(1 + \frac{K_{mA} b}{K_{eA} K_{mB}} + \frac{K_{mQ} p}{K_{eQ} K_{mP}} \right) I_1$$

$$+ \left(\frac{a}{K_{eA}} + \frac{K_{mQ} ap}{K_{eA} K_{mP} K_{eQ}} + \frac{K_{mA} pq}{K_{eA}^2 K_{mB} K'_{eq}} \right) I_2$$

$$+ \left(\left[\frac{1}{K_{eA} K_{mB}} - \frac{K_{mQ} K'_{eq}}{K_{eQ}^2 K_{mP}} \right] ab + \left[\frac{1}{K_{mP} K_{eQ}} - \frac{K_{mA}}{K_{eA}^2 K_{mB} K_{eq}} \right] pq \right.$$

$$\left. + \frac{K_{mQ} abp}{K_{eA} K_{eB} K_{mP} K_{eQ}} + \frac{K_{mA} bpq}{K_{eA} K_{mB} K_{eP} K_{eQ}} \right) I_3$$

$$+ \left(\frac{q}{K_{eQ}} + \frac{K_{mQ} K'_{eq} ab}{K_{eQ}^2 K_{mP}} + \frac{K_{mA} bq}{K_{eA} K_{mB} K_{eQ}} \right) I_4.$$ (4.55)

The kinetic constants in Equation (4.55) are defined:

$$V_m = \frac{E_o k_{+3} k_{+4}}{k_{+3} + k_{+4}}$$

$$K_{mA} = \frac{k_{+3} k_{+4}}{k_{+1}(k_{+3} + k_{+4})}$$

$$K_{mB} = \frac{k_{+4}(k_{-2} + k_{+3})}{k_{+2}(k_{+3} + k_{+4})}$$

$$K_{mP} = \frac{k_{-1}(k_{-2} + k_{+3})}{k_{-3}(k_{-2} + k_{-1})}$$

$$K_{mQ} = \frac{k_{-1} k_{-2}}{k_{-4}(k_{-1} + k_{-2})}$$

$$K_{eA} = \frac{k_{-1}}{k_{+1}}$$

$$K_{eB} = \frac{k_{-2}}{k_{+2}}$$

$$K_{eP} = \frac{k_{+3}}{k_{-3}}$$

$$K_{eQ} = \frac{k_{+4}}{k_{-4}}.$$ (4.56)

Expressing the steady state kinetics in terms of these parameters, only the V_m parameter, which has units of mass per unit time per unit volume, has units that include time. All other parameters have units of concentration (mass per unit volume). In addition, the eight concentration parameters cannot vary independently. For example we can compute K_{eQ} in terms of the other parameters if the equilibrium constant of the reaction is known:

$$K_{eQ} = \frac{K_{eq} K_{eA} K_{eB}}{K_{eP}}. \tag{4.57}$$

4.4.3.2 Overview of other bimolecular mechanisms

The "random bi-uni" mechanism (random bi-substrate binding order with single product) has the form:

$$E + A \underset{k_{-1}}{\overset{k_{+1}}{\rightleftharpoons}} E \cdot A, \quad E + B \underset{k_{-2}}{\overset{k_{+2}}{\rightleftharpoons}} E \cdot B,$$

$$E \cdot A + B \underset{k_{-3}}{\overset{k_{+3}}{\rightleftharpoons}} E \cdot AB, \quad E \cdot B + A \underset{k_{-4}}{\overset{k_{+4}}{\rightleftharpoons}} E \cdot AB,$$

$$E \cdot AB \underset{k_{-5}}{\overset{k_{+5}}{\rightleftharpoons}} E + P.$$

The random bi-uni quasi-steady flux is expressed (in terms of rate constants) [183]

$$J = \frac{k_{+5} \left(k_{+1}k_{+3}(k_{-2} + k_{+4}[A]) + k_{+2}k_{+4}(k_{-1} + k_{+3}[B]) \right) ([A][B] - K_d[P])}{D} \tag{4.58}$$

where

$$\begin{aligned} D = &\left([A][B] + \frac{k_{-3}[A]}{k_{+3}} + \frac{k_{-3}k_{-1}}{k_{+3}k_{+1}} + \frac{k_{-3}k_{-1}k_{+2}[B]}{k_{+3}k_{+1}k_{-2}} \right) \\ &\times (k_{+1}k_{+3}(k_{-2} + k_{+4}[A]) + k_{+2}k_{+4}(k_{-1} + k_{+3}[B])) \\ &+ (k_{+5} + k_{-5}[P])(k_{-2} + k_{+4}[A])(k_{-1} + k_{+3}[B]) \\ &+ (k_{-1} + k_{+3}[B])(k_{+2}k_{+5}[B] + k_{-4}k_{-5}[P]) \\ &+ (k_{-2} + k_{+4}[A])(k_{+1}k_{+5}[A] + k_{-3}k_{-5}[P]) \end{aligned}$$

and the constant K_d is

$$K_d = \frac{k_{-1}k_{-3}k_{-5}}{k_{+1}k_{+3}k_{+5}} = \frac{k_{-2}k_{-4}k_{-5}}{k_{+2}k_{+4}k_{+5}}.$$

The so-called "ping-pong" mechanism has the form

$$E + A \underset{k_{-1}}{\overset{k_{+1}}{\rightleftharpoons}} E \cdot A \underset{k_{-2}}{\overset{k_{+2}}{\rightleftharpoons}} E^* + P, \quad E^* + B \underset{k_{-3}}{\overset{k_{+3}}{\rightleftharpoons}} E^* \cdot B \underset{k_{-4}}{\overset{k_{+4}}{\rightleftharpoons}} E + Q.$$

In this case the enzyme first reacts with one substrate, leading to a covalently modified (unbound) enzyme E^* and the release of P. The enzyme in state E^* then reacts with a second substrate Q and is converted back to E. An example of an enzyme that displays this mechanism is phosphoglycerate mutase [105].

The flux expression for the ping-pong mechanism is [183]

$$J = \frac{k_{+1}k_{+2}k_{+3}k_{+4}[A][B] - k_{-1}k_{-2}k_{-3}k_{-4}[P][Q]}{D} \tag{4.59}$$

where

$$\begin{aligned} D = &\ k_{+1}k_{+2}(k_{-3} + k_{+4})[A] + k_{+3}k_{+4}(k_{+2} + k_{-1})[B] + k_{-1}k_{-2}(k_{-3} + k_{+4})[P] \\ &+ k_{-3}k_{-4}(k_{-1} + k_{+2})[Q] + k_{+1}k_{+3}(k_{+2} + k_{+4})[A][B] \\ &+ k_{+1}k_{-2}(k_{-3} + k_{+4})[A][P] + k_{-2}k_{-4}(k_{-1} + k_{-3})[P][Q] \\ &+ k_{+3}k_{-4}(k_{-1} + k_{+2})[B][Q]. \end{aligned}$$

J. J. Hopfield has proposed a so-called *energy relay model* [101], in which the ping-pong mechanism operates on two identical substrate molecules and releases two identical product molecules. The model introduces the novel idea of *dynamic cooperativity* by which biomolecular processes in living cells, such as DNA replication and protein biosynthesis, can achieve high fidelity. The model was developed as an alternative to the kinetic proofreading mechanism which we shall discuss in Chapter 5.

4.4.4 Example: enzyme kinetics of citrate synthase

As an example to illustrate analysis of kinetic data to characterize the mechanism of a real enzyme, here we apply the general compulsory-order ternary mechanism introduced above to citrate synthase to determine kinetic parameters for several isoforms of this enzyme and to elucidate the mechanisms behind inhibition by products and other species not part of the overall chemical reaction.

Citrate is a key intermediate of the tricarboxylic acid (TCA) cycle, also known as the Krebs cycle, in the central metabolism of cells. (The set of reactions of the TCA cycle will be considered in some detail in Chapter 6.) One reaction in the cycle is the combination of oxaloacetate (OAA) and the acetyl group from acetyl coenzyme A (ACCOA), in the presence of H_2O, to form citrate (CIT), thiol coenzyme A (COASH), and hydrogen ion (H^+). The chemical reference reaction for this aldol condensation-hydrolysis reaction catalyzed by citrate synthase is:

$$ACCOA^0 + OAA^{2-} + H_2O \rightleftharpoons CIT^{3-} + COAS^- + 2H^+ \tag{4.60}$$

Table 4.1 *Thermodynamic and cation binding parameter values for reactants involved in citrate synthase reaction*

Reactant	Abbr.	Ref. species	$\Delta_f G^o$ (kJ·mol^{-1})	Ion-bound species	pK
water	H$_2$O	H$_2$O	-235.74	$-$	$-$
coenzyme A	COASH	COAS$^-$	-0.2		8.13
acetyl-coenzyme A	ACCOA	ACCOA^0	-178.19	$-$	$-$
oxaloacetate	OAA	OAA^{2-}	-794.41	MgOAA^0	0.0051[a]
citrate	CIT	CIT^{3-}	-1165.59	HCIT^{2-}	5.63
				MgCIT$^-$	3.37[a]
				KCIT^{2-}	0.339[a]
adenosine triphosphate	ATP	ATP^{4-}	-2771.00	HATP^{3-}	6.59
				MgATP^{2-}	3.82[a]
				KATP^{3-}	1.87[a]
adenosine diphosphate	ADP	ADP^{3-}	-1903.96	HADP^{2-}	6.42
				MgADP$^-$	2.79[a]
				KADP^{2-}	1.53[a]
adenosine monophosphate	AMP	AMP^{2-}	-1034.66	HAMP$^-$	6.22
				MgAMP^0	1.86[a]
				KAMP$^-$	1.05[a]
succinyl-coenzyme A	SCOA	SCOA$^-$	-507.55	HSCOA^0	3.96

All values from [4] unless otherwise noted.
[a]NIST database 46: Critical Stability Constants [134].

where the biochemical species and related thermodynamic data are listed in Table 4.1. Data for additional species that act as inhibitors of the citrate synthase enzymes are also listed in the table.

The standard Gibbs free energy is computed

$$\Delta_r G^o_{cts} = \Delta_f G^o_{CIT} + \Delta_f G^o_{COASH} - \Delta_f G^o_{ACCOA} - \Delta_f G^o_{OAA} - \Delta_f G^o_{H_2O}$$
$$= 42.36 \,\text{kJ} \cdot \text{mol}^{-1} \tag{4.61}$$

where the basic thermodynamic data are listed in Table 4.1. The equilibrium constant for reaction is computed from the standard Gibbs free energy

$$K^o_{eq.cts} = \frac{1}{h^2} \exp\left(-\frac{\Delta_r G^o_{cts}}{RT}\right) \tag{4.62}$$

where we have introduced the definition $h = 10^{-pH}$ and this equilibrium constant explicitly accounts for pH. Therefore this constant represents the equilibrium ratio of $[COAS^-][CIT^{3-}]/[ACCOA^0][OAA^{2-}]$ at given pH. The relationships between species and reactant concentrations, which depend on the pH and concentration of metal ions that reversibly bind to biochemical species, are expressed in terms of the binding polynomials:

$$P_{OAA} = 1 + \frac{[Mg^{2+}]}{K_{M,OAA}}$$

$$P_{ACCOA} = 1$$

$$P_{CIT} = 1 + \frac{h}{K_{H,CIT}} + \frac{[K^+]}{K_{K,CIT}} + \frac{[Mg^{2+}]}{K_{M,CIT}}$$

$$P_{COASH} = 1 + \frac{h}{K_{H,COASH}}$$

$$P_{ATP} = 1 + \frac{h}{K_{H,ATP}} + \frac{[K^+]}{K_{K,ATP}} + \frac{[Mg^{2+}]}{K_{M,ATP}}$$

$$P_{ADP} = 1 + \frac{h}{K_{H,ADP}} + \frac{[K^+]}{K_{K,ADP}} + \frac{[Mg^{2+}]}{K_{M,ADP}}$$

$$P_{AMP} = 1 + \frac{h}{K_{H,AMP}} + \frac{[K^+]}{K_{K,AMP}} + \frac{[Mg^{2+}]}{K_{M,AMP}}$$

$$P_{SCOA} = 1 + \frac{h}{K_{H,SCOA}}. \tag{4.63}$$

Note that only states that are expected to be significant in the pH and ionic range studied are included in these calculations. Therefore some binding polynomials do not include terms for all possible cation-bound states. Given these forms of the binding polynomials, the relationships between the reference species concentrations and the reactant concentrations take the usual form:

$$[OAA^{2-}] = [OAA]/P_{OAA}$$

$$[ACCOA^0] = [ACCOA]/P_{ACCOA}$$

$$[CIT^{3-}] = [CIT]/P_{CIT}$$

$$[COAS^-] = [COASH]/P_{COASH}$$

$$[ATP^{4-}] = [ATP]/P_{ATP}$$

$$[ADP^{3-}] = [ADP]/P_{ADP}$$

$$[AMP^{2-}] = [AMP]/P_{AMP}$$

$$[SCOA^-] = [SCOA]/P_{SCOA}. \tag{4.64}$$

The apparent equilibrium constant for the biochemical reaction is computed as a function of pH, $[K^+]$, and $[Mg^{2+}]$:

$$K_{eq.cts} = K^o_{eq.cts} \frac{P_{COASH} P_{CIT}}{P_{OAA} P_{ACCOA}}. \qquad (4.65)$$

Although investigations have led to proposing more complex behavior, involving cooperativity and random order and dead-end binding of substrates [135], here we propose that the standard compulsory-order ternary-complex mechanism derived above can explain the kinetic behavior of citrate synthase with substrate and products identified as: $a = [OAA^{2-}]$, $b = [ACCOA^0]$, $p = [COAS^-]$, $q = [CIT^{3-}]$. The specific mechanism is:

$$\mathrm{E} + \mathrm{OAA}^{2-} \underset{k_{-1}}{\overset{k_{+1}}{\rightleftharpoons}} \mathrm{E} \cdot \mathrm{OAA}^{2-}$$

$$\mathrm{E} \cdot \mathrm{OAA}^{2-} + \mathrm{ACCOA}^0 \underset{k_{-2}}{\overset{k_{+2}}{\rightleftharpoons}} \mathrm{E} \cdot \mathrm{OAA}^{2-} \cdot \mathrm{ACCOA}^0$$

$$\mathrm{E} \cdot \mathrm{OAA}^{2-} \cdot \mathrm{ACCOA}^0 + \mathrm{H_2O} \underset{k_{-3}}{\overset{k_{+3}}{\rightleftharpoons}} \mathrm{E} \cdot \mathrm{CIT}^{3-} + \mathrm{COAS}^- + 2\,\mathrm{H}^+$$

$$\mathrm{E} \cdot \mathrm{CIT}^{3-} \underset{k_{-4}}{\overset{k_{+4}}{\rightleftharpoons}} \mathrm{E} + \mathrm{CIT}^{3-}. \qquad (4.66)$$

In this mechanism only the third reaction (in which hydrogen ion explicitly appears) depends on pH. Since K_{eP}, the equilibrium constant for the third reaction, depends on pH while the others do not, we compute K_{eP} as a function of the equilibrium constant for the reference reaction

$$K_{eP} = \frac{K^o_{eq.cts} K_{eA} K_{eB}}{K_{eQ}}. \qquad (4.67)$$

The rate constant k_{-3} is assumed to depend on pH according to the formula $k_{-3} = (h/10^{-7})^2 k'_{43}$ where k'_{43} is independent of pH. Therefore the kinetic constant K_{mP} is defined to depend on pH as

$$K_{mP} = \left(\frac{10^{-7}}{h} \right)^2 K'_{mP}, \qquad (4.68)$$

where K'_{mP} is a kinetic constant that is independent of pH. In addition to the pH dependency of the kinetic constants, the overall enzyme activity depends on pH, with the numerator of the flux expression taking the form:

$$n = \frac{V_m}{K_{eA} K_{mB}} \frac{ab - pq/K^o_{eq.cts}}{1 + h/K_{iH}} \qquad (4.69)$$

Enzyme-catalyzed reactions

Table 4.2 *Kinetic parameter values for citrate synthase*

Parameter	Rat kidney	Rat liver	Bovine heart
V_m (μmol·min^{-1}·μg^{-1})	0.3355	–	–
K_{mA} (μM)	8.227	1.347	1.775
K_{mB} (μM)	7.402	13.62	8.041
K'_{mP} (μM)	0.156	–	–
K_{mQ} (mM)	4.548	–	–
K_{eA} (μM)	0.8879	1.936	1.143
K_{eB} (μM)	30.05	–	–
K_{eQ} (mM)	3.618	–	–
K_{iATP} (μM)	–	38.43	70.11
K_{iADP} (μM)	–	139.4	–
K_{iAMP} (μM)	–	1023	–
K_{iSCOA} (μM)	–	–	70.06
K_{iH} (μM)	–	0.055	–

which is used to reproduce the pH dependency observed by Shepherd and Garland [63]. Equation (4.69) assumes that the enzyme is a monobasic acid, with dissociation constant K_{iH}.

Studies have revealed that a number of substances, including succinyl-coenzyme A and adenine nucleotides, act as inhibitors of citrate synthase. Analysis of kinetic data on citrate synthase from rat liver and bovine heart indicate a model involving ATP, ADP, and AMP inhibiting the enzyme by forming unproductive complexes with enzyme state 2 is consistent with the observed data. This study was not able to elucidate the site of SCOA binding: models assuming binding at either state 1 or state 2 are equally well able to explain the observed data. Since the adenine nucleotide inhibition was determined to occur at enzyme state 2, a model was developed assuming that SCOA binds to this complex as well. Based on this formulation of the model, the inhibition term I_2 is

$$I_2 = 1 + \frac{[\text{ATP}^{4-}]}{K_{iATP}} + \frac{[\text{ADP}^{3-}]}{K_{iADP}} + \frac{[\text{AMP}^{2-}]}{K_{iAMP}} + \frac{[\text{SCOA}^-]}{K_{iSCOA}} \qquad (4.70)$$

and inhibition at other complexes is not considered: $I_1 = I_3 = I_4 = 1$.

Kinetic parameter values for citrate synthase for several isoforms of the model may be estimated for the general analysis of the compulsory-order ternary mechanism outlined in Section 4.4.3.1 based on observed data. The parameter estimates for citrate synthase obtained from rat kidney, rat liver, and bovine heart are listed in Table 4.2.

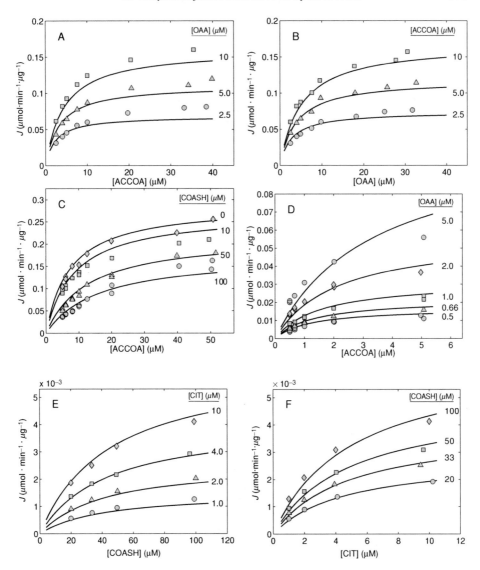

Figure 4.10 Fits to kinetic data from [135] on the operation of citrate synthase from rat kidney. Data (flux as a function of substrate concentrations) were obtained from Figures 2, 3, 4, 5, 6, 7, and 9 of [135]. Initial fluxes (μmol of COASH (or CIT) synthesized per minute per μg of enzyme) measured at the substrate concentrations indicated are plotted in A, B, C, and D. For A, B, and D, the initial product (CIT and COASH) concentrations are zero. In C, flux was measured with COASH added in various concentrations to investigate the kinetics of product inhibition. E and F show fits to kinetic data on the reverse operation of kidney enzyme, with product concentrations indicated in the figure. All data were obtained at pH = 8.1 at 28 °C. Model fits in all cases are plotted as solid lines.

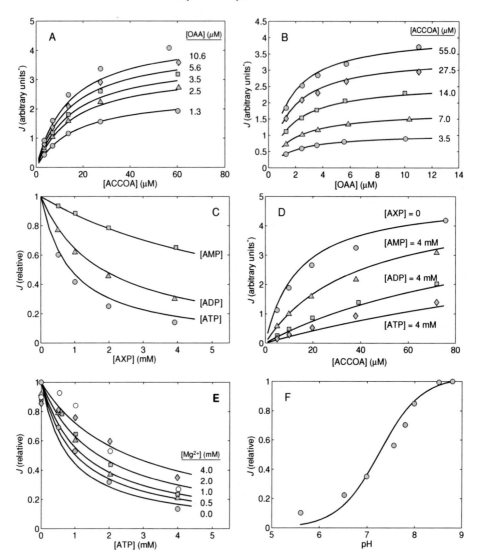

Figure 4.11 Fits to kinetic data from [63] on the forward operation of liver enzyme. Measured flux in arbitrary units was obtained from Figures 1, 2, 5, 6, 13, and 14 of [63]. For all cases the product (CIT and COASH) concentrations are zero and total substrate and inhibitor concentrations are indicated in the figure. Data obtained with no inhibitors present are plotted in A and B. In C the relative activity (normalized to its maximum) of the enzyme is plotted as functions of [ATP], [ADP], and [AMP] measured at [ACCOA] = 11 μM and [OAA] = 1.9 μM. D. The measured flux is plotted as a function of [ACCOA] at [OAA] = 34 μM with ATP, ADP, and AMP present as indicated in the figure. In E the relative activity of the enzyme is plotted as functions of [ATP] at [Mg^{2+}] = 0 mM (shaded circles), 0.5 mM (shaded triangles), 1.0 mM (shaded squares), 2.0 mM (open circles), and 4.0 mM (diamonds). In F relative activity is plotted as a function of pH. Substrate concentrations are [ACCOA] = 21 μM and [OAA] = 8.6 μM. All data were obtained at 25 °C. pH is fixed a 7.4 for A. Model fits are plotted as solid lines.

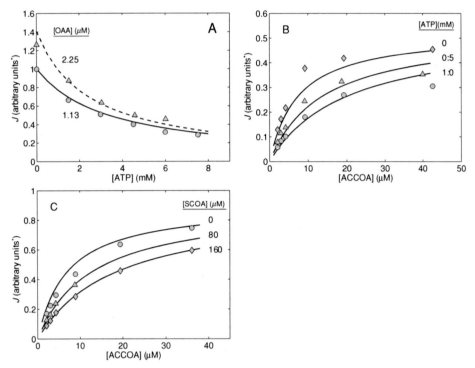

Figure 4.12 Analysis of data from Smith and Williamson [188] on inhibition of cardiac enzyme. Measured flux in arbitrary units was obtained from Figures 1 and 2 of [188]. A. Flux is plotted as a function inhibitor ATP concentration for [ACCOA] = 16 μM and [OAA] = 1.13 and 2.25 μM. B. Flux is plotted as a function of [ACCOA] at [OAA] = 5 μM at three different concentrations of ATP indicated in figure. C. Flux is plotted as a function of [ACCOA] at [OAA] = 3.1 μM at three different concentrations of SCOA indicated in figure. All data were obtained at pH = 7.4 at 21 °C. Model fits are plotted as solid lines.

Note that data sensitive to all parameter values are not available for all isoforms of the enzyme. Therefore estimates are not provided for all parameters for all tissue types.

The data and associated model fits used to obtain these kinetic constants are shown in Figures 4.10 through 4.12. These data on quasi-steady reaction flux as functions of reactant and inhibitor concentrations are obtained from a number of independent sources, as described in the figure legends. Note that the data sets were obtained under different biochemical states. In fact, it is typical that data on biochemical kinetics are obtained under non-physiological pH and ionic conditions. Therefore the reported kinetic constants are not necessarily representative of the biochemical states obtained in physiological systems.

In this analysis we have addressed and corrected this problem by posing the reaction mechanism in terms of species and ensuring that mechanisms properly account for thermodynamics. This basic approach was first introduced by Frieden and Alberty [61], yet has received little attention.

Concluding remarks

Predicting and understanding how and why concentrations and states of biomolecules in living systems change with time represents the central goal of quantitative research in biological systems. Since enzymes catalyze nearly all of the chemical transformations occurring in a cell, enzyme kinetics represents the heart and soul of this endeavor. While we have not yet reached the point where all of the processes occurring in a cell can be simulated in a convincing way, a great deal of individual cellular processes are realistically represented by the techniques introduced in this chapter. Integrating individual processes together, we are able to build computer simulations of biological systems. While systems of progressively increasing complexity are considered in Chapters 5, 6, and 7 of this book, we hope that readers will find the tools of enzyme kinetics useful in building simulations of scale and realism beyond anything described in this book or elsewhere.

Exercises

4.1 Show that the non-dimensional Equations (4.23) follow from $u = [S]/S_o$, $v = [ES]/E_o$, and Equations (4.22). Substitute Equation (4.24) into Equation (4.23) to yield Equation (4.25).

4.2 Equation (4.25) has an analytic solution in which τ can be expressed as an explicit function of $u(\tau)$. What is this function? Plot $u(\tau)$ versus τ based on this equation.

4.3 Determine the constants a, b, c, and d in terms of K_1, K_2, k_{f1}, and k_{f2} for the example in Section 4.3.2.

4.4 What is the Hill coefficient for the function $y(x) = x^n/(x^n + x_o^n)$? What is the value of the Hill coefficient for the curves plotted in Figure 4.6? [Hint: the identity $\frac{x}{y}\frac{dy}{dx} = \frac{d \ln x}{d \ln y}$ is useful.]

4.5 Show that the relationship $\lambda_1\lambda_2 = \alpha\gamma$, used in Section 4.3.4, is satisfied by the eigenvalues of Equation (4.17).

4.6 Derive the steady state rate expression of Equations (4.38) and (4.39) from the kinetic mechanism of Equation (4.37). What is the apparent Michaelis–Menten constant for this mechanism?

4.7 Show that the reverse cycle time for the catalytic cycle illustrated in Figure 4.2B is given by $T_{\text{cycle}}^- = E_o/J^-$ for this mechanism.

5

Biochemical signaling modules

Overview

The central dogma of molecular biology describes how one form of biological information (an organism's genetic sequence) is processed in terms of DNA replication, RNA transcription, and protein synthesis. However, a related mystery is yet to be worked out in sufficient detail: how is the information encoded in the DNA (i.e., genotypes) related to cellular functions (i.e., phenotypes)? How do different signals tell different cells to synthesize different proteins?

To tackle these questions we adopt a view of the cell as a machine that processes diverse information [206, 166]. The hardware for cellular information processing consists of specialized biochemical reactions and their associated molecules, forming so-called *signal transduction networks*. As we have discussed in the previous chapter, the majority of biochemical reactions involve proteins acting as enzymatic catalysts. Reactions in signaling systems are no exception. In fact it is a common motif in signaling systems for enzymes to carry information via regulations of their biochemical activities; activities are modulated by covalent modification or allosteric binding by effector molecules.

A central question in cellular biology is now to elucidate (meaning to develop models with reliable predictive power) the mechanisms by which the cells transduce information and perform their functions. Cellular biochemical signaling systems are customarily visualized as "logic circuits"; the components for the circuitry, now popularly called "modules" [78], consist of molecules and biochemical reactions. Hence they can be subjected to kinetic and thermodynamic analysis as we have introduced in the previous chapters. In this chapter, we study several such modules that occur widely in cellular biology.

5.1 Kinetic theory of the biochemical switch

Biochemical switches inside a cell are usually based on the conformational transition of a protein: the protein can have little or no biological activity in one state but high biological activity in another. Conformational transition between the two states, thus, constitutes a molecular switch. The biological activity of interest is often the catalytic activity of an enzyme and the transitions are effected via enzyme-mediated reactions. Thus often enzymes serve as substrates for other enzymes in biochemical signaling pathways.

There are essentially two types of control mechanisms for biochemical switching: allosteric cooperative transition and reversible chemical modification. Allosteric cooperativity, which was discussed in Chapter 4, was discovered in 1965 by Jacques Monod, Jefferies Wyman, and Jean-Pierre Changeux [143], and independently by Daniel Koshland, George Némethy and David Filmer [116]. The molecular basis of this phenomenon, which is well understood in terms of three-dimensional protein crystal structures and protein–ligand interaction, is covered in every biochemistry textbook [147] as well as special treatises [215].

Reversible chemical modification of enzymes, which was discovered in 1955 by Edmond Fischer and Edwin Krebs [58], is a more prevalent mechanism for cellular signaling switching. Fischer and Krebs showed that enzymes can be turned from an inactive form to an active form via phosphorylation of certain residues of the protein. Enzymes that catalyze phosphorylation (addition of a phosphate group coupled with ATP or GTP hydrolysis) are called protein *kinases*. Enzymes that catalyze dephosphorylation (which is not the reverse reaction of the phosphorylation) are called *phosphatases*. For example, a protein tyrosine phosphatase is an enzyme that catalyzes the removal of a phosphate group from a tyrosine residue in a phosphorylated protein [57].

Complex signaling networks can result from having a number of interacting enzymes catalyzing the activation and deactivation (or switching-on and switching-off) of one another. Often non-protein messenger molecules are thrown into the mix as well, as is shown in Figure 5.1, which illustrates the signaling network associated with phosphatidylinositol-3,4,5-triphosphate (PIP_3). The PIP_3 pathway is involved in regulating a number of processes in a number of cell types. One of the tasks of computational cell biology is to translate cartoon illustrations such as Figure 5.1 into quantitative biochemical kinetic models.

If we concentrate on one particular component of this map – the phosphorylation of $PI(4,5)P_2$ to $PI(3,4,5)P_3$ by PI3K and the dephosphorylation of $PI(3,4,5)P_3$ to $PI(4,5)P_2$ by PTEN, we can study the detailed enzyme kinetic scheme of this so-called phosphorylation–dephosphorylation cycle, which is illustrated in Figure 5.2. This illustrated cycle represents a ubiquitous module in biochemical signaling. It could, for example, represent the phosphorylation of mitogen-activation protein kinase (MAPK) by MAPK kinase (MAPKK) and dephosphorylation of MAPK by MAPK phosphatase (MKP).

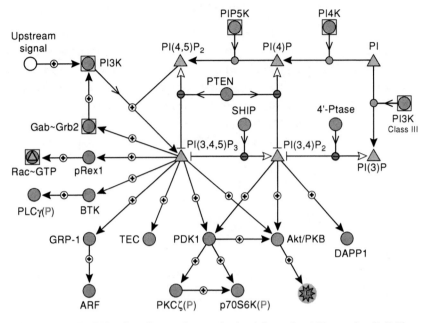

Figure 5.1 The PIP$_3$ signaling pathway obtained from the Alliance for Cell Signaling [177]. Proteins are illustrated as circles and non-protein molecules as triangles. Arrows indicate the direction of information flow in the network, including phosphorylation reactions, such as the phosphorylation of PI to PI(4)P, which is catalyzed by phosphoinositide 4-kinase (PI4K), and allosteric activations indicated by arrows with "+" signs.

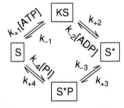

Figure 5.2 A typical cellular biochemical switch consisting of a phosphorylation–dephosphorylation cycle. The substrate molecule S may be a protein or other signaling molecule. If S is a protein then the phosphorylation of S is catalyzed by a protein kinase (K) and the dephosphorylation is catalyzed by a protein phosphatase (P). The entire cycle is accompanied by the reaction ATP $\rightleftharpoons$ ADP+PI. In the context of mitogen-activation protein kinase pathway, S, K, and P correspond to MAPK, MAPKK, and MKP, respectively. In the context of the example from the PIP$_3$ pathway, the kinase is PI3K and the phosphatase is PTEN.

5.1.1 The phosphorylation–dephosphorylation cycle

The simplest kinetic model for the phosphorylation–dephosphorylation cycle assumes that the concentration of substrate S is sufficiently lower than the Michaelis–Menten constants: $[S] \ll K_1$ and K_2, where K_1 and K_2 are the effective Michaelis–Menten constants for S for the kinase and phosphatase, respectively. Similarly, $[S^*] \ll K_1^*$ and K_2^*, where K_1^* and K_2^* are the effective Michaelis–Menten constants for S^*. Applying the formulas from Equation (4.9) to the mechanism of Figure 5.2, we have the following expressions for the Michaelis–Menten parameters:

$$K_1 = \frac{k_{-1} + k_{+2}}{k_{+1}[\text{ATP}]}, \quad K_2 = \frac{k_{-3} + k_{+4}}{k_{-4}[\text{PI}]}$$

$$K_1^* = \frac{k_{-1} + k_{+2}}{k_{-2}[\text{ADP}]}, \quad K_2^* = \frac{k_{-3} + k_{+4}}{k_{+3}} \tag{5.1}$$

when ATP, ADP, and PI concentrations are held fixed. Under the assumption that the enzymes remain unsaturated the reversible Michaelis–Menten flux of Equation (4.10) varies linearly with reactant concentrations. For this system, we have

$$J_k = \frac{k_{+2} K_o}{K_1}[\text{S}] - \frac{k_{-1} K_o}{K_1^*}[\text{S}^*]$$

$$J_p = \frac{k_{+4} P_o}{K_2^*}[\text{S}^*] - \frac{k_{-3} P_o}{K_2}[\text{S}], \tag{5.2}$$

where J_k and J_p are the net kinase and phosphatase fluxes and K_o and P_o are the concentrations of the kinase and phosphatase enzymes, respectively.

Equivalently, we model the system with mass action kinetics as:

$$\text{S} + \text{ATP} \underset{a_- K_o}{\overset{a_+ K_o}{\rightleftharpoons}} \text{S}^* + \text{ADP}, \quad \text{S}^* \underset{b_- P_o}{\overset{b_+ P_o}{\rightleftharpoons}} \text{S} + \text{PI}, \tag{5.3}$$

with rate constants for phosphorylation

$$a_+ = \frac{k_{+1} k_{+2}}{k_{-1} + k_{+2}}, \quad a_- = \frac{k_{-1} k_{-2}}{k_{-1} + k_{+2}},$$

and for dephosphorylation

$$b_+ = \frac{k_{+3} k_{+4}}{k_{-3} + k_{+4}}, \quad b_- = \frac{k_{-3} k_{-4}}{k_{-3} + k_{+4}}. \tag{5.4}$$

The kinetics is governed by

$$\frac{d[\text{S}]}{dt} = \left(a_-[\text{S}^*][\text{ADP}] - a_+[\text{S}][\text{ATP}]\right) K_o + \left(b_+[\text{S}^*] - b_-[\text{S}][\text{PI}]\right) P_o, \tag{5.5}$$

in which $[S] + [S^*] = S_o$ is the total substrate concentration. The steady state population fraction in the phosphorylated state is

$$f^* = \frac{[S^*]^{ss}}{[S_o]} = \frac{a_+[ATP]K_o + b_-[PI]P_o}{a_+[ATP]K_o + b_-[PI]P_o + a_-[ADP]K_o + b_+P_o},$$

which can be rewritten as [166]

$$f^* = \frac{[S^*]^{ss}}{S_o} = \frac{\theta + \mu}{\theta + \mu + \theta/(\mu\gamma) + 1}, \tag{5.6}$$

with the three parameters

$$\theta = \frac{a_+[ATP]K_o}{b_+P_o} = \frac{k_{+1}k_{+2}(k_{-3} + k_{+4})[ATP]K_o}{(k_{-1} + k_{+2})k_{+3}k_{+4}P_o}, \tag{5.7a}$$

$$\mu = \frac{b_-[PI]}{b_+} = \frac{k_{-3}k_{-4}[PI]}{k_{+3}k_{+4}}, \tag{5.7b}$$

$$\gamma = \frac{a_+b_+[ATP]}{a_-b_-[ADP][PI]} = \frac{k_{+1}k_{+2}k_{+3}k_{+4}[ATP]}{k_{-1}k_{-2}k_{-3}k_{-4}[ADP][PI]}. \tag{5.7c}$$

Equation (5.6) is the fundamental equation for a phosphorylation–dephosphorylation switch. The parameter θ is the control parameter that represents the ratio of the apparent kinase activity to that of phosphatase; K catalyzes phosphorylation and P catalyzes dephosphorylation. The parameter μ characterizes the level of S^* in the absence of the kinase K (when $\theta = 0$); μ determines the basal activity and is usually very small.

The parameter γ represents the amount of available free energy for ATP hydrolysis; specifically, $-RT \ln \gamma$ is the free energy change associated with ATP hydrolysis. To see this, consider the four reactions of Figure 5.2:

$$S + ATP + K \underset{k_{-1}}{\overset{k_{+1}}{\rightleftharpoons}} KS$$

$$KS \underset{k_{-2}}{\overset{k_{+2}}{\rightleftharpoons}} S^* + ADP + K$$

$$S^* + P \underset{k_{-3}}{\overset{k_{+3}}{\rightleftharpoons}} S^*P$$

$$S^*P \underset{k_{-4}}{\overset{k_{+4}}{\rightleftharpoons}} S + PI + P,$$

which sum to the overall reaction:

$$ATP \rightleftharpoons ADP + PI,$$

with equilibrium constant $K_{eq} = (k_{+1}k_{+2}k_{+3}k_{+4})/(k_{-1}k_{-2}k_{-3}k_{-4})$. Thus $\Delta G_{ATP} = -RT \ln \gamma$.

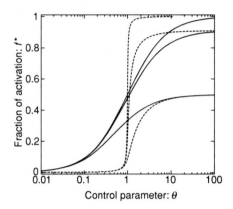

Figure 5.3 Phosphorylation–dephosphorylation cycle activation as a function of the activating signal θ and available free energy γ. The solid lines and dashed lines are without and with enzyme saturation, i.e., Equations (5.6) and (5.18), respectively. In both cases, from top to bottom: $\gamma = 10^{10}$, 10^4, and 10^3. All computations are done with $\mu = 0.001$, and for the dashed lines $\frac{K_1}{S_o} = \frac{K_2}{S_t} = 0.01$. If $\gamma = 1$, then both the solid and dashed lines will be strictly horizontal.

Equation (5.6) indicates that if there is no available free energy for ATP hydrolysis ($\gamma = 1$) then

$$\frac{[S^*]^{ss}}{S_o} = \frac{\mu}{1 + \mu}, \tag{5.8}$$

which is independent of θ altogether. From this we draw an important lesson: a biochemical switch cannot function without free energy input. No energy, no switch. The solid lines in Figure 5.3 show how f^* changes as a function of both θ and γ according to Equation (5.6).

We can further characterize the amplitude of the switch, AOS:

$$AOS = \left(\frac{[S^*]^{ss}}{S_o}\right)_{\theta=\infty} - \left(\frac{[S^*]^{ss}}{S_o}\right)_{\theta=0}$$

$$= \frac{\mu(\gamma - 1)}{(\mu\gamma + 1)(\mu + 1)} \leq \frac{\gamma - 1}{1 + \gamma + 2\sqrt{\gamma}}. \tag{5.9}$$

The inequality indicates that the optimal μ for the maximal AOS is when $\mu = \frac{1}{\sqrt{\gamma}}$. Substituting $\Delta G_{ATP} = -RT \ln \gamma$, we have the optimal AOS

$$\text{optimal } AOS = \frac{\sqrt{\gamma} - 1}{\sqrt{\gamma} + 1} = \tanh\left(\frac{-\Delta G}{4RT}\right). \tag{5.10}$$

As a function of θ, f^* in Equation (5.6) increases with θ. Hence, we can characterize the sharpness of the transition in terms of the Hill coefficient introduced in

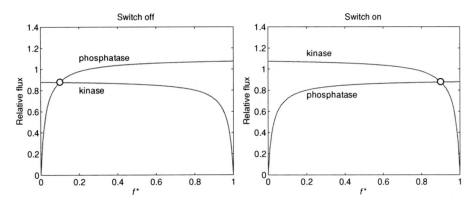

Figure 5.4 Switch-like behavior of the phosphorylation–dephosphorylation cycle. The left panel illustrates the off position (unphosphorylated) of the biochemical switch, in which the phosphatase activity is higher than the kinase activity. When the kinase activity exceeds the phosphatase activity, as in the right panel, the biochemical switch is in the opposite state.

Equation (4.34):[1]

$$n_h = 2 \left(\frac{d \ln f^*}{d \ln \theta} \right)_{f^*=0.5} = \frac{(1 - \mu)(\mu \gamma - 1)}{\mu(\gamma - 1)}. \qquad (5.11)$$

Thus $n_h \leq 1$ and equals 1 when $\mu \ll 1$ and $\mu \gamma \gg 1$. That is,

$$f^* = \frac{\theta}{1 + \theta}, \qquad (5.12)$$

which is the maximum obtainable sensitivity for the phosphorylation–dephosphorylation kinetics given in Equation (5.3).

5.1.2 Ultrasensitivity and the zeroth-order phosphorylation–dephosphorylation cycle

Can a phosphorylation–dephosphorylation switch be more sensitive to the level of kinase concentration than $n_h = 1$ as given in Equation 5.12? We note that the kinetic scheme in Equation (4.7) is obtained under the assumption of no Michaelis–Menten saturation. Since this assumption may not be realistic, let us move on to study the enzyme kinetics in Figure (5.2) in terms of saturable Michaelis–Menten kinetics. The mechanism by which saturating kinetics of the kinase and phosphatase leads to sensitive switch-like behavior is illustrated in Figure 5.4. The reaction fluxes as a function of f^* (the ratio $[S^*]/S_o$) for two cases are plotted. The first case (switch off)

---

[1] See exercise 1.

represents the situation where the phosphatase activity is slightly higher than the kinase activity. The value of f^* for which the reaction fluxes are equal (the steady state) is close to zero and the biochemical switch is in the unphosphorylated state (here denoted as the "off" state). When kinase activity is higher than phosphatase activity, the switch moves to the "on" state ($f^* \approx 1$).

To analyze this system, we assume that the enzymes obey the flux expression of Equation (4.10). Using this Michaelis–Menten expression, we replace Equation (5.5) with the following

$$\frac{d[S]}{dt} = -\frac{\frac{V_1[S]}{K_1} - \frac{V_1^*[S^*]}{K_1^*}}{1 + \frac{[S]}{K_1} + \frac{[S^*]}{K_1^*}} + \frac{\frac{V_2^*[S^*]}{K_2^*} - \frac{V_2[S]}{K_2}}{1 + \frac{[S^*]}{K_2^*} + \frac{[S]}{K_2}}. \tag{5.13}$$

In Equation (5.13),

$$V_1 = k_{+2}K_o, \quad V_1^* = k_{-1}K_o \tag{5.14}$$

are forward and reverse V_{max}'s for the kinase enzyme. Again, the constant K_o is the total concentration of the kinase. Similarly,

$$V_2^* = k_{+4}P_o, \quad V_2 = k_{-3}P_o \tag{5.15}$$

are parameters for the phosphatase enzyme, where P_o is the total concentration of the phosphatase.

Again examining the steady state, the fraction of phosphorylated $f^* = [S^*]^{ss}/S_o$ satisfies

$$\theta = \frac{\mu\gamma[\mu - (\mu + 1)f^*]\left(f^* - \frac{K_1^*(S_o + K_1)}{(K_1^* - K_1)S_o}\right)K_2K_2^*(K_1^* - K_1)}{[\mu\gamma - (\mu\gamma + 1)f^*]\left(f^* + \frac{K_2^*(S_o + K_2)}{(K_2 - K_2^*)S_o}\right)K_1K_1^*(K_2 - K_2^*)}, \tag{5.16}$$

where

$$\theta = \frac{V_1 K_2^*}{K_1 V_2^*} = \frac{k_{+1}k_{+2}(k_{-3} + k_{+4})[ATP]K_o}{(k_{-1} + k_{+2})k_{+3}k_{+4}P_o}, \tag{5.17a}$$

$$\mu = \frac{V_2 K_2^*}{K_2 V_2^*} = \frac{k_{-3}k_{-4}[PI]}{k_{+3}k_{+4}}, \tag{5.17b}$$

$$\gamma = \frac{V_1 K_1^* V_2^* K_2}{V_1^* K_1 V_2 K_2^*} = \frac{k_{+1}k_{+2}k_{+3}k_{+4}[ATP]}{k_{-1}k_{-2}k_{-3}k_{-4}[ADP][PI]}, \tag{5.17c}$$

are the same parameters in Equation (5.7). Equation (5.16) can be solved for an explicit equation for f^*. However, such an expression is terribly messy and it is more convenient to deal with the implicit function of Equation (5.16).

If we assume that affinity of kinase for S^* and affinity of phosphatase for S are low, that is $K_1^* \gg K_1$, $K_2 \gg K_2^*$, and $K_2 \gg S_o$, then Equation (5.16) is simplified into

$$\sigma = \frac{V_1}{V_2^*} = \frac{\mu\gamma[\mu - (\mu + 1)f^*]\left(f^* - 1 - \frac{K_1}{S_o}\right)}{[\mu\gamma - (\mu\gamma + 1)f^*]\left(f^* + \frac{K_2^*}{S_o}\right)}. \tag{5.18}$$

Equation (5.18) is the general equation for a phosphorylation–dephosphorylation switch with Michaelis–Menten kinetics [163]. Two limiting cases are particularly worth mentioning. First, if the kinase and the phosphatase are not saturated with respect to the substrate, i.e., $S_o \ll K_1, K_2^*$, then we have

$$\theta = \frac{\sigma K_2^*}{K_1} = \frac{\mu\gamma[(\mu + 1)f^* - \mu]}{\mu\gamma - (\mu\gamma + 1)f^*}. \tag{5.19}$$

As expected, solving f^* in Equation (5.19) reduces to the Equation (5.6).

Second, if the free energy is infinite, i.e., $\gamma \to \infty$, and the basal level of phosphorylation is zero, i.e., $\mu = 0$, then we have

$$\sigma = \frac{f^*\left(1 - f^* + \frac{K_1}{S_o}\right)}{(1 - f^*)\left(f^* + \frac{K_2^*}{S_o}\right)}. \tag{5.20}$$

This is the famous Goldbeter–Koshland formula for zeroth-order ultrasensitivity [70]. The most important feature of this equation for f^* as function of σ, the control parameter, is that the transition can be very sharp. In fact, the Hill coefficient is

$$n_h = \frac{\left(1 + 2\frac{K_1}{S_o}\right)\left(1 + 2\frac{K_2^*}{S_o}\right)}{\frac{K_1}{S_o} + \frac{K_2^*}{S_o} + 4\frac{K_1 K_2^*}{S_o^2}}. \tag{5.21}$$

For example, for $\frac{K_1}{S_o} = \frac{K_2^*}{S_o} = 0.01$, $n_h = 51$, indicating an *ultrasensitive* switch! The coefficient n_h is large if K_1 and $K_2^* \ll S_o$. That is, the enzyme reactions are highly saturated. This means the rates for the phosphorylation and dephosphorylation reactions $S \rightleftharpoons S^*$ are independent of the respective substrate concentrations [S] and [S*]. Hence, both reactions are effectively zeroth order. Ultrasensitivity arises from this zeroth-order behavior. In comparison with the case where there is no enzyme saturation, the dashed lines in Figure 5.3 show the sharp transitions. Figure 5.3 also shows that the ultrasensitivity does not change the amplitude of the switch.

5.1.3 Substrate selectivity of the phosphorylation–dephosphorylation switch

We now study how, as a kinetic circuit, a phosphorylation–dephosphorylation switch can amplify the affinity difference between two competing substrate

proteins, thus achieving higher specificity than expected from in vitro measurements. Our study here is based on the simple unsaturated kinetics of Section 5.1.1.
Focusing on the ratio of the concentrations of the phosphorylated to the dephosphorylated forms of a substrate protein we have: $\frac{f^*}{1-f^*}$, where f^* is given in Equation (5.6):

$$\frac{f^*}{1 - f^*} = \frac{\theta^{(i)} + \mu^{(i)}}{\theta^{(i)}/(\mu^{(i)}\gamma) + 1},$$

(5.22)

in which the superscripts $i = 1, 2$ are for the two competing substrate proteins.
As before, γ is related to the free energy of ATP hydrolysis that is independent
of the substrate proteins or the kinetic constants. In order to make a comparison,
let us assume that the two substrate proteins are structurally homologous in the
S form but identical in the S* form, with differences only in the dissociation rate
constants k_{-1} and k_{+4}. Because the equilibrium constant for ATP hydrolysis $K_{eq} =$
$(k_{+1}k_{+2}k_{+3}k_{+4})/(k_{-1}k_{-2}k_{-3}k_{-4})$ is fixed, we have

$$\frac{k_{-1}^{(1)}}{k_{-1}^{(2)}} = \frac{k_{+4}^{(1)}}{k_{+4}^{(2)}} = \xi.$$

(5.23)

Without the loss of generality, we assume $\xi < 1$. This means the affinity of the
kinase for the protein 1 is greater than that of the protein 2.

The ratio of the ratios,

$$\eta = \frac{\left(\frac{f^*}{1-f^*}\right)^{(1)}}{\left(\frac{f^*}{1-f^*}\right)^{(2)}} = \frac{\mu^{(1)}\left(\frac{\theta^{(1)}}{\mu^{(1)}} + 1\right)\left(\frac{\theta^{(2)}}{\mu^{(2)}\gamma} + 1\right)}{\mu^{(2)}\left(\frac{\theta^{(1)}}{\mu^{(1)}\gamma} + 1\right)\left(\frac{\theta^{(2)}}{\mu^{(2)}} + 1\right)},$$

(5.24)

characterizes the difference in the affinities of the phosphorylation–dephosphorylation cycle for proteins 1 and 2. In a test tube with $\gamma = 1$ (in the absence of ATP free
energy), $\eta = \mu^{(1)}/\mu^{(2)} = 1/\xi$. However, with ATP hydrolysis driving the system,

$$\eta = \frac{(\omega + \xi)(\omega + \gamma)}{\xi(\omega + \xi\gamma)(\omega + 1)},$$

(5.25)

in which $\omega = k_{+1}k_{+2}[\text{ATP}]K_o/(k_{-1}^{(2)}k_{-4}[\text{PI}]P_o)$ is a constant. Equation (5.25) is
obtained if $k_{-1} \gg k_{+2}$ and $k_{-3} \gg k_{+4}$, leading to $\theta^{(1)}/\theta^{(2)} = 1/\xi^2$. Equation (5.25)
gives $\eta = 1/\xi$ for $\omega = 0$ and ∞. The maximal selectivity occurs when $\omega = \sqrt{\xi\gamma}$
and we have

$$\eta = \frac{\left(\sqrt{\xi} + \sqrt{\gamma}\right)^2}{\xi\left(1 + \sqrt{\xi\gamma}\right)^2}.$$

(5.26)

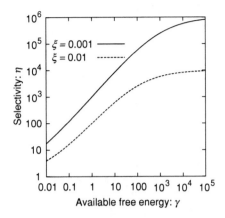

Figure 5.5 Biomolecule selectivity, η, as a function of the intrinsic affinity, ξ, and the available free energy $\Delta G = k_B T \ln \gamma$, according to Equation (5.26).

Equation (5.26) gives the selectivity as a function of the available free energy in ATP hydrolysis [164]. In the limit of $\gamma \gg 1$, we have $\eta = 1/\xi^2$. This is the celebrated result of John Hopfield and Jacques Ninio [100, 150], who independently discovered in the mid 1970s the kinetic proofreading mechanism for biosynthesis. They showed that biomolecular specificity can be amplified in living cells through pure kinetic means (at the expense of consuming cellular free energy) without altering molecular structures and equilibrium affinities. Figure 5.5 shows how the selectivity increases with increasing free energy. We also note that if the hydrolysis reaction were to be thermodynamically driven in the reverse direction, i.e., $\gamma \leq 1$, then the same mechanism can diminish the difference between the affinities of two substrates.

5.1.4 The GTPase signaling module

The transmembrane GTPase system, which was discovered by Martin Rodbell and Alfred Gilman in the 1970s [175, 176], is another important cellular signaling module. The GTPase signaling system involves a GTP-hydrolyzing protein (here called GTPase, or G-protein) that acts with a protein called GTPase activating protein (GAP) and a protein called guanine-nucleotide exchange factor (GEF). The "wiring diagram" of the GTPase system is remarkably isomorphic to that of the phosphorylation–dephosphorylation cycle, as shown in Figure 5.6. Usually, the GTP-bound GTPase is the active form of the biological molecule. The hydrolysis of GTP in the protein–nucleotide complex is catalyzed by GAP, generating a GTPase–GDP complex. The GTPase–GDP complex exchanges its nucleotide to form GTPase–GTP via a reaction catalyzed by the GEF, as illustrated in Figure 5.6.

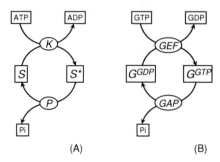

Figure 5.6 "Wiring diagrams" for two most important cellular signaling modules (small networks). (A) Phosphorylation–dephosphorylation cycle as shown in Figure (5.2): substrate protein that can be in either phosphorylated and dephosphorylated states, S and S^*. K is a protein kinase and P is a protein phosphatase. The entire switching on and off cycle hydrolyzes one ATP. (B) Essentially isomorphic topology for the GTPase transmembrane signaling system. G^{GTP} and G^{GDP} are GTP- and GDP-bound GTPase. GEF is guanine-nucleotide exchange factor and GAP is GTPase activating protein.

The reactions of the GTPase system are

$$G^{GDP} + GEF \quad \underset{k^o_{-1}}{\overset{k^o_{+1}}{\rightleftharpoons}} \quad G^{GDP} \cdot GEF$$

$$G^{GDP} \cdot GEF + GTP \quad \underset{k^o_{-2}}{\overset{k^o_{+2}}{\rightleftharpoons}} \quad G^{GTP} + GDP + GEF$$

$$G^{GTP} + GAP \quad \underset{k^o_{-3}}{\overset{k^o_{+3}}{\rightleftharpoons}} \quad G^{GTP} \cdot GAP$$

$$G^{GTP} \cdot GAP \quad \underset{k^o_{-4}}{\overset{k^o_{+4}}{\rightleftharpoons}} \quad G^{GDP} + GAP + PI. \tag{5.27}$$

In the previous section, we obtained the steady state population of the phosphorylated protein in a phosphorylation–dephosphorylation network that consists of a kinase and a phosphatase. Based on the homologous reaction networks, the mathematics for the phosphorylation network translates immediately to the GTPase system.

If we consider a single GTPase protein, the chemical physics of macromolecules dictates that it fluctuates between the GDP-bound and GTP-bound states. In this case f^* then can be interpreted as the probability of a single substrate protein in the GTP state. In addition to this steady state probability, we would also like to know how frequently the signaling protein switches between the on and off states. The rate of cycling per unit time is directly related to the mean dwell time in the activated state.

Denoting T as the time in the GTPase–GTP state, then the number of switching cycles per unit time is simply f^*/T. More precisely, each time the protein is activated, the dwell time is stochastic; the T defined above is the mean dwell time. In addition to the mean time, we are interested in the probability distribution of dwell times. This distribution, which was first studied in [128], can be biologically important since the GTPase has been widely considered as a timer in cellular signaling: the amount of time a GTPase is in the GTP-bound state is directly related to the "amount of signal" transmitted to the down-stream biochemical event.

5.1.5 Duration of switch activation and a biochemical timer

In analyzing the temporal behavior of a biochemical switching molecule, we can study either of the equivalent models of the phosphorylation–dephosphorylation cycle or the GTPase signaling module. In particular, we are interested in the duration of each activation event at the single-molecule level.

For the GTPase system, every GTPase molecule can be in one of the four states. Using 1, 2, 3, and 4 to represent the G^{GDP}, $G^{GDP} \cdot GEF$, G^{GTP}, and $G^{GTP} \cdot GAP$, we have from Equation (5.27)

$$1 \underset{k_{-1}}{\overset{k_{+1}}{\rightleftharpoons}} 2, \quad 2 \underset{k_{-2}}{\overset{k_{+2}}{\rightleftharpoons}} 3, \quad 3 \underset{k_{-3}}{\overset{k_{+3}}{\rightleftharpoons}} 4, \quad 4 \underset{k_{-4}}{\overset{k_{+4}}{\rightleftharpoons}} 1, \tag{5.28}$$

with pseudo-first order rate constants defined as $k_{+1} = k_{+1}^{o}[GEF]$, $k_{-1} = k_{-1}^{o}$, $k_{+2} = k_{+2}^{o}[GTP]$, $k_{-2} = k_{-2}^{o}[GDP][GEF]$, $k_{+3} = k_{+3}^{o}[GAP]$, $k_{-3} = k_{-3}^{o}$, $k_{+4} = k_{+4}^{o}$, $k_{-4} = k_{-4}^{o}[PI][GAP]$. Again, the overall system is driven by free energy dissipation, in this case through hydrolysis of GTP to GDP and PI rather than ATP to ADP and PI as in the phosphorylation–dephosphorylation system.

States 3 and 4 correspond to GTPase with GTP bound. Hence we are interested in the dwell time in state 3 and 4: $(3 \cup 4)$. If we can determine the probability of a molecule remaining in state $3 \cup 4$ at time t, given that the system was in state $3 \cup 4$ at $t = 0$, then we will have the distribution of dwell times in the state. From the kinetics, the probabilities satisfy the equations

$$\frac{dp_3}{dt} = -(k_{-2} + k_{+3})p_3 + k_{-3}p_4 \tag{5.29a}$$

$$\frac{dp_4}{dt} = k_{+3}p_3 - (k_{-3} + k_{+4})p_4. \tag{5.29b}$$

This is a system of two linear ordinary differential equations. The eigenvalues of the system are

$$\lambda_{1,2} = \frac{1}{2}\Big[(k_{-2} + k_{+3} + k_{-3} + k_{+4})$$
$$\pm \sqrt{(k_{-2} + k_{+3} + k_{-3} + k_{+4})^2 - 4(k_{-2}k_{-3} + k_{+3}k_{+4} + k_{-2}k_{+4})}\Big].$$
$$\tag{5.30}$$

The solution to the Equation (5.29), with an initial condition of $p_3(0) + p_4(0) = 1$, is

$$p_3(t) + p_4(t) = \alpha e^{-\lambda_1 t} + (1 - \alpha)e^{-\lambda_2 t}, \quad (\lambda_2 > \lambda_1 > 0), \tag{5.31}$$

where the amplitude depends on the value of $p_3(0) \in (0, 1)$:

$$\alpha = \frac{(k_{+4} - k_{-2})p_3(0) - k_{+4} + \lambda_2}{\lambda_2 - \lambda_1}. \tag{5.32}$$

If the system is in state $3 \cup 4$ at $t = 0$, then the probability of remaining in the state at time t, $p_{3\cup4}(t)$ in Equation (5.31), is the probability that the dwell time is greater than t. That is, $p_{3\cup4}(t) = \Pr\{T > t\}$. The probability density function for the dwell time, T, then is

$$f_T(t) = \frac{d}{dt}(1 - p_{3\cup4}(t)) = \alpha\lambda_1 e^{-\lambda_1 t} + (1 - \alpha)\lambda_2 e^{-\lambda_2 t}, \tag{5.33}$$

which has mean and variance

$$\langle T \rangle = \frac{\alpha}{\lambda_1} + \frac{1 - \alpha}{\lambda_2} \tag{5.34}$$

$$\langle (\Delta T)^2 \rangle = \frac{2\alpha}{\lambda_1^2} + \frac{2(1 - \alpha)}{\lambda_2^2} - \langle T \rangle^2. \tag{5.35}$$

To see how broad is the probability distribution for T, we consider the relative variance (r.v.):

$$\boxed{\frac{\langle (\Delta T)^2 \rangle}{\langle T \rangle^2} = \frac{2\alpha\lambda_2^2 + 2(1 - \alpha)\lambda_1^2}{(\alpha\lambda_2 + (1 - \alpha)\lambda_1)^2} - 1. \qquad (5.36)}$$

We see that if $\alpha = 1$ or $\alpha = 0$, the r.v. is 1. In fact, for $0 \le \alpha \le 1$, the r.v. is always greater than 1. If $\alpha \gg 1$, then the r.v. can be very small. One can verify, however, that the α in Equation (5.33) has to be $\le \frac{\lambda_2}{\lambda_2 - \lambda_1}$ in order for the $f_T(t)$ to be positive for all t – a necessary condition for the $f_T(t)$ to be a meaningful probability density function. Hence, when $\alpha = \frac{\lambda_2}{\lambda_2 - \lambda_1}$, the $f_T(t)$ achieves the minimal r.v. of

$$\frac{\lambda_1^2 + \lambda_2^2}{(\lambda_1 + \lambda_2)^2}. \tag{5.37}$$

This minimal value of r.v. corresponds to the maximal accuracy of the timer. In the limit that r.v. is small, the dwell time at the single-molecule level is consistent between individual molecules. Yet the minimal value that this r.v. can obtain is 1/2, which occurs in the limit that $\lambda_1 \approx \lambda_2$.

We note that Equation (5.32) gives the minimal r.v. obtained in the limit that $[GDP] \to 0$ (and thus $k_{-2} \to 0$) and $p_3(0) = 1$. Indeed, near irreversibility is provided by the free energy of GTP hydrolysis, which drives the cycle in a living cell. When GDP and PI go to zero, k_{-4} and k_{-2} approach zero, and $\gamma \to \infty$. Thus the optimal timer is obtained (as we have seen for the cases of optimal sensitivity and specificity), when the free energy GTP (or ATP) hydrolysis is maximized.

In a closed chemical system in which the GTP is in chemical equilibrium with GDP and PI, there is no net GTP hydrolysis. In this case

$$\frac{k_{+1}k_{+2}k_{+3}k_{+4}}{k_{-1}k_{-2}k_{-3}k_{-4}} = \frac{k^o_{+1}k_{+2}k_{+3}k_{+4}[\text{GTP}]}{k_{-1}k^o_{-2}k_{-3}k^o_{-4}[\text{GDP}][\text{PI}]} = 1,$$

and

$$p_3(0) = \frac{k_{+1}k_{+2}}{k_{+1}k_{+2} + k_{-1}k_{-4}}. \tag{5.38}$$

Substituting this result into Equation (5.32), we obtain

$$\alpha = \frac{k_{+1}k_{+2}(\lambda_2 - k_{-2}) + k_{-1}k_{-4}(\lambda_2 - k_{+4})}{(k_{+1}k_{+2} + k_{-1}k_{-4})(\lambda_2 - \lambda_1)}. \tag{5.39}$$

We note that both k_{+4} and $k_{-2} \geq \lambda_1$, and both $\leq \lambda_2$.[2] Hence

$$\frac{k_{+1}k_{+2}k_{-2} + k_{-1}k_{-4}k_{+4}}{k_{+1}k_{+2} + k_{-1}k_{-4}} \geq \lambda_1.$$

This leads to $0 \leq \alpha \leq 1$. Thus, without free energy from GTP hydrolysis, the probability density function for the dwell time is monotonic with r.v. greater than 1, resulting in a non-ideal and inaccurate timer.

The results of this mathematical analysis are perhaps most easily illustrated based on an example with model parameters assigned specific values. Using the parameter values $k^o_{-2} = 5 \times 10^6\,\text{M}^{-2}{\cdot}\text{sec}^{-1}$, $k^o_{+3} = 3.30 \times 10^4\,\text{M}^{-1}{\cdot}\text{sec}^{-1}$, $k^o_{-3} = 3 \times 10^{-5}\,\text{sec}^{-1}$, and $k^o_{+4} = 0.0333\,\text{sec}^{-1}$, and physiologically realistic values of the concentrations, $[\text{GEF}] = 1\,\mu\text{M}$, $[\text{GAP}] = 1\,\mu\text{M}$, $[\text{GTP}] = 5\,\text{mM}$, $[\text{GDP}] = 0.1\,\text{mM}$, and $[\text{PI}] = 1\,\text{mM}$, we obtain $\lambda_1 = 0.0324\,\text{sec}^{-1}$ and $\lambda_2 = 0.0344\,\text{sec}^{-1}$. Note that concentrations of $[\text{GTP}]$, $[\text{GDP}]$, and $[\text{PI}]$ yield $\Delta G'_{GTP} = -60\,\text{kJ}{\cdot}\text{mol}^{-1}$ for the reaction $\text{GTP} \rightleftharpoons \text{GDP} + \text{PI}$ and correspond to a physiologically reasonable cellular phosphorylation potential. With these values and setting $p_3(0) = 1$, Equation (5.32) yields $\alpha = 16.98$. The functions $p_{3\cup4}(t)$ and $f_T(t)$ for these values are plotted as solid lines in Figure 5.7.

To treat the case where the reaction $\text{GTP} \rightleftharpoons \text{GDP} + \text{PI}$ is in equilibrium, we compute the concentrations $[\text{GTP}] = 4.1 \times 10^{-8}\,\text{mM}$, $[\text{GDP}] = 5.1\,\text{mM}$, and

---

[2] The square-root term in Equation (5.30) can be rewritten as $\sqrt{(k_{+4} + k_{-3} - k_{-2} - k_{+3})^2 + 4k_{+3}k_{-3}}$. Hence, $\lambda_1 \leq \frac{1}{2}[(k_{-2} + k_{+3} + k_{-3} + k_{+4}) - |k_{+4} + k_{-3} - k_{-2} - k_{+3}|]$ and $\lambda_2 \geq \frac{1}{2}[(k_{-2} + k_{+3} + k_{-3} + k_{+4}) + |k_{+4} + k_{-3} - k_{-2} - k_{+3}|]$. That is, $\lambda_1 \leq$ both $(k_{+4} + k_{-3})$ and $(k_{-2} + k_{+3}) \leq \lambda_2$.

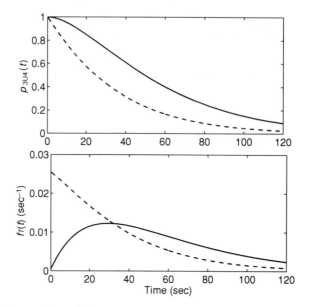

Figure 5.7 Comparison of GTPase timing for physiological (non-equilibrium) and non-physiological (equilibrium) cases. Results at cellular phosphorylation potential ($\Delta G'_{GTP} = -60\,\text{kJ·mol}^{-1}$) are plotted as solid lines; results at equilibrium ($\Delta G'_{GTP} = 0$) are plotted as dashed lines. The top panel plots $p_{3 \cup 4}(t)$, the probability that the G protein is in the GTP-bound state given that it is in the state G^{GTP} at $t = 0$. The bottom panel plots $f_T(t)$ the probability distribution of dwell time in the GTP-bound state. See text for details and parameter values.

[PI] = 6 mM, based on the equilibrium constant value $K_{eq} = 7.5 \times 10^5$ M. Assuming that all rate constants remain the same as for the physiological example, these concentrations yield $\lambda_1 = 0.0333\,\text{sec}^{-1}$ and $\lambda_2 = 0.0585\,\text{sec}^{-1}$. In this case, setting $p_3(0) = 1$, Equation (5.32) yields $\alpha = 1.31$. The functions $p_{3 \cup 4}(t)$ and $f_T(t)$ for the equilibrium case are plotted as dashed lines in Figure 5.7.

Note that the relative variance of the timing probability distribution is considerably larger in the equilibrium case than in the physiological case. However, even in the physiological case, the system behavior is far from that of a perfect timer. In this near ideal case, r.v. $\approx 1/2$, which is the minimal value obtained by Equation (5.37) when $\lambda_1 \approx \lambda_2$.

In cells, improved timing accuracy arises from cascades of phosphorylation events. This insight is in fact the theoretical basis for a recent kinetic model for phototransduction signaling in vertebrate rod cells published by Hamer *et al.* [76]. It is proposed that multiple, successive phosphorylations of rhodopsin by rhodopsin kinase lead to high accuracy in the single-photon responses of a rod cell. Briefly, if we assume n identical, irreversible phosphorylation steps with only the fully

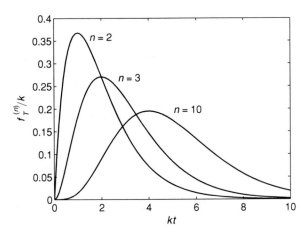

Figure 5.8 Dwell time probability distributions predicted by Equation (5.40) for several values of n.

phosphorylated rhodopsin being active, then the probability distribution for the dwell time will be simply

$$f_T^{(n)} = \frac{k^n t^{n-1}}{(n-1)!} e^{-kt} \tag{5.40}$$

in which the k is the rate constant associated with a single step. Equation (5.40) is known as the Gamma distribution in statistics. The r.v. for the Gamma distribution is $\frac{1}{n}$. Our previous example has $n = 2$. In the model by Hamer *et al.*, n may be as large as 7. Figure 5.8 illustrates dwell time probability distributions predicted by Equation (5.40) for several values of n.

5.1.6 Synergistic action of kinases and phosphatases and the phosphorylation energy hypothesis

We have seen that the behaviors of biochemical signaling modules, in terms of their sensitivity (Section 5.1.2), specificity (Section 5.1.3), and timing (Section 5.1.5), are all intimately tied to the available free energy for ATP and GTP hydrolysis. The previous sections of this chapter have shown that, as a function of the level of kinases and phosphatases, the fractions of the signaling molecules in various states can display either a graded transition, if the kinases and phosphatases are not saturated, or an all-or-none transition, if the kinases and phosphatases are highly saturated. (See Figure 5.3.) Both types of responses are employed in cellular regulations and signaling.

The interplay of kinases and phosphatases can go beyond the simple activation switching. Even though it is widely accepted that the state of activity of a signaling protein depends on the ratio of its phosphorylated to dephosphorylated forms ($\frac{f^*}{1-f^*}$), other mechanisms have been suggested, in particular for the tyrosine kinases and phosphatases. Fischer and his colleagues [57] have suggested a scenario in which dephosphorylated proteins have temporal memory: after dephosphorylation by a protein tyrosine phosphatase (PTP), the substrate protein can retain its activity for a finite period of time, before it returns to the dephosphorylated, inactive state.

As we have seen, phosphatase-catalyzed protein dephosphorylation reactions do not merely counteract the kinase catalyzed protein phosphorylation. The hydrolysis cycle is not "futile." Free energy from the ATP and GTP hydrolysis is essential in the functioning of biochemical switches. Thus, it is not surprising that biological organisms have evolved to use phosphorylation as one of the dominant mechanisms for signal transduction regulations. The *phosphorylation hypothesis*, recently proposed by one of the authors, suggests that free energy derived from cellular phosphorylation is necessary for overcoming intrinsic biochemical "noise" from thermal agitations, small copy numbers, and limited affinities, guaranteeing precise and robust cell signaling and functions [166].

5.2 Biochemical regulatory oscillations

Engineers have long been fascinated by oscillations. Oscillatory dynamics in mechanics and electronics provides our modern lives with clocks and radios among many other devices. Yet it is arguable that chemical oscillations are yet to find an application apparent in daily life. In biology, oscillations are everywhere, but the governing molecular mechanisms are often elusive. In recent years, bioengineers have started to study biological circuits that oscillate at the cellular level. This approach, now known as *synthetic biology* [9], has revealed a host of interesting dynamics a cellular system can exhibit. Through these efforts we have improved our level of understanding and confidence in cellular modeling predictions.

5.2.1 Gene regulatory networks and the repressilator

As we have seen in Section 3.1.4.2, oscillations can arise from chemical kinetic systems. As a general rule, in order for such a kinetic system to oscillate, it must exhibit both activation and inhibition. Such a feedback loop in fact exists in one of the central reactions of molecular biology: gene expression, producing mRNA and leading to protein synthesis. If the synthesized protein serves as a transcription factor and is a repressor of the gene expression, then there is a simple feedback loop. This idea motivated several researchers to construct such feedback transcriptional regulatory networks in living cells. In independent laboratories in 2000, Becskei and Serrano [20] constructed such a loop involving one repressor protein; Gardner, Cantor and Collins [62] constructed such a loop with two repressor proteins; and Elowitz and Leibler [50] designed a system consisting of three repressor proteins. These investigators showed that by increasing the delay in the feedback loop, the dynamics of the synthetic transcription system becomes more complex. Becskei and Serrano observed on and off transitions, Gardner *et al.* were able to show bistability in the on and off states. Elowitz and Leibler indeed saw oscillations, which they named a *repressilator*.

Elowitz and Leibler modeled their system in terms of the concentrations of three repressor proteins, *lacI*, *tetR*, and *cI*, and their corresponding mRNA concentrations. If we use subscripts 1, 2, and 3 to denote these three, their generic equations for the mRNA and protein concentrations m_i and p_i ($i = 1, 2, 3$) are

$$\frac{dm_i}{dt} = -\gamma_i m_i + \alpha_{i0} + \frac{\alpha_{i1}}{1 + \left(\frac{p_{i-1}}{K_i}\right)^2}, \tag{5.41a}$$

$$\frac{dp_i}{dt} = \lambda_i m_i - \beta_i p_i, \tag{5.41b}$$

where $p_0 = p_3$. The rationale for these kinetic equations is as follows. The ith mRNA has associated with it an intrinsic degradation rate γ_i. The parameter α_{i0} is the rate of mRNA transcription in the presence of saturating repressor $i - 1$, representing promotor "leakiness." The repressors act via cooperative binding with an affinity K_i and Hill coefficient of 2. The sum $(\alpha_{i1} + \alpha_{i0})$ is the maximal rate transcription of the ith mRNA in the absence of the repressor. The parameter λ_i is the rate of synthesis of protein i, and β_i is the rate of the protein degradation. Mathematical analysis of models like that in Equation (5.41) have a rich history. The model was first proposed by Brian Goodwin in 1965 for studying oscillatory enzyme control processes [71]. (For a recent discussion of the model, see [51].)

There are six ODEs and eighteen parameters in the full model of Elowitz and Leibler. To simplify the analysis, Elowitz and Leibler assumed that all the parameters are independent of the i in their subscripts. That leaves six parameters, which can be further simplified by introducing unitless variables

$$u_i = \frac{\lambda}{\beta K} m_i, \ v_i = \frac{1}{K} p_i, \ \tau = \gamma t \tag{5.42}$$

$$\frac{du_i}{d\tau} = -u_i + a_0 + \frac{a_1}{1 + v_{i-1}^2}, \tag{5.43a}$$

$$\frac{dv_i}{d\tau} = b(u_i - v_i), \tag{5.43b}$$

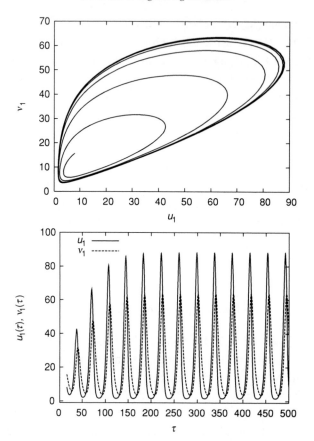

Figure 5.9 Biochemical oscillation of an engineered signaling system, "repressi-lator." u_1 and v_1: *lacI* mRNA and protein concentrations. The behavior of the other two genes, *tetR* and *cI*, is essentially the same, with a time delay. See [50] for more details.

where $i = 1, 2, 3$, $v_0 = v_3$, and

$$a_0 = \frac{\lambda \alpha_0}{\beta \gamma K}, \ a_1 = \frac{\lambda \alpha_1}{\beta \gamma K}, \ b = \frac{\beta}{\gamma}. \tag{5.44}$$

The set of parameter values $\alpha_0 = 5 \times 10^{-4} \sec^{-1}$, $\alpha_1 = 0.5 \sec^{-1}$, $\lambda = 20 \beta K$, $\beta = 1.67 \times 10^{-3} \sec^{-1}$, $\gamma = 0.83 \times 10^{-2} \sec^{-1}$, $K = 40$, yields $a_0 = 1.2$, $a_1 = 1200$, and $b = 0.2$ for Equation (5.43). Simulation of the oscillatory dynamics of this system is illustrated in Figure (5.9).

To obtain the steady state(s) of Equation (5.43), we have by repeated substitution:

$$u_1 = f(u_3) = f(f(u_2)) = f(f(f(u_1))), \tag{5.45}$$

in which the function

$$f(x) = a_0 + \frac{a_1}{1 + x^2} \tag{5.46}$$

is a strictly monotonic decreasing function of x for $x > 0$. Hence, $f(f(f(x)))$ is a strictly monotonic decreasing function of x, and Equation (5.45) has a single root x^*. The system of ODEs, thus, has a unique steady state: $u_i = v_i = x^*$ $(i = 1, 2, 3)$, which satisfies

$$(1 + x^{*2})(x^* - a_0) + a_1 = 0. \tag{5.47}$$

How can the result of unique steady state be consistent with the observed oscillation in Figure 5.9? The answer is that the steady state, which mathematically exists, is physically impossible since it is *unstable*. By unstable, we mean that no matter how close the system comes to the unstable steady state, the dynamics leads the system away from the steady state rather than to it. This is analogous to the situation of a simple pendulum, which has an unstable steady state when the weight is suspended at exactly at $180°$ from its resting position. (Stability analysis, which is an important topic in model analysis and in differential equations in general, is discussed in detail in a number of texts, including [146].)

5.2.2 Biochemical oscillations in cell biology

The 1952 Hodgkin–Huxley model for membrane electrical potential is perhaps the oldest and the best known cellular kinetic model that exhibits temporal oscillations. The phenomenon of the nerve action potential, also known as excitability, has grown into a large interdisciplinary area between biophysics and neurophysiology, with quite sophisticated mathematical modeling. See [103] for a recent treatise.

Derived from Hodgkin–Huxley's celebrated theory and inspired by the experimental observations, cellular calcium dynamics, either stimulated via inositol 1,4,5-trisphosphate (IP_3) receptor in many non-muscle cells [69, 139], or via the ryanodine receptor in muscle cells [108], is another extensively studied oscillatory system. Both receptors are themselves Ca^{2+} channels, and both can be activated by Ca^{2+}, leading to calcium-induced calcium release from endoplasmic reticulum.

The cell cycle oscillator is one of the best studied cellular signaling networks in terms of kinetic models. Readers are encouraged to consult the very readable paper [198] and book chapter [197] by John Tyson and his colleagues. For a succinct review of recent studies of various oscillations in cell biology, see [120].

Concluding remarks

Traditional cellular biochemical research, focusing on either metabolism or signal transduction, primarily deals with pathways. By pathways, we mean the biochemical reaction schemes as defined by a set of chemical reactions that define a set of species (or *players* in layman's terminology), and a set of reactions that connect the species. One important research focus is to determine schematics like that in Figure (5.1) for important pathways and networks. Over a period of time from approximately the 1940s to the 1970s, the studies on metabolic pathways focused on the production and utilization of important biochemicals inside cells. These metabolic pathways include the biosynthesis of DNA, proteins, and a host of other macro- and small molecules that are essential for cell survival. More recently, the diverse functional aspects of cells have led molecular biology to focus on regulation of the pathways. The cellular regulations are themselves achieved through biochemical networks consisting of reactions and regulatory proteins. These issues underlie the importance of studying biochemical reactions and processes from an integrated systems perspective since studying biochemical processes as isolated phenomena is often not sufficient for understanding the behavior of cells [113].

One of the possible approaches to current research in systems biology of cells is to quantify the biochemical pathways and develop integrative mathematical models for the pathways. Of course mathematical models come in different forms. The focus of this chapter (and the following chapter on modeling reaction networks) is on models developed in terms of chemical reaction kinetics. While this approach has the advantage of a concrete foundation in physical chemistry, it does not necessarily mean that models developed through such an approach are automatically valid. The difficulty is in the incompleteness of the chemical knowledge we have in biochemical reactions, and in the impossibility of computations from first principles. Early emphasis on kinetics and thermodynamics in biochemistry was greatly diminished in recent decades, perhaps resulting in less than optimal progress on the basic chemical knowledge. We anticipate that the current emphasis on quantitative model building, combined with technologies for high-throughput measurement and databasing, will result in a dramatic increase in the available kinetic and thermodynamic data for biochemical systems modeling.

Nevertheless one still needs to proceed in physicochemical-based modeling with caution. It is easy to build models of great size and complexity that involve biochemical reactions with many more unknown mechanisms and parameter values than are available from experimental observation. The value in studying small biochemical modules is that such models can be experimentally tested, through efforts such as in synthetic biology. These small systems (or modules) serve intermediate entities in our quantitative thinking and actual modeling of large systems.

Exercises

5.1 Show that when $\theta = \frac{\mu\gamma(1-\mu)}{\mu\gamma-1}$ in Equation (5.6) for the non-saturating phosphorylation–dephosphorylation cycle of Section (5.1.1), $f^* = \frac{1}{2}$.

5.2 For Equation (5.12), show that if one wants to increase f^* from 0.1 to 0.9 by changing θ, one needs to increase θ by a factor of 81. That is, it takes about a one-hundred-fold increase in the kinase activity to turn on the switch.

5.3 Consider a phosphorylation–dephosphorylation cycle for a substrate protein (S) with saturated kinase (K) but unsaturated, first-order, phosphatase (P). For simplicity, we neglect the cofactors such as ATP, ADP, and PI:

$$S + K \underset{k_2}{\overset{k_1}{\rightleftharpoons}} KS \xrightarrow{k_3} S^* + K, \ S^* \xrightarrow{k_4 P} S,$$

in which the concentration of phosphatase, [P], is assumed to be constant.

(a) Establish the kinetic equations for the concentrations [S], [K], [KS], and [S*]. Note that the total concentrations for the substrate protein and kinase are constants:

$$S_t = [S] + [KS] + [S^*], \quad K_t = [K] + [KS].$$

Show that the steady state concentration of the phosphorylated $[S^*]^{ss} = x$ satisfies the quadratic equation

$$\left(1 + \frac{K_t}{\theta S_t}\right) x^2 - (\theta S_t + K_M + K_t + S_t)x + \theta S_t^2 = 0,$$

in which $\theta = \frac{k_3 K_t}{k_4 [P] S_t}$, and $K_M = \frac{k_2 + k_3}{k_1}$ is the Michaelis–Menten constant for the kinase.

(b) Solve for x in (a) as a function of θ. This function is called the *activation curve*. Explore the dependence of the activation curve on the parameters of the model: S_t, K_t, and K_M.

(c) If we introduce $f^* = \frac{[S^*]}{S_t}$, then one can compute the Hill coefficient defined in Equation (5.11). Try to show numerically that n_h is never greater than 2. What is the condition under which the $n_h = 2$?

5.4 Consider the biochemical reaction system with four species:

$$A \xrightarrow{k_1} X, \ X + Y \xrightarrow{k_2} 2Y, \ Y \xrightarrow{k_3} B.$$

(a) Assume that the species A and B are with fixed concentrations a and b, while the concentrations of X and Y vary. Write down the kinetic equations for [X] and [Y]. Show that the concentrations of X and Y oscillate.

(b) Next consider the case in which all reactions are reversible, with backward reaction rate constants k_{-1}, k_{-2}, and k_{-3}, for the three reactions. Show that the equilibrium constant for the ratio of b/a is $K_{AB} = \frac{k_1 k_2 k_3}{k_{-1} k_{-2} k_{-3}}$. Write down the kinetic equations for [X] and [Y] for the reversible system. Show that if $b/a = K_{AB}$, there will be no oscillation.

6

Biochemical reaction networks

Overview

Previous chapters have introduced methods for simulating the kinetics of relatively simple chemical systems, such as the phosphorylation–dephosphorylation system of Section 5.1 or the model of glycolysis illustrated in Section 3.1.4.2. However, the essential fact that biochemical reactants in solution exist as sums of rapidly interconverting species, as described in Chapter 2, is not explicitly taken into account in these simple models. As a result, influences of the binding of hydrogen and metal ions to reactants on thermodynamic driving forces and reactions' kinetics are not taken into account in these simulations.

Since a great deal of information is available regarding the thermodynamic and ion-binding properties of biochemical reactants [4, 5], it is possible to construct simulations of biochemical systems that properly incorporate these data. Specifically, realistic simulations of biochemical systems require combining the following concepts into the simulations.

(i) A formal treatment of biochemical reactants as sums of distinct species formed by different hydrogen and metal ion binding states.
(ii) Conservation of mass based on reaction stoichiometry and multiple equilibria of biochemical reactions.
(iii) pH and ionic dependence on enzyme kinetics and apparent equilibria and thermodynamic driving forces for biochemical reactions.

This chapter will present a general methodology for incorporating these elements into a simulation of a biochemical system and illustrate the concepts based on the specific example of a kinetic model of the tricarboxylic acid (TCA) cycle.

The final section of this chapter develops various methods and theories to explore control and regulation in biochemical networks.

6.1 Formal approach to biochemical reaction kinetics

The essential components of a methodology for biochemical network kinetic simulation formally treating reactants of sums of rapidly inter-converting species were synthesized by Vinnakota *et al.* in a computer model of glycogenolysis in skeletal muscle [205]. The methods, which follow from that primary work, are presented here in four steps: (1) establishing the components of the biochemical network model; (2) determining the expressions for the biochemical fluxes; (3) determining the differential equations for biochemical reactants, pH, and binding ions; and (4) computational implementation and testing.

6.1.1 Establishing the components of the biochemical network model

The first step is to determine what reactions and associated reactants are to be considered in a given model, and to compile all of the information necessary to specify the thermodynamic properties of the system. The necessary information for each reactant and each reaction in the system is specified below.

For each *reactant* we compile the following information.

(i) Definition of a reference species for a given reactant. The choice of reference species is arbitrary. However, it is convenient to choose the minimum-proton-bound state. For example, for the reactant ATP, we choose ATP^{4-} as the reference species.

(ii) List of all species making up a given reactant. This list includes all H^+-, K^+-, and Mg^{2+}-bound states (and any other significant states that are to be considered in the system). For example, for ATP, the significant species considered may be ATP^{4-}, $HATP^{3-}$, $MgATP^{2-}$, and $KATP^{3-}$. The state H_2ATP^{2-}, which is present in a significant fraction for pH values below 6, may or may not be important to include, depending on the pH range to be treated by the model.

(iii) Binding constants for H^+ and other ions that are associated with the species of each reactant.

(iv) Thermodynamic information. If available, it is convenient to compile the apparent free energy of formation $\Delta_f G^o$ for each individual species of a given reactant for the appropriate ionic strength and temperature that are to be simulated. The apparent equilibrium constants for each reaction in the system may be calculated based on the $\Delta_f G^o$ of the reference species of each reactant, the binding constants, and the pH and the free concentrations of other binding ions.

For example, for the reactants ATP, ADP, and PI, the following information is compiled:

Reactant	Reference species	$\Delta_f G^o$ (I = 0.17 M)	Ion-bound species	pK
ATP	ATP^{4-}	-2771.00	$HATP^{3-}$	6.59
			H_2ATP^{2-}	3.83
			$MgATP^{2-}$	3.82
			$KATP^{3-}$	1.87
ADP	ADP^{3-}	-1903.96	$HADP^{2-}$	6.42
			H_2ADP^-	3.79
			$MgADP^-$	2.79
			$KADP^{2-}$	1.53
PI	HPO_4^{2-}	-1098.27	$H_2PO_4^-$	6.71
			$MgHPO_4$	1.69
			$K_2PO_4^-$	0.0074
H_2O	H_2O	-235.74		

where free energies are given in kJ mol^{-1} and the pK values are the negatives of the base-10 logarithms of the binding constants for H^+, Mg^{2+}, and K^+ in units of Molar. Apparent equilibrium free energies and values of $\Delta_f G^o$ vary with total ionic strength of the solution. The above values are tabulated for ionic strength I = 0.17 M, which is a suitable value to represent the intracellular medium.

We define each *reaction* in terms of the reference reaction stoichiometry, which conserves mass (in terms of all elements) and charge. Therefore the reference reactions include explicit proton and H_2O stoichiometry. For example, the reaction of ATP hydrolysis is given by

$$ATP^{4-} + H_2O \rightleftharpoons ADP^{3-} + HPO_4^{2-} + H^+ \qquad (6.1)$$

as described in Section 2.2.2.

Mass- and charge-balanced reference reactions are the basis for computing the apparent thermodynamic properties of a given reaction as functions of pH, $[Mg^{2+}]$, and $[K^+]$, as described in Chapter 2. For example, for the ATP hydrolysis reaction, the ΔG^o for the reference reaction is computed

$$\Delta G^o = +\Delta_f G^o_{ADP^{3-}} + \Delta_f G^o_{HPO_4^{2-}} + \Delta_f G^o_{H^+}$$
$$- \Delta_f G^o_{ATP^{4-}} - \Delta_f G^o_{H_2O} = 4.51 \text{ kJ mol}^{-1} \qquad (6.2)$$

and the equilibrium constant for the reference reaction is

$$K_{eq} = e^{-\Delta G^o/RT} \approx 0.16. \qquad (6.3)$$

The apparent equilibrium constant and equilibrium free energy for the reaction are computed

$$K'_{eq} = e^{-\Delta G'^o/RT} = K_{eq} \frac{P_{ADP}([H^+], [Mg^{2+}], [K^+]) \, P_{Pi}([H^+], [Mg^{2+}], [K^+])}{[H^+] \, P_{ATP}([H^+], [Mg^{2+}], [K^+])}.$$

(6.4)

The functions $P_{ATP}([H^+], [Mg^{2+}], [K^+])$, $P_{ADP}([H^+], [Mg^{2+}], [K^+])$, and $P_{Pi}([H^+], [Mg^{2+}], [K^+])$ are the binding polynomials for the three reactants (see Chapter 2) that account for H^+, Mg^{2+}, and K^+ binding:

$$P_{ATP} = 1 + \frac{[H^+]}{K_{H-ATP_1}} + \frac{[H^+]^2}{K_{H-ATP_1} K_{H-ATP_2}} + \frac{[Mg^{2+}]}{K_{Mg-ATP}} + \frac{[K^+]}{K_{K-ATP}}$$

$$P_{ADP} = 1 + \frac{[H^+]}{K_{H-ADP_1}} + \frac{[H^+]^2}{K_{H-ADP_1} K_{H-ADP_2}} + \frac{[Mg^{2+}]}{K_{Mg-ADP}} + \frac{[K^+]}{K_{K-ADP}}$$

$$P_{Pi} = 1 + \frac{[H^+]}{K_{H-Pi_1}} + \frac{[Mg^{2+}]}{K_{Mg-Pi}} + \frac{[K^+]}{K_{K-Pi}}.$$

(6.5)

Given the binding constant and $\Delta_f G^o$ values listed above for the example network of Section 6.2, setting pH = 7, $[Mg^{2+}] = 1$ mM, and $[K^+] = 150$ mM yields

$$\Delta G'^o = -34.3 \text{ kJ mol}^{-1}$$
$$K'_{eq} = 1.004 \times 10^6$$

(6.6)

for ATP hydrolysis.

6.1.2 Determining expressions for biochemical fluxes for the reactions

Appropriate expressions for the fluxes of each of the reactions in the system must be determined. Typically, biochemical reactions proceed through multiple-step catalytic mechanisms, as described in Chapter 4, and simulations are based on the quasi-steady state approximations for the fluxes through enzyme-catalyzed reactions. (See Section 3.1.3.2 and Chapter 4 for treatments on the kinetics of enzyme catalyzed reactions.)

The quasi-steady approximation is strictly valid when the rate of change of enzyme-bound intermediate concentrations is small compared to the rate of change of reactant concentrations. This is the case either when a given reaction remains in an approximately steady state (reactant concentrations remain nearly constant) or when total reactant concentrations are significantly higher than total enzyme concentration, as illustrated in Figure 3.4.

For a given reaction, the expression used to model the flux must be constrained based on the apparent equilibrium constant and overall transformed thermodynamic driving force. Specifically it is required that the flux goes to zero when the reaction reaches equilibrium and that the forward and reverse fluxes satisfy the relationship $\Delta G' = -RT \ln(J^+/J^-)$ introduced in Section 3.1.2.

As an example of a flux expression derived from the quasi-steady approximation consider the reversible Michaelis–Menten flux arrived at in Section 3.1.3.2:

$$J_{MM}([A], [B]) = J^+ - J^- = \frac{k_f[A] - k_r[B]}{1 + [A]/K_a + [B]/K_b} \tag{6.7}$$

for the reaction $A \rightleftharpoons B$. It is straightforward to show that the relationship $\Delta G' = -RT \ln(J^+/J^-)$ is satisfied when the ratio k_f/k_r is equal to the equilibrium constant for the reaction. Thus when the equilibrium constant for the biochemical reaction $A \rightleftharpoons B$ is a function of pH and free binding ion concentrations, then either or both of k_f and k_r must be appropriate functions of pH and binding ion concentrations to ensure that the constraint $k_f/k_r = K'_{eq}$ remains satisfied. Examples of flux expressions that satisfy this thermodynamic criterion are given below in Section 6.2.3.

6.1.3 Determining the differential equations

The strategy for determining the differential equations for biochemical reactants, pH, and binding ions is to express the equations for reactants based on the stoichiometry of the reference reactions and to determine the kinetics of pH and binding ions based on mass balance.

For example, if the ATP hydrolysis flux (for the reaction of Equation (6.1)) were denoted J_{ATPase}, then the differential equations for concentrations of *reactants* ATP, ADP, and PI would follow from the stoichiometry of the *reference species* in the reference reaction:

$$d[ATP]/dt = -J_{ATPase}$$
$$d[ADP]/dt = +J_{ATPase}$$
$$d[PI]/dt = +J_{ATPase}. \tag{6.8}$$

However, the differential equations for the pH, $[Mg^{2+}]$, and $[K^+]$ (and concentrations of any other binding ions) are not as straightforward to determine as the equations for the biochemical reactants.

The kinetics of pH are governed by proton binding and unbinding as well as the consumption and generation of protons via chemical reactions. For a general system of N_r reactants, and considering $[H^+]$, $[Mg^{2+}]$, and $[K^+]$ binding, the total concentration of protons bound to reactants is calculated

$$[H_{bound}] = \sum_{i=1}^{N_r} ([L_i H_1] + 2[L_i H_2] + \cdots), \tag{6.9}$$

where i indexes the reactants denoted by L_i; the concentrations of the one- and two-proton bound species of reactant i are denoted by $[L_i H_1]$ and $[L_i H_2]$, respectively. If we ignore species with two or more protons bound, then

$$[H_{bound}] = \sum_{i=1}^{N_r} [L_i H_1] = \sum_{i=1}^{N_r} [L_i] \frac{[H^+]/K_i^H}{P_i([H^+], [Mg^{2+}], [K^+])}, \tag{6.10}$$

where $[L_i]$ is the total concentration for reactant i, K_i^H, and P_i are the proton binding constant and binding polynomial for the reactant. (Equation (6.10) follows from Equation (2.11), where Equation (2.11) includes only proton binding in the binding polynomial while Equation (6.10) includes H^+-, Mg^{2+}-, and K^+-bound species.) The binding polynomials are calculated

$$P_i([H^+], [Mg^{2+}], [K^+]) = 1 + [H^+]/K_i^H + [Mg^{2+}]/K_i^{Mg} + [K^+]/K_i^K, \tag{6.11}$$

where K_i^{Mg} and K_i^K are the binding constants for Mg^{2+} and K^+. Equation (6.10) expresses the total concentration of reversibly bound protons as a function of the total reactant concentrations in the system, and the pH and free concentrations of Mg^{2+} and K^+.

If the system is closed, then the rate of change of free $[H^+]$ can be calculated based on mass conservation. The total proton concentration in the system is given by

$$H_o = [H^+] + [H_{bound}] + [H_{reference}], \tag{6.12}$$

where $[H_{reference}]$ is the concentration of hydrogen in reference species and $[H^+]$ is the free hydrogen ion concentration. If the system is closed $dH_o/dt = 0$ and

$$d[H^+]/dt = -d[H_{bound}]/dt - d[H_{reference}]/dt. \tag{6.13}$$

The term $d[H_{reference}]/dt$ is computed from the proton fluxes of the reference reactions:

$$\frac{d[H_{reference}]}{dt} = -\sum_k n_k J_k. \tag{6.14}$$

The stoichiometric number of protons generated by the kth reference reaction and the flux through the reaction are denoted by n_k and J_k. Summation in Equation (6.14) is over all reactions.

The term $d[H_{bound}]/dt$ is expressed via the chain rule:

$$\frac{d[H_{bound}]}{dt} = +\frac{\partial[H_{bound}]}{\partial[H^+]}\frac{d[H^+]}{dt} + \frac{\partial[H_{bound}]}{\partial[Mg^{2+}]}\frac{d[Mg^{2+}]}{dt}$$
$$+\frac{\partial[H_{bound}]}{\partial[K^+]}\frac{d[K^+]}{dt} + \sum_{i=1}^{N_r}\frac{\partial[H_{bound}]}{\partial[L_i]}\frac{d[L_i]}{dt}. \qquad (6.15)$$

Combining Equations (6.13)–(6.15) yields the following equation for the free H^+ concentration:

$$\frac{d[H^+]}{dt} = \frac{-\frac{\partial[H_{bound}]}{\partial[Mg^{2+}]}\frac{d[Mg^{2+}]}{dt} - \frac{\partial[H_{bound}]}{\partial[K^+]}\frac{d[K^+]}{dt} - \sum_{i=1}^{N_r}\frac{\partial[H_{bound}]}{\partial[L_i]}\frac{d[L_i]}{dt} + \sum_k n_k J_k}{1 + \frac{\partial[H_{bound}]}{\partial[H^+]}},$$

$$(6.16)$$

where the above partial derivatives can be expressed:

$$\frac{\partial[H_{bound}]}{\partial[Mg^{2+}]} = -\sum_{i=1}^{N_r}\frac{[L_i][H^+]/K_i^H}{K_i^{Mg}(P_i([H^+],[Mg^{2+}],[K^+]))^2},$$

$$\frac{\partial[H_{bound}]}{\partial[K^+]} = -\sum_{i=1}^{N_r}\frac{[L_i][H^+]/K_i^H}{K_i^K(P_i([H^+],[Mg^{2+}],[K^+]))^2},$$

$$\frac{\partial[H_{bound}]}{\partial[L_i]} = \frac{[H^+]/K_i^H}{P_i([H^+],[Mg^{2+}],[K^+])},$$

$$\frac{\partial[H_{bound}]}{\partial[H^+]} = \sum_{i=1}^{N_r}\frac{[L_i]\left(1 + [Mg^{2+}]/K_i^{Mg} + [K^+]/K_i^K\right)}{K_i^H(P_i([H^+],[Mg^{2+}],[K^+]))^2}. \qquad (6.17)$$

Notice that the partial derivatives $\frac{\partial[H_{bound}]}{\partial[Mg^{2+}]}$ and $\frac{\partial[H_{bound}]}{\partial[K^+]}$ are always negative since magnesium and potassium compete with hydrogen ion for binding with the reactants in the system. The partial derivatives $\frac{\partial[H_{bound}]}{\partial[L_i]}$ and $\frac{\partial[H_{bound}]}{\partial[H^+]}$ are always positive because as either the reactant concentration or the free hydrogen ion concentration increases, the amount of hydrogen ion bound to reactants increases.

Also notice that Equation (6.16) assumes that all of the proton binding species in the system are included in the N_r reactants. The validity of this assumption is application specific. We will see in Section 7.4.1.5 of the following chapter that it is straightforward to modify these equations to account for additional buffers in the system.

If the system is open and hydrogen ion is transported in or out, then Equation (6.16) becomes

$$\frac{d[\mathrm{H^+}]}{dt} = \frac{-\frac{\partial[\mathrm{H}_{bound}]}{\partial[\mathrm{Mg^{2+}}]}\frac{d[\mathrm{Mg^{2+}}]}{dt} - \frac{\partial[\mathrm{H}_{bound}]}{\partial[\mathrm{K^+}]}\frac{d[\mathrm{K^+}]}{dt} + \Phi^H}{1 + \frac{\partial[\mathrm{H}_{bound}]}{\partial[\mathrm{H^+}]}}, \tag{6.18}$$

where

$$\Phi^H = -\sum_{i=1}^{N_r}\frac{\partial[\mathrm{H}_{bound}]}{\partial[\mathrm{L}_i]}\frac{d[\mathrm{L}_i]}{dt} + \sum_k n_k J_k + J_t^H$$

and J_t^H is the flux of $[\mathrm{H^+}]$ into the system.

The equations for free $\mathrm{Mg^{2+}}$ and $\mathrm{K^+}$ are developed in a manner similar to that for hydrogen ion, with the major difference that neither $\mathrm{Mg^{2+}}$ nor $\mathrm{K^+}$ are involved in the biochemical reference reactions. The kinetic equation for magnesium ion is derived from the expression for the total bound magnesium in the system:

$$[\mathrm{Mg}_{bound}] = \sum_{i=1}^{N_r}[\mathrm{L}_i]\frac{[\mathrm{Mg^{2+}}]/K_i^{Mg}}{P_i([\mathrm{H^+}],[\mathrm{Mg^{2+}}],[\mathrm{K^+}])}. \tag{6.19}$$

The rate of change of free magnesium ion is given by:

$$\frac{d[\mathrm{Mg^{2+}}]}{dt} = \frac{-\frac{\partial[\mathrm{Mg}_{bound}]}{\partial[\mathrm{H^+}]}\frac{d[\mathrm{H^+}]}{dt} - \frac{\partial[\mathrm{Mg}_{bound}]}{\partial[\mathrm{K^+}]}\frac{d[\mathrm{K^+}]}{dt} + \Phi^M}{1 + \frac{\partial[\mathrm{Mg}_{bound}]}{\partial[\mathrm{Mg^{2+}}]}}, \tag{6.20}$$

where

$$\Phi^M = -\sum_{i=1}^{N_r}\frac{\partial[\mathrm{Mg}_{bound}]}{\partial[\mathrm{L}_i]}\frac{d[\mathrm{L}_i]}{dt} + J_t^M$$

and J_t^M is the flux of $[\mathrm{Mg^{2+}}]$ into the system. The partial derivatives are expressed:

$$\frac{\partial[\mathrm{Mg}_{bound}]}{\partial[\mathrm{H^+}]} = -\sum_{i=1}^{N_r}\frac{[\mathrm{L}_i][\mathrm{Mg^{2+}}]/K_i^{Mg}}{K_i^H(P_i([\mathrm{H^+}],[\mathrm{Mg^{2+}}],[\mathrm{K^+}]))^2},$$

$$\frac{\partial[\mathrm{Mg}_{bound}]}{\partial[\mathrm{K^+}]} = -\sum_{i=1}^{N_r}\frac{[\mathrm{L}_i][\mathrm{Mg^{2+}}]/K_i^{Mg}}{K_i^K(P_i([\mathrm{H^+}],[\mathrm{Mg^{2+}}],[\mathrm{K^+}]))^2},$$

$$\frac{\partial[\mathrm{Mg}_{bound}]}{\partial[\mathrm{L}_i]} = \frac{[\mathrm{Mg^{2+}}]/K_i^{Mg}}{P_i([\mathrm{H^+}],[\mathrm{Mg^{2+}}],[\mathrm{K^+}])},$$

$$\frac{\partial[\mathrm{Mg}_{bound}]}{\partial[\mathrm{Mg^{2+}}]} = \sum_{i=1}^{N_r}\frac{[\mathrm{L}_i]\left(1 + [\mathrm{H^+}]/K_i^H + [\mathrm{K^+}]/K_i^K\right)}{K_i^{Mg}(P_i([\mathrm{H^+}],[\mathrm{Mg^{2+}}],[\mathrm{K^+}]))^2}. \tag{6.21}$$

The magnesium transport flux is denoted J_t^M. Similarly, for potassium we have

$$[\text{K}_{bound}] = \sum_{i=1}^{N_r} [\text{L}_i] \frac{[\text{K}^+]/K_i^K}{P_i([\text{H}^+], [\text{Mg}^{2+}], [\text{K}^+])}, \tag{6.22}$$

$$\frac{d[\text{K}^+]}{dt} = \frac{-\frac{\partial[\text{K}_{bound}]}{\partial[\text{H}^+]} \frac{d[\text{H}^+]}{dt} - \frac{\partial[\text{K}_{bound}]}{\partial[\text{Mg}^{2+}]} \frac{d[\text{Mg}^{2+}]}{dt} + \Phi^K}{1 + \frac{\partial[\text{K}_{bound}]}{\partial[\text{K}^+]}}, \tag{6.23}$$

$$\Phi^K = -\sum_{i=1}^{N_r} \frac{\partial[\text{K}_{bound}]}{\partial[\text{L}_i]} \frac{d[\text{L}_i]}{dt} + J_t^K,$$

and

$$\frac{\partial[\text{K}_{bound}]}{\partial[\text{H}^+]} = -\sum_{i=1}^{N_r} \frac{[\text{L}_i][\text{K}^+]/K_i^K}{K_i^H (P_i([\text{H}^+], [\text{Mg}^{2+}], [\text{K}^+]))^2},$$

$$\frac{\partial[\text{K}_{bound}]}{\partial[\text{Mg}^{2+}]} = -\sum_{i=1}^{N_r} \frac{[\text{L}_i][\text{K}^+]/K_i^K}{K_i^{Mg} (P_i([\text{H}^+], [\text{Mg}^{2+}], [\text{K}^+]))^2},$$

$$\frac{\partial[\text{K}_{bound}]}{\partial[\text{L}_i]} = \frac{[\text{K}^+]/K_i^K}{P_i([\text{H}^+], [\text{Mg}^{2+}], [\text{K}^+])},$$

$$\frac{\partial[\text{K}_{bound}]}{\partial[\text{K}^+]} = \sum_{i=1}^{N_r} \frac{[\text{L}_i]\left(1 + [\text{H}^+]/K_i^H + [\text{Mg}^{2+}]/K_i^{Mg}\right)}{K_i^K (P_i([\text{H}^+], [\text{Mg}^{2+}], [\text{K}^+]))^2}. \tag{6.24}$$

Since the time derivatives $d[\text{H}^+]/dt$, $d[\text{Mg}^{2+}]/dt$, and $d[\text{K}^+]/dt$ are expressed as functions of $d[\text{H}^+]/dt$, $d[\text{Mg}^{2+}]/dt$, and $d[\text{K}^+]/dt$ in Equations (6.18), (6.20), and (6.23), these derivatives cannot be calculated as explicit functions directly from these expressions. However, this system of three equations can be solved for three explicit functions for $d[\text{H}^+]/dt$, $d[\text{Mg}^{2+}]/dt$, and $d[\text{K}^+]/dt$, yielding the following unavoidably complex expressions.

$$\frac{d[\text{H}^+]}{dt} = \left[\left(\frac{\partial[\text{K}_{bound}]}{\partial[\text{Mg}^{2+}]} \cdot \frac{\partial[\text{Mg}_{bound}]}{\partial[\text{K}^+]} - \alpha_M \alpha_K \right) \Phi^H \right.$$

$$+ \left(\alpha_K \frac{\partial[\text{H}_{bound}]}{\partial[\text{Mg}^{2+}]} - \frac{\partial[\text{H}_{bound}]}{\partial[\text{K}^+]} \cdot \frac{\partial[\text{K}_{bound}]}{\partial[\text{Mg}^{2+}]} \right) \Phi^M$$

$$+ \left. \left(\alpha_M \frac{\partial[\text{H}_{bound}]}{\partial[\text{K}^+]} - \frac{\partial[\text{H}_{bound}]}{\partial[\text{Mg}^{2+}]} \cdot \frac{\partial[\text{Mg}_{bound}]}{\partial[\text{K}^+]} \right) \Phi^K \right] \Big/ D, \tag{6.25}$$

$$\frac{d[\text{Mg}^{2+}]}{dt} = \left[\left(\alpha_K \frac{\partial[\text{Mg}_{bound}]}{\partial[\text{H}^+]} - \frac{\partial[\text{K}_{bound}]}{\partial[\text{H}^+]} \cdot \frac{\partial[\text{Mg}_{bound}]}{\partial[\text{K}^+]} \right) \Phi^H \right.$$

$$+ \left(\frac{\partial[\text{K}_{bound}]}{\partial[\text{H}^+]} \cdot \frac{\partial[\text{H}_{bound}]}{\partial[\text{K}^+]} - \alpha_H \alpha_K \right) \Phi^M$$

$$\left. + \left(\alpha_H \frac{\partial[\text{Mg}_{bound}]}{\partial[\text{K}^+]} - \frac{\partial[\text{H}_{bound}]}{\partial[\text{K}^+]} \cdot \frac{\partial[\text{Mg}_{bound}]}{\partial[\text{H}^+]} \right) \Phi^K \right] \bigg/ D, \quad (6.26)$$

and

$$\frac{d[\text{K}^+]}{dt} = \left[\left(\alpha_M \frac{\partial[\text{K}_{bound}]}{\partial[\text{H}^+]} - \frac{\partial[\text{K}_{bound}]}{\partial[\text{Mg}^{2+}]} \cdot \frac{\partial[\text{Mg}_{bound}]}{\partial[\text{H}^+]} \right) \Phi^H \right.$$

$$+ \left(\alpha_H \frac{\partial[\text{K}_{bound}]}{\partial[\text{Mg}^{2+}]} - \frac{\partial[\text{K}_{bound}]}{\partial[\text{H}^+]} \cdot \frac{\partial[\text{H}_{bound}]}{\partial[\text{Mg}^{2+}]} \right) \Phi^M$$

$$\left. + \left(\frac{\partial[\text{Mg}_{bound}]}{\partial[\text{H}^+]} \cdot \frac{\partial[\text{H}_{bound}]}{\partial[\text{Mg}^{2+}]} - \alpha_H \alpha_M \right) \Phi^K \right] \bigg/ D, \quad (6.27)$$

where

$$D = \alpha_H \frac{\partial[\text{K}_{bound}]}{\partial[\text{Mg}^{2+}]} \cdot \frac{\partial[\text{Mg}_{bound}]}{\partial[\text{K}^+]} + \alpha_K \frac{\partial[\text{H}_{bound}]}{\partial[\text{Mg}^{2+}]} \cdot \frac{\partial[\text{Mg}_{bound}]}{\partial[\text{H}^+]}$$

$$+ \alpha_M \frac{\partial[\text{H}_{bound}]}{\partial[\text{K}^+]} \cdot \frac{\partial[\text{K}_{bound}]}{\partial[\text{H}^+]} - \alpha_M \alpha_K \alpha_H$$

$$- \frac{\partial[\text{H}_{bound}]}{\partial[\text{K}^+]} \cdot \frac{\partial[\text{K}_{bound}]}{\partial[\text{Mg}^{2+}]} \cdot \frac{\partial[\text{Mg}_{bound}]}{\partial[\text{H}^+]}$$

$$- \frac{\partial[\text{H}_{bound}]}{\partial[\text{Mg}^{2+}]} \cdot \frac{\partial[\text{Mg}_{bound}]}{\partial[\text{K}^+]} \cdot \frac{\partial[\text{K}_{bound}]}{\partial[\text{H}^+]}, \quad (6.28)$$

and

$$\alpha_H = 1 + \frac{\partial[\text{H}_{bound}]}{\partial[\text{H}^+]},$$

$$\alpha_M = 1 + \frac{\partial[\text{Mg}_{bound}]}{\partial[\text{Mg}^{2+}]},$$

$$\alpha_K = 1 + \frac{\partial[\text{K}_{bound}]}{\partial[\text{K}^+]}. \quad (6.29)$$

6.1.4 Computational implementation and testing

Once the differential equations for a given system are determined, it is left to construct a computer simulation based on the equations. This step requires supplying a computer code to compute the time derivatives of the state variables and using a simulation package to integrate the derivatives, as outlined in Chapter 3. In general a numerical solver will introduce some degree of numerical error into the solution and it is prudent to check that the error involved in one's simulation of a given system is within an acceptable tolerance.

A convenient check on the accuracy of a simulation is based on checking conserved quantities in the system. The equations of Section 6.1.3 conserve overall elemental mass balance. Thus, were the equations to be solved exactly, then the total proton, magnesium, and potassium concentrations would remain constant in time.

As an example system we return to the simple ATP hydrolysis reaction with reactants ATP, ADP, and PI. For this reaction, the time derivatives of the reactants are given by Equation (6.8) and the time derivatives of the binding ions are given by Equations (6.25), (6.26), and (6.27). To complete the system of equations, we use the following form for the reaction flux: $J^+ = k[\text{ATP}]$, $J^- = J^+ e^{\Delta G'/RT}$, where k is a constant (set equal to 0.1 sec^{-1}), and $\Delta G'$ is the apparent free energy for the reaction.

Figure 6.1 illustrates results from a simulation of this system obtained from the following initial conditions: $[\text{ATP}] = 10$ mM; $[\text{ADP}] = 100\,\mu$M; $[\text{PI}] = 1$ mM; $[\text{H}^+] = 1 \times 10^{-7}$ M; $[\text{Mg}^{2+}] = 1$ mM; and $[\text{K}^+] = 150$ mM. Solutions are obtained using the Matlab solver 'ode45', which is based on an explicit fourth-order Runge–Kutta algorithm, using the default numerical settings (Matlab version 7 Release 14).

Also shown are the total protons H_o, total magnesium M_o, and total potassium K_o computed from the simulation:

$$H_o = [\text{H}^+] + [\text{H}_{bound}] + (0.010\,\text{M} - [\text{ATP}])$$
$$M_o = [\text{Mg}^{2+}] + [\text{Mg}_{bound}]$$
$$K_o = [\text{K}^+] + [\text{K}_{bound}]. \tag{6.30}$$

The third term on the right-hand side of the equation for H_o accounts for protons released from the reference species ATP^{4-} as the ATP concentration drops below the initial value of 10 mM. It is apparent from the figure that the H_o, M_o, and K_o remain essentially constant over the course of the simulation. The numerical noise in these variables is less than 10^{-4} times of their mean values. The numerical accuracy may be increased based on adjusting the tolerances used in the simulation and/or using different algorithms. Of course one must keep in mind that in general increases in accuracy cost additional computational time.

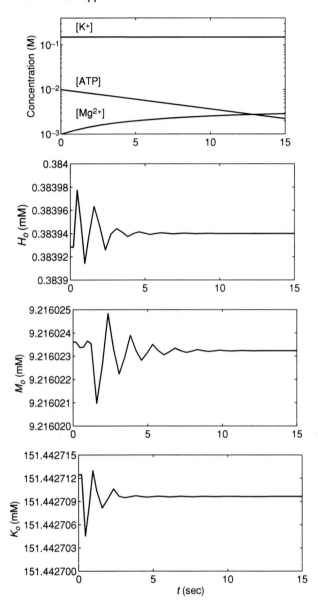

Figure 6.1 Simulation of ATP hydrolysis reaction described in Section 6.1.4. Upper panel shows the predicted concentrations of ATP, Mg^{2+}, and K^+ as functions of time. In the lower panel the quantities H_o, M_o, and K_o, which report the conservation of protons, magnesium, and potassium in system simulation, are plotted versus time.

6.2 Kinetic model of the TCA cycle

6.2.1 Overview

The tricarboxylic acid (TCA) cycle (also known as the citric acid cycle and the Krebs cycle) is a collection of biochemical reactions that oxidize certain organic molecules, generating CO_2 and reducing the cofactors NAD and FAD to NADH and $FADH_2$ [147]. In turn, NADH and $FADH_2$ donate electrons in the electron transport chain, an important component of oxidative ATP synthesis. The TCA cycle also serves to feed precursors to a number of important biosynthetic pathways, making it a critical "hub in metabolism" [147] for aerobic organisms. Its ubiquity and importance make it a useful example for the development of a kinetic network model.

6.2.2 Components of the TCA cycle reaction network

In prokaryotic animals, the TCA cycle reactions are confined to occur within the intracellular organelle called the mitochondrion. The reactions are illustrated in Figure 6.2, with a total of eight enzyme-catalyzed reactions responsible for oxidizing citrate to oxaloacetate (OAA) and synthesizing citrate from oxaloacetate and acetyl-coenzyme A (ACCOA). In the kinetic model we consider one source of ACCOA, the pyruvate dehydrogenase reaction (reaction number 1 in the figure), which generates ACCOA from pyruvate. In addition, the aspartate aminotransferase reaction (reaction 11), which is important in several cell types, is included in the diagram. The biochemical reactants in this system, with the abbreviations used here and the reference species definitions, are given in Table 6.1. The biochemical reference reactions are listed in Table 6.2.

Charges for all of the references species in Table 6.1 are indicated using superscripts, even when the charge is zero. This notation distinguishes biochemical reactants (for example ACCOA) from references species (for example $ACCOA^0$).

The Gibbs free energies of formation and pKs for cation-bound species are listed in Table 6.1 for ionic strength $I = 0.17\,M$. (See Section 2.5.) Here we have considered species of each reactant that make significant contributions in the region of pH = 7. All cation-species that make significant contributions at the pH and concentrations considered in the model are listed in the table. Some reactants, such as ACCOA, do not significantly bind hydrogen ions or other cations. Therefore no bound states are listed for these reactants.

We have chosen to use $CO_3{}^{2-}$ as a reference species in reaction numbers 1, 4, and 5. Therefore the apparent thermodynamic properties of these reactions will be calculated in terms of the biochemical reactant ΣCO_2. (See Section 2.6 for a discussion on the treatment of CO_2 in biochemical reactions.)

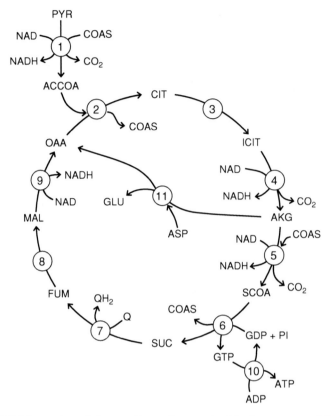

Figure 6.2 Illustration of the reactions of pyruvate dehydrogenase and the TCA cycle. The abbreviations for the biochemical reactants are listed in Table 6.1 and the stoichiometries of the 11 biochemical reference reactions are listed in Table 6.2.

The reference reactions tabulated in Table 6.2 correspond to the 11 reactions illustrated in Figure 6.2. The first ten reactions sum to an overall reaction for the TCA cycle of:

$$PYR^- + 4\,NAD^- + CoQ^0 + ADP^{3-} + PI^{2-} + 5\,H_2O$$
$$\rightleftharpoons 4\,NADH^{2-} + CoQH_2{}^0 + ATP^{4-} + 3\,CO_3{}^{2-} + 9\,H^+. \qquad (6.31)$$

The eleventh reaction (aspartate aminotransferase) serves as a shortcut through the cycle, generating oxaloacetate and glutamate directly from α-ketoglutarate and glutamate.

The source of ACCOA for citrate synthase (reaction 2) is the oxidation of carbohydrate, fatty acid, and amino acid molecules. The COAS generated by citrate synthase is used in the oxidation of substrates (such as pyruvate via pyruvate

Table 6.1 *Biochemical reactants of the TCA cycle*[a]

Reactant	Abbrev.	Reference species	$\Delta_f G^o$	Additional species	pK
adenosine triphosphate	ATP	ATP^{4-}	-2771.00	$HATP^{3-}$	6.59
				$MgATP^{2-}$	3.82
				$KATP^{3-}$	1.87
adenosine diphosphate	ADP	ADP^{3-}	-1903.96	$HADP^{2-}$	6.42
				$MgADP^{-}$	2.79
				$KADP^{2-}$	1.53
inorganic phosphate	PI	$HPO_4{}^{2-}$	-1098.27	$H_2PO_4{}^{-}$	6.71
				$MgHPO_4$	1.69
				$KHPO_4{}^{-}$	0.0074
guanosine triphosphate	GTP	GTP^{4-}	-2771.00	$HGTP^{3-}$	6.59
				$MgGTP^{2-}$	3.82
				$KGTP^{3-}$	1.87
guanosine diphosphate	GDP	GDP^{3-}	-1903.96	$HGDP^{2-}$	6.42
				$MgGDP^{-}$	2.79
				$KGDP^{2-}$	1.53
acetyl-coenzyme A	ACCOA	$ACCOA^{0}$	-178.19		
coenzyme A-SH	COAS	$COAS^{-}$	-0.72	$COASH^{0}$	8.13
oxalo-acetate	OAA	OAA^{2-}	-794.41	$MgOAA^{0}$	0.0051
citrate	CIT	CIT^{3-}	-1165.59	$HCIT^{2-}$	5.63
				$MgCIT^{-}$	3.37
				$KCIT^{2-}$	0.339
isocitrate	ICIT	$ICIT^{3-}$	-1158.94	$HCIT^{2-}$	5.64
				$MgCIT^{-}$	2.46
NAD^{-} (oxidized)	NAD^{-}	NAD^{-}	18.10		
NAD^{-} (reduced)	NADH^{2-}	$NADH^{2-}$	39.31		
α-keto-glutarate	AKG	AKG^{2-}	-793.41		
carbon dioxide	ΣCO_2	$CO_3{}^{2-}$	-530.71	$HCO_3{}^{-}$	9.75
succinyl-coenzyme A	SCOA	$SCOA^{-}$	-507.55	$HSCOA^{0}$	3.96

Table 6.1 *(continued)*

Reactant	Abbrev.	Reference species	$\Delta_f G^o$	Additional species	pK
succinate	SUC	SUC^{2-}	−690.44	$HSUC^-$	5.13
				$MgSUC^0$	1.17
				$KSUC^-$	0.503
fumarate	FUM	FUM^{2-}	−603.32	$HFUM^-$	4.10
malate	MAL	MAL^{2-}	−842.66	$HMAL^-$	4.75
				$MgMAL^0$	1.55
				$KMAL^-$	−0.107
ubiquinol (oxidized)	CoQ	CoQ^0	65.17		
ubiquinol (reduced)	$CoQH_2$	$CoQH_2{}^0$	−23.30		
aspartate	ASP	ASP^-	−692.26	$HASP^0$	3.65
				$MgASP^+$	2.32
glutamate	GLU	GLU^-	−692.40	$HGLU^0$	4.06
				$MgGLU^+$	1.82
water	H_2O	H_2O	−235.74		

[a] Values computed for $T = 298.15$ K and $I = 0.17$ M based on data in Alberty [3, 4] and NIST database of stability constants [134] as described in [213].

dehydrogenase), while the reduced cofactors NADH and $CoQH_2$ generated by the remaining reactions of the TCA cycle donate electrons to the electron transport system. The GTP synthesized is converted to ATP via the nucleoside diphosphate kinase (NDK) reaction; ATP is transported out of the mitochondria to the cytoplasm via the andenylate translocase (ANT) transporter. (The kinetics of ANT and of other transporters are treated in Chapter 7.)

6.2.3 Flux expressions for TCA cycle reaction network

As outlined in Section 6.1, the next step in building a computational model of the TCA cycle is determining an expression for the biochemical fluxes in the system. Flux expressions used here are adopted from Wu *et al.* [213], who developed thermodynamically balanced flux expressions for the reactions illustrated in Figure 6.2 and listed in Table 6.2. Here we describe in detail the mechanistic model and the associated rate law for one example enzyme (pyruvate dehydrogenase) from Wu *et al.*'s model. For all other enzymes we simply list the flux expression and refer readers to the supplementary material to [213] for further details.

Table 6.2 *Reference reactions of pyruvate dehydrogenase and the TCA cycle*

Enzyme name and reference reaction

---

1. pyruvate dehydrogenase:

$$PYR^- + COAS^- + NAD^- + H_2O \rightleftharpoons CO_3{}^{2-} + ACCOA^0 + NADH^{2-} + H^+$$

2. citrate synthase:

$$ACCOA^0 + OAA^{2-} + H_2O \rightleftharpoons CIT^{3-} + COAS^- + 2H^+$$

3. aconitase:

$$CIT^{3-} \rightleftharpoons ICIT^{3-}$$

4. isocitrate dehydrogenase:

$$ICIT^{3-} + NAD^- + H_2O \rightleftharpoons AKG^{2-} + NADH^{2-} + CO_3{}^{2-} + 2H^+$$

5. α-ketoglutarate dehydrogenase:

$$AKG^{2-} + COAS^- + NAD^- + H_2O \rightleftharpoons SCOA^- + NADH^{2-} + CO_3{}^{2-} + H^+$$

6. succinyl-CoA synthetase:

$$SCOA^- + GDP^{3-} + PI^{2-} \rightleftharpoons SUC^{2-} + GTP^{4-} + COAS^- + H^+$$

7. succinate dehydrogenase:

$$SUC^{2-} + CoQ^0 \rightleftharpoons FUM^{2-} + CoQH_2{}^0$$

8. fumarase:

$$FUM^{2-} + H_2O \rightleftharpoons MAL^{2-}$$

9. malate dehydrogenase:

$$MAL^{2-} + NAD^- \rightleftharpoons OAA^{2-} + NADH^{2-} + H^+$$

10. nucleoside diphosphokinase:

$$GTP^{4-} + ADP^{3-} \rightleftharpoons ATP^{4-} + GDP^{3-}$$

11. aspartate aminotransferase:

$$AKG^- + ASP^{2-} \rightleftharpoons OAA^{2-} + GLU^-$$

6.2.3.1 Pyruvate dehydrogenase

The standard Gibbs free energy for the chemical reference reaction for pyruvate dehydrogenase is computed

$$\Delta_r G_{pdh}^o = \Delta_f G_{\Sigma CO_2}^o + \Delta_f G_{ACCOA}^o + \Delta_f G_{NADH}^o$$
$$- \Delta_f G_{PYR}^o - \Delta_f G_{COAS}^o - \Delta_f G_{NAD}^o - \Delta_f G_{H_2O}^o$$
$$= 19.59 \text{ kJ} \cdot \text{mol}^{-1}. \tag{6.32}$$

The equilibrium constant for the reference reaction is

$$K_{pdh}^o = \exp\left(-\frac{\Delta_r G_{pdh}^o}{RT}\right) = 5.02 \times 10^{-4} \text{ M}^{-1}. \tag{6.33}$$

The apparent equilibrium constant for the biochemical reaction is

$$K'_{pdh} = K^o_{pdh} \frac{1}{[\text{H}^+]} \frac{P_{\Sigma CO_2} P_{ACCOA} P_{NADH}}{P_{PYR} P_{COAS} P_{NAD}}, \tag{6.34}$$

where the binding polynomials are functions of pH, $[\text{K}^+]$, and $[\text{Mg}^{2+}]$ as detailed in Chapter 2.

The reaction mechanism is assumed to be *hexa-uni-ping-pong* [34] when the reactant water is ignored with the following steps [114]:

$$\text{E} + \text{PYR}^- \rightleftharpoons \text{E} \cdot \text{PYR}^-$$
$$\text{E} \cdot \text{PYR}^- \rightarrow \text{E} \cdot \text{CHOCH}_3^- + \text{CO}_2$$
$$\text{E} \cdot \text{CHOCH}_3^- + \text{COASH}^0 \rightleftharpoons \text{E} \cdot \text{ACCOA}^-$$
$$\text{E} \cdot \text{ACCOA}^- \rightarrow \text{E}^- \cdot \text{ACCOA}^0$$
$$\text{E}^- + \text{NAD}^- \rightleftharpoons \text{E} + \text{NAD}^{2-}$$
$$\text{E} \cdot \text{NAD}^{2-} \rightarrow \text{E} + \text{NADH}^{2-}. \tag{6.35}$$

The flux expression for this mechanism follows from application of the method of King and Altman [112]:

$$J^+_{pdh} = \frac{V_{mf}[\text{A}][\text{B}][\text{C}]}{K_{mC}[\text{A}][\text{B}] + K_{mB}[\text{A}][\text{C}] + K_{mA}[\text{B}][\text{C}] + [\text{A}][\text{B}][\text{C}]}, \tag{6.36}$$

where we have used the notation $[\text{A}] = [\text{PYR}]$, $[\text{B}] = [\text{COAS}]$, $[\text{C}] = [\text{NAD}]$; V_{mf}, K_{mA}, K_{mB}, and K_{mC} are kinetic constants.

The model of Wu *et al.* [213] assumes that ACCOA and NADH bind to the enzyme in competition for COASH and NAD with binding constants K_{iACCOA} and K_{iNADH}, respectively. Incorporating these competitive inhibitions into the model, the flux expression becomes

$$J^+_{pdh} = \frac{V_{mf}[\text{A}][\text{B}][\text{C}]}{K_{mC}\alpha_{i2}[\text{A}][\text{B}] + K_{mB}\alpha_{i1}[\text{A}][\text{C}] + K_{mA}[\text{B}][\text{C}] + [\text{A}][\text{B}][\text{C}]}, \tag{6.37}$$

where

$$\alpha_{i1} = 1 + \frac{[\text{ACCOA}]}{K_{iACCOA}}$$

and

$$\alpha_{i2} = 1 + \frac{[\text{NADH}]}{K_{iNADH}}.$$

Assuming that the expression for J_{pdh}^+ represents the flux for the one-way mechanism of Equation 6.35, we can compute the net flux as:

$$J_{pdh} = J_{pdh}^+ - J_{pdh}^- = J_{pdh}^+ \left(1 - \frac{1}{K'_{pdh}} \frac{[P][Q][R]}{[A][B][C]}\right), \qquad (6.38)$$

where $[P] = [\Sigma CO_2]$, $[Q] = [ACCOA]$, and $[R] = [NADH]$.

The kinetic parameters for pyruvate dehydrogenase are assigned values $V_{mf} = 0.122$ mol sec^{-1} (1 mito)$^{-1}$, $K_{mA} = 38.3$ μM, $K_{mB} = 9.9$ μM, $K_{mC} = 60.7$ μM, $K_{iACCOA} = 40.2$ μM, and $K_{iNADH} = 40.0$ μM; estimation and assignment of parameter values is discussed in [213]. The flux and V_{mf} are expressed in units of mass per unit time per unit mitochondrial volume.

6.2.3.2 Citrate synthase

The flux expression for citrate synthase is

$$J_{cits} = J_{cits}^+ \left(1 - \frac{1}{K'_{cits}} \frac{[P][Q]}{[A][B]}\right), \qquad (6.39)$$

where $[A] = [OAA]$, $[B] = [ACCOA]$, $[P] = [COAS]$, and $[Q] = [CIT]$ and K'_{cits} is computed according to the usual procedures. The forward flux J_{cits}^+ is computed

$$J_{cits}^+ = \frac{V_{mf}[A][B]}{K_{ia}K_{mB}\alpha_{i1} + K_{mA}\alpha_{i1}[B] + K_{mB}\alpha_{i2}[A] + [A][B]} \qquad (6.40)$$

where

$$\alpha_{i1} = 1 + \frac{[fCIT]}{K_{iCIT}}$$

and

$$\alpha_{i2} = 1 + \frac{[fATP]}{K_{iATP}} + \frac{[fADP]}{K_{iADP}} + \frac{[fAMP]}{K_{iAMP}} + \frac{[COAS]}{K_{iCOAS}} + \frac{[SCOA]}{K_{iSCOA}}.$$

The concentrations [fCIT], [fATP], [fADP], [fAMP] refer to unchelated (non-magnesium-bound) citrate, ATP, ADP, and AMP.

The parameter values are $V_{mf} = 11.6$ mol sec^{-1} (1 mito)$^{-1}$, $K_{mA} = 4$ μM, $K_{mB} = 14$ μM, $K_{ia} = 3.33$ μM, $K_{iCIT} = 1600$ μM, $K_{iATP} = 900$ μM, $K_{iADP} = 1800$ μM, $K_{iAMP} = 6000$ μM, $K_{iCOAS} = 67$ μM, and $K_{iSCOA} = 140$ μM.

6.2.3.3 Aconitase

The flux expression for aconitase is

$$J_{acon} = \frac{\frac{V_{mf}}{K_{mA}}[A] - \frac{V_{mf}}{K_{mP}}[P]}{1 + [A]/K_{mA} + [P]/K_{mP}} \qquad (6.41)$$

where $[A] = [CIT]$ and $[P] = [ICIT]$.

The parameter values are $V_{mf} = 3.21 \times 10^{-2} \, \text{mol sec}^{-1} \, (1 \, \text{mito})^{-1}$, $K_{mA} = 1161 \, \mu\text{M}$, and $K_{mP} = 434 \, \mu\text{M}$. The value of V_{mr} is computed

$$V_{mr} = \frac{V_{mf} K_{mP}}{K'_{acon} K_{mA}}, \tag{6.42}$$

where K'_{acon} is the apparent equilibrium constant for the biochemical reaction.

6.2.3.4 Isocitrate dehydrogenase

The flux expression for isocitrate dehydrogenase is

$$J_{isod} = J^+_{isod} \left(1 - \frac{1}{K'_{isod}} \frac{[P][Q][R]}{[A][B]} \right), \tag{6.43}$$

where $[A] = [NAD]$, $[B] = [ICIT]$, $[P] = [AKG]$, $[Q] = [NADH]$, and $[R] = [\Sigma CO_2]$ and K'_{isod} is the apparent equilibrium constant. The forward flux J^+_{isod} is computed

$$J^+_{isod} = \frac{V_{mf}[A][B]^{n_H}}{[A][B]^{n_H} + K^{n_H}_{mB}\alpha_i[A] + K_{mA}\left([B]^{n_H} + K^{n_H}_{ib}\alpha_i + \alpha_i[Q][B]^{n_H}/K_{iq}\right)} \tag{6.44}$$

where

$$\alpha_i = 1 + \frac{K_{aADP}}{[fADP]}\left(1 + \frac{[fATP]}{K_{iATP}}\right).$$

The concentrations [fATP] and [fADP] refer to unchelated (non-magnesium-bound) ATP and ADP.

The parameter values are $V_{mf} = 0.425 \, \text{mol sec}^{-1} \, (1 \, \text{mito})^{-1}$, $n_H = 3.0$, $K_{mA} = 74 \, \mu\text{M}$, $K_{mB} = 183 \, \mu\text{M}$, $K_{ib} = 23.8 \, \mu\text{M}$, $K_{iq} = 29 \, \mu\text{M}$, $K_{iATP} = 91 \, \mu\text{M}$, and $K_{aADP} = 50 \, \mu\text{M}$.

6.2.3.5 α-Ketoglutarate dehydrogenase

The flux expression for α-ketoglutarate dehydrogenase is

$$J_{akgd} = J^+_{akgd} \left(1 - \frac{1}{K'_{akgd}} \frac{[P][Q][R]}{[A][B][C]} \right), \tag{6.45}$$

where $[A] = [AKG]$, $[B] = [COAS]$, $[C] = [NAD]$, $[P] = [\Sigma CO_2]$, $[Q] = [SCOA]$, and $[R] = [NADH]$ and K'_{akgd} is the apparent equilibrium constant. The forward flux J^+_{akgd} is computed

$$J^+_{akgd} = \frac{V_{mf}}{\left[1 + \alpha_i \frac{K_{mA}}{[A]} + \frac{K_{mB}}{[B]}\left(1 + \frac{[Q]}{K_{iq}}\right) + \frac{K_{mC}}{[C]}\left(1 + \frac{[R]}{K_{ir}}\right)\right]} \tag{6.46}$$

where

$$\alpha_i = 1 + \frac{K_{aADP}}{[\text{fADP}]}\left(1 + \frac{[\text{fATP}]}{K_{iATP}}\right).$$

The concentrations [fATP] and [fADP] refer to unchelated (non-magnesium-bound) ATP and ADP.

The parameter values are $V_{mf} = 7.70 \times 10^{-2}$ mol sec^{-1} (1 mito)$^{-1}$, $K_{mA} = 80\,\mu\text{M}$, $K_{mB} = 55\,\mu\text{M}$, $K_{mC} = 21\,\mu\text{M}$, $K_{iq} = 6.9\,\mu\text{M}$, $K_{ir} = 0.60\,\mu\text{M}$, $K_{iATP} = 50\,\mu\text{M}$, and $K_{aADP} = 100\,\mu\text{M}$.

6.2.3.6 Succinyl-CoA synthetase

The flux expression for succinyl-CoA synthetase is

$$J_{scs} = \frac{V_{mf}V_{mr}\left([\text{A}][\text{B}][\text{C}] - \frac{[\text{P}][\text{Q}][\text{R}]}{K'_{scs}}\right)}{denom}, \tag{6.47}$$

where [A] = [GDP], [B] = [SCOA], [C] = [PI], [P] = [COAS], [Q] = [SUC], [R] = [GTP], K'_{scs} is the apparent equilibrium constant, and the denominator in this expression is

$$
\begin{aligned}
denom =\ & V_{mr}K_{ia}K_{ib}K_{mC} + V_{mr}K_{ib}K_{mc}[\text{A}] + V_{mr}K_{ia}K_{mB}[\text{C}] \\
& + V_{mr}K_{mC}[\text{A}][\text{B}] + V_{mr}K_{mB}[\text{A}][\text{C}] + V_{mr}K_{mA}[\text{B}][\text{C}] + V_{mr}[\text{A}][\text{B}][\text{C}] \\
& + \frac{1}{K'_{scs}}(V_{mf}K_{ir}K_{mQ}[\text{P}] + V_{mf}K_{iq}K_{mP}[\text{R}] + V_{mf}K_{mR}[\text{P}][\text{Q}] \\
& + V_{mf}K_{mQ}[\text{P}][\text{R}] + V_{mf}K_{mP}[\text{Q}][\text{R}] + V_{mf}[\text{P}][\text{Q}][\text{R}]) \\
& + \frac{1}{K_{ia}K'_{scs}}(V_{mf}K_{mQ}K_{ir}[\text{A}][\text{P}] + V_{mf}K_{mR}[\text{A}][\text{P}][\text{Q}]) \\
& + \frac{1}{K_{ir}}(V_{mr}K_{ia}K_{mB}[\text{C}][\text{R}] + V_{mr}K_{mA}[\text{B}][\text{C}][\text{R}]) \\
& + \frac{1}{K_{ia}K_{ib}K'_{scs}}(V_{mf}K_{mQ}K_{ir}[\text{A}][\text{B}][\text{Q}] + V_{mf}K_{mR}[\text{A}][\text{B}][\text{P}][\text{Q}]) \\
& + \frac{1}{K_{iq}K_{ir}}(V_{mr}K_{ia}K_{mB}[\text{C}][\text{Q}][\text{R}] + V_{mf}K_{mA}[\text{B}][\text{C}][\text{Q}][\text{R}]) \\
& + \frac{1}{K_{ip}K_{iq}K_{ir}}(V_{mr}K_{mA}K_{ic}[\text{B}][\text{P}][\text{Q}][\text{R}] + V_{mr}K_{ia}K_{mB}[\text{C}][\text{P}][\text{Q}][\text{R}] \\
& + V_{mf}K_{mA}[\text{B}][\text{C}][\text{P}][\text{Q}][\text{R}]) \\
& + \frac{1}{K_{ia}K_{ib}K_{ic}K'_{scs}}(V_{mf}K_{ir}K_{mQ}[\text{A}][\text{B}][\text{C}][\text{P}] \\
& + V_{mf}K_{ip}K_{mR}[\text{A}][\text{B}][\text{C}][\text{Q}] + V_{mf}K_{mR}[\text{A}][\text{B}][\text{C}][\text{P}][\text{Q}]). \tag{6.48}
\end{aligned}
$$

The parameter values are $V_{mf} = 0.582 \, \text{mol sec}^{-1} \, (1 \, \text{mito})^{-1}$, $K_{mA} = 16 \, \mu\text{M}$, $K_{mB} = 55 \, \mu\text{M}$, $K_{mC} = 660 \, \mu\text{M}$, $K_{mP} = 20 \, \mu\text{M}$, $K_{mQ} = 880 \, \mu\text{M}$, $K_{mR} = 11.1 \, \mu\text{M}$, $K_{ia} = 5.5 \, \mu\text{M}$, $K_{ib} = 100 \, \mu\text{M}$, $K_{ic} = 2000 \, \mu\text{M}$, $K_{ip} = 20 \, \mu\text{M}$, $K_{iq} = 3000 \, \mu\text{M}$, and $K_{ir} = 11.1 \, \mu\text{M}$. The value of V_{mr} is computed

$$V_{mr} = \frac{V_{mf} K_{mP} K_{iq} K_{ir}}{K'_{scs} K_{ia} K_{ib} K_{mC}}. \tag{6.49}$$

6.2.3.7 Succinate dehydrogenase

The flux expression for succinate dehydrogenase is

$$J_{sdh} = \frac{V_{mf} V_{mr} \left([\text{A}][\text{B}] - \frac{[\text{P}][\text{Q}]}{K'_{sdh}} \right)}{denom}, \tag{6.50}$$

where $[\text{A}] = [\text{SUC}]$, $[\text{B}] = [\text{CoQ}]$, $[\text{P}] = [\text{CoQH}_2]$, $[\text{Q}] = [\text{FUM}]$, and K'_{sdh} is the apparent equilibrium constant. The denominator of the flux expression is

$$denom = V_{mr} K_{ia} K_{mB} \alpha_i + V_{mr} K_{mB}[\text{A}] + V_{mr} K_{mA} \alpha_i [\text{B}] + \frac{V_{mf} K_{mQ} \alpha_i}{K'_{sdh}}[\text{P}]$$

$$+ \frac{V_{mf} K_{mP}}{K'_{sdh}}[\text{Q}] + V_{mr}[\text{A}][\text{B}] + \frac{V_{mf} K_{mQ}}{K'_{sdh} K_{ia}}[\text{A}][\text{P}] + \frac{V_{mr} K_{mA}}{K_{iq}}[\text{B}][\text{Q}]$$

$$+ \frac{V_{mf}}{K'_{sdh}}[\text{P}][\text{Q}], \tag{6.51}$$

where

$$\alpha_i = \left(1 + \frac{[\text{OAA}]}{K_{iOAA}} + \frac{[\text{SUC}]}{K_{aSUC}} + \frac{[\text{FUM}]}{K_{aFUM}} \right) \Big/ \left(1 + \frac{[\text{SUC}]}{K_{aSUC}} + \frac{[\text{FUM}]}{K_{aFUM}} \right).$$

The parameter values are $V_{mf} = 6.23 \times 10^{-2} \, \text{mol sec}^{-1} \, (1 \, \text{mito})^{-1}$, $K_{mA} = 467 \, \mu\text{M}$, $K_{mB} = 480 \, \mu\text{M}$, $K_{mP} = 2.45 \, \mu\text{M}$, $K_{mQ} = 1200 \, \mu\text{M}$, $K_{ia} = 120 \, \mu\text{M}$, $K_{iq} = 1275 \, \mu\text{M}$, $K_{iOAA} = 1.5 \, \mu\text{M}$, $K_{aSUC} = 450 \, \mu\text{M}$, and $K_{aFUM} = 375 \, \mu\text{M}$. The value of V_{mr} is computed

$$V_{mr} = \frac{V_{mf} K_{mP} K_{iq}}{K'_{sdh} K_{ia} K_{mB}}. \tag{6.52}$$

6.2.3.8 Fumarase

The flux expression for fumarase is

$$J_{fum} = \frac{V_{mf} V_{mr} \left([\text{A}] - \frac{[\text{P}]}{K'_{fum}} \right)}{K_{mA} V_{mr} \alpha_i + V_{mr}[\text{A}] + \frac{V_{mf}[\text{P}]}{K'_{fum}}}, \tag{6.53}$$

where $[A] = [FUM]$, $[P] = [MAL]$, and K'_{fum} is the apparent equilibrium constant. The factor α_i accounts for competitive inhibition by citrate and unchelated ATP, ADP, GTP, and GDP:

$$\alpha_i = 1 + \frac{[CIT]}{K_{iCIT}} + \frac{[fATP]}{K_{iATP}} + \frac{[fADP]}{K_{iADP}} + \frac{[fGTP]}{K_{iGTP}} + \frac{[fGDP]}{K_{iGDP}}.$$

The concentrations [fATP], [fADP], [fGTP], and [fGDP] refer to unchelated (non-magnesium-bound) ATP, ADP, GTP, GDP.

The parameter values are $V_{mf} = 7.12 \times 10^{-3}$ mol sec^{-1} (1 mito)$^{-1}$, $K_{mA} = 44.7$ μM, $K_{mP} = 197.7$ μM, $K_{iCIT} = 3500$ μM, $K_{iATP} = 40$ μM, $K_{iADP} = 400$ μM, $K_{iGTP} = 80$ μM, and $K_{iGDP} = 330$ μM. The value of V_{mr} is computed

$$V_{mr} = \frac{V_{mf} K_{mP}}{K'_{fum} K_{mA}}. \tag{6.54}$$

6.2.3.9 Malate dehydrogenase

The flux expression for malate dehydrogenase is

$$J_{mdh} = \frac{V_{mf} V_{mr} \left([A][B] - \frac{[P][Q]}{K'_{mdh}}\right)}{denom}, \tag{6.55}$$

where $[A] = [NAD]$, $[B] = [MAL]$, $[P] = [OAA]$, $[Q] = [NADH]$, and K'_{mdh} is the apparent equilibrium constant. The denominator of the flux expression is

$$denom = V_{mr} K_{ia} K_{mB} \alpha_i + V_{mr} K_{mB}[A] + V_{mr} K_{mA} \alpha_i [B] + \frac{V_{mf} K_{mQ} \alpha_i}{K'_{mdh}}[P]$$

$$+ \frac{V_{mf} K_{mP}}{K'_{mdh}}[Q] + V_{mr}[A][B] + \frac{V_{mf} K_{mQ}}{K'_{mdh} K_{ia}}[A][P] + \frac{V_{mr} K_{mA}}{K_{iq}}[B][Q]$$

$$+ \frac{V_{mf}}{K'_{mdh}}[P][Q] + \frac{V_{mr}}{K_{ip}}[A][B][P] + \frac{V_{mf}}{K_{ib} K'_{mdh}}[B][P][Q] \tag{6.56}$$

where

$$\alpha_i = \left(1 + \frac{[fATP]}{K_{iATP}} + \frac{[fADP]}{K_{iADP}} + \frac{[fAMP]}{K_{iAMP}}\right).$$

The parameter values are $V_{mf} = 6.94$ mol sec^{-1} (1 mito)$^{-1}$, $K_{mA} = 90.55$ μM, $K_{mB} = 250$ μM, $K_{mP} = 6.13$ μM, $K_{mQ} = 2.58$ μM, $K_{ia} = 279$ μM, $K_{ib} = 360$ μM, $K_{ip} = 5.5$ μM, $K_{iq} = 3.18$ μM, $K_{iATP} = 183.2$ μM, $K_{iADP} = 394.4$ μM, and $K_{iAMP} = 420$ μM. The value of V_{mr} is computed

$$V_{mr} = \frac{V_{mf} K_{mP} K_{iq}}{K'_{mdh} K_{ia} K_{mB}}. \tag{6.57}$$

6.2.3.10 Nucleoside diphosphokinase

The flux expression for nucleoside diphosphokinase is

$$J_{ndk} = \frac{V_{mf} V_{mr} \left([A][B] - \frac{[P][Q]}{K'_{ndk}} \right) \big/ \alpha_i}{denom},$$
(6.58)

where $[A] = [GTP]$, $[B] = [ADP]$, $[P] = [GDP]$, $[Q] = [ATP]$, K'_{ndk} is the apparent equilibrium constant, and

$$\alpha_i = \left(1 + \frac{[fAMP]}{K_{iAMP}} \right).$$

The denominator of the flux expression is

$$denom = V_{mr} K_{mB}[A] + V_{mr} K_{mA}[B] + V_{mr}[A][B] + \frac{V_{mf} K_{mQ}}{K'_{ndk}}[P]$$

$$+ \frac{V_{mf} K_{mP}}{K'_{ndk}}[Q] + \frac{V_{mf}}{K'_{ndk}}[P][Q] + \frac{V_{mr} K_{mA}}{K_{iq}}[B][Q]$$

$$+ \frac{V_{mf} K_{mQ}}{K'_{ndk} K_{ia}}[A][P].$$
(6.59)

The parameter values are $V_{mf} = 2.65 \times 10^{-2}$ mol sec^{-1} (1 mito)$^{-1}$, $K_{mA} = 111$ μM, $K_{mB} = 100$ μM, $K_{mP} = 260$ μM, $K_{mQ} = 278$ μM, $K_{ia} = 170$ μM, $K_{ib} = 143.6$ μM, $K_{ip} = 146.6$ μM, $K_{iq} = 156.5$ μM, $K_{iAMP} = 650$ μM. The value of V_{mf} is computed

$$V_{mr} = \frac{V_{mf} K_{mQ} K_{ip}}{K'_{ndk} K_{mB} K_{ia}}.$$
(6.60)

6.2.3.11 Aspartate aminotransferase

The flux expression for aspartate aminotransferase is

$$J_{aat} = \frac{V_{mf} V_{mr} \left([A][B] - \frac{[P][Q]}{K'_{aat}} \right)}{denom},$$
(6.61)

where $[A] = [ASP]$, $[B] = [AKG]$, $[P] = [OAA]$, $[Q] = [GLU]$, and K'_{ndk} is the apparent equilibrium constant. The denominator of the flux expression is

$$denom = V_{mr} K_{mB}[A] + V_{mr} K_{mA}[B] + V_{mr}[A][B] + \frac{V_{mf} K_{mQ}}{K'_{aat}}[P]$$

$$+ \frac{V_{mf} K_{mP}}{K'_{aat}}[Q] + \frac{V_{mf}}{K'_{aat}}[P][Q] + \frac{V_{mr} K_{mA}}{K_{iq}}[B][Q]$$

$$+ \frac{V_{mf} K_{mQ}}{K'_{aat} K_{ia}}[A][P].$$
(6.62)

The parameter values are $V_{mf} = 7.96 \, \text{mol} \, \text{sec}^{-1} \, (1 \, \text{mito})^{-1}$, $K_{mA} = 3900 \, \mu\text{M}$, $K_{mB} = 430 \, \mu\text{M}$, $K_{mP} = 88 \, \mu\text{M}$, $K_{mQ} = 8900 \, \mu\text{M}$, $K_{ia} = 3480 \, \mu\text{M}$, $K_{ib} = 50 \, \mu\text{M}$, $K_{ip} = 710 \, \mu\text{M}$, and $K_{iq} = 3480 \, \mu\text{M}$. The value of V_{mr} is computed

$$V_{mr} = \frac{V_{mf} K_{mQ} K_{ip}}{K'_{aat} K_{mB} K_{ia}}. \tag{6.63}$$

6.2.4 Differential equations for TCA cycle reaction network

The next step in building a biochemical network model for the TCA cycle is determining the governing differential equations. Since we are not treating transport of material into or out of the mitochondrion, the reactants of the overall reaction of Equation (6.31) are held clamped in this model. In addition, ASP and GLU are held fixed because there are no sources or sinks for the metabolites other than the aspartate aminotransferase reaction built into the model at this stage. Since the electron transport system is not modeled, proton transport is not included and pH is held fixed.

The fixed concentrations are set at $[\text{PYR}] = 0.076 \, \text{mM}$, $[\text{CoQ}] = 0.97 \, \text{mM}$, $[\text{CoQH}_2] = 0.38 \, \text{mM}$, $[\text{PI}] = 1.8 \, \text{mM}$, $[\text{ASP}] = 0.06 \, \text{nM}$, $[\text{GLU}] = 0.06 \, \text{mM}$, $[\Sigma\text{CO}_2] = 21.4 \, \text{mM}$, and $[\text{H}^+] = 10^{-7.2} \, \text{M}$, which are physiologically reasonable values. (In Chapter 7 this model is integrated with a detailed model of oxidative phosphorylation and mitochondrial substrate and ion transport in which none of these species or reactants are held at fixed concentrations.) The concentrations of NAD, NADH, ATP, and ADP are varied in order to examine the biochemical mechanisms regulation flux of the TCA cycle.

For the remaining (non-clamped) reactants, we have the following differential equations that arise from the stoichiometry of the reference reactions:

$$
\begin{aligned}
d[\text{ACCOA}]/dt &= (-J_{cits} + J_{pdh})/W_x \\
d[\text{CIT}]/dt &= (+J_{cits} - J_{acon})/W_x \\
d[\text{ICIT}]/dt &= (+J_{acon} - J_{isod})/W_x \\
d[\text{AKG}]/dt &= (+J_{isod} - J_{akgd} - J_{aat})/W_x \\
d[\text{SCOA}]/dt &= (+J_{akgd} - J_{scs})/W_x \\
d[\text{COAS}]/dt &= (-J_{pdh} - J_{akgd} + J_{scs} + J_{cits})/W_x \\
d[\text{SUC}]/dt &= (+J_{scs} - J_{sdh})/W_x \\
d[\text{FUM}]/dt &= (+J_{sdh} - J_{fum})/W_x \\
d[\text{MAL}]/dt &= (+J_{fum} - J_{mdh})/W_x \\
d[\text{OAA}]/dt &= (-J_{cits} + J_{mdh} + J_{aat})/W_x \\
d[\text{GTP}]/dt &= (+J_{scs} - J_{ndk})/W_x \\
d[\text{GDP}]/dt &= (-J_{scs} + J_{ndk})/W_x,
\end{aligned}
\tag{6.64}
$$

where $W_x = 0.6514$ (l water)$\cdot$(l mito)$^{-1}$ is the mitochondrial matrix water space per unit mitochondrial volume [13]. Since pH is held fixed, the equations for binding ions are simplified and we have, for Mg^{2+} and K^+ kinetics:

$$\frac{d[Mg^{2+}]}{dt} = \frac{\alpha_K \Phi^M - \left(\frac{\partial[Mg_{bound}]}{\partial[K^+]}\right) \Phi^K}{\alpha_M \alpha_K - \frac{\partial[Mg_{bound}]}{\partial[K^+]} \frac{\partial[K_{bound}]}{\partial[Mg^{2+}]}} \qquad (6.65)$$

and

$$\frac{d[K^+]}{dt} = \frac{\alpha_M \Phi^K - \left(\frac{\partial[K_{bound}]}{\partial[Mg^{2+}]}\right) \Phi^M}{\alpha_M \alpha_K - \frac{\partial[Mg_{bound}]}{\partial[K^+]} \frac{\partial[K_{bound}]}{\partial[Mg^{2+}]}}. \qquad (6.66)$$

6.2.5 Simulation of TCA cycle kinetics

We start our analysis of the TCA cycle kinetics by examining the predicted steady state production of NADH as a function of the NAD and ADP concentrations. From Equation (6.31) we see that there can be no net flux through the TCA cycle when concentration of either NAD or ADP, which serve as substrates for reactions in the cycle, is zero. Thus when the ratios [ATP]/[ADP] and [NADH]/[NAD] are high, we expect the TCA cycle reaction fluxes to be inhibited by simple mass action. In addition, the allosteric inhibition of several enzymes (for example inhibition of pyruvate dehydrogenase by NADH and ACCOA) has important effects.

The overall control of integrated system behavior by NAD and ADP can be understood based on simulation of the model as follows. We define the rate of NADH production as $J_{DH} = J_{pdh} + J_{isod} + J_{akgd} + J_{mdh}$, and compute the predicted steady state J_{DH} as a function of $[ADP]/A_o$ and $[NAD]/N_o$, where $A_o = [ATP] + [ADP] = 10.0$ mM and $N_o = [NADH] + [NAD] = 2.97$ mM are the total concentrations of adenine nucleotide and NAD nucleotide [213].

We can see from Figure 6.3 that the relative NADH concentration is the more important controller of steady state TCA cycle flux, in agreement with experimental observations [124]. When ADP concentration is low, a variation in $[NAD]/N_o$ from 0 to 1 produces a change in J_{DH} from 0 to nearly 1.59 mmol $\cdot$ sec^{-1} $\cdot$ (l mito)$^{-1}$. When NAD concentration is near zero, the rate of NADH production is not sensitive to ADP. Yet the flux is by no means insensitive to ADP. Neither NAD nor ADP represents a sole independent controller of the system.

To understand how the TCA cycle responds kinetically to changes in demand, we can examine the predictions in time-dependent reaction fluxes in response to changes in the primary controlling variable NAD. Figure 6.4 plots predicted reaction fluxes for pyruvate dehydrogenase, aconitase, fumarase, and malate dehydrogenase in response to an instantaneous change in NAD. The initial steady state is obtained

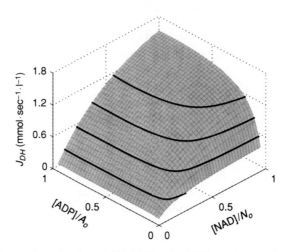

Figure 6.3 Rate of production of NADH by the TCA cycle as a function of the relative concentrations of NAD and ADP. The flux J_{DH} is given in units of mmole NADH generated per second per unit mitochondrial volume. Contours are drawn at $J_{DH} = 0.30, 0.60, 0.90,$ and $1.20 \, \text{mmol} \cdot \text{sec}^{-1} \cdot (\text{l mito})^{-1}$.

at $[\text{ADP}]/A_o = 0.5$ and $[\text{NAD}]/N_o = 0.20$, for which the predicted net rate of NADH generation is $J_{DH} = 0.58 \, \text{mmol} \cdot \text{sec}^{-1} \cdot (\text{l mito})^{-1}$. At time $t = 5 \, \text{seconds}$, the $[\text{NAD}]/N_o$ ratio is suddenly increased to 0.80, which produces a near maximal steady state $J_{DH} = 1.43 \, \text{mmol} \cdot \text{sec}^{-1} \cdot (\text{l mito})^{-1}$.

The four fluxes plotted in the figure correspond to different locations on the TCA cycle diagram. (The reaction numbers from the diagram of Figure 6.2 are indicated in Figure 6.4.) We can see that the predicted fluxes reach peak values within one second following the perturbation before settling down to the new steady state. As expected, the pyruvate dehydrogenase flux (reaction 1) responds sharply to an increase in NAD. (The peak is not shown in the plot for pyruvate dehydrogenase.) After an initial increase from 0.15 to approximately $2.0 \, \text{mmol} \cdot \text{sec}^{-1} \cdot (\text{l mito})^{-1}$, the flux approaches the new steady state value of approximately $0.33 \, \text{mmol} \cdot \text{sec}^{-1} \cdot (\text{l mito})^{-1}$.

The aconitase flux (reaction 3) and the fumarase flux (reaction 8) display overshoots that are small compared to pyruvate dehydrogenase. The aconitase reaction flux reaches a peak value that is only a few percent greater than the final steady state value. The fumarase flux does overshoot the final steady state, but the overshoot is too small to be observed on the scale plotted. This behavior occurs because these reactions are downstream of any reactions directly using NAD as a substrate. Therefore their response is muted compared to reactions that are directly controlled by NAD/NADH.

The final dehydrogenase in the system (malate dehydrogenase, reaction 11) uses NAD as a substrate and, like pyruvate dehydrogenase, responds sharply to

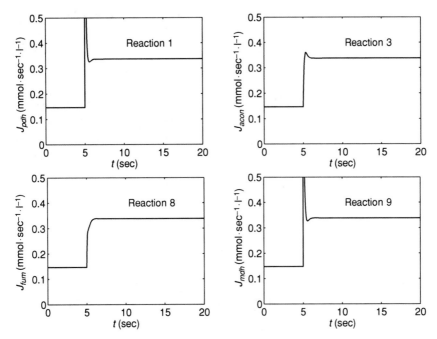

Figure 6.4 Predicted response in flux of several reactions in the TCA cycle following an instantaneous increase in NAD. At time $t = 5$ seconds the ratio $[NAD]/N_o$ is increased from 0.20 to 0.80.

the global increase in NAD. Similar to the pyruvate dehydrogenase reaction, the predicted peak flux following the perturbation (not shown) is approximately 2.1 $mmol \cdot sec^{-1} \cdot (1\ mito)^{-1}$.

One important shortcoming of the model and associated results reported in this section is the fact that input and output concentrations of Equation (6.31) are held fixed, and that the reactions that consume NADH and $CoQH_2$ and regenerate NAD and CoQ are not included. Modeling these processes requires integrating the reaction system studied here with mitochondrial transport processes. Such a study is the subject of an example model developed in Chapter 7, which is devoted to integrating reaction and transport systems. Based on simulating the integrated system of the TCA cycle and oxidative phosphorylation, it is demonstrated that ADP and PI act as the primary controllers of oxidative ATP synthesis [213]. Since PI is a TCA cycle substrate, PI concentration is a significant determinant of both oxidative phosphorylation activity and also TCA cycle activity.

6.3 Control and stability in biochemical networks

In the previous section we explored the impact of changes in certain concentrations on TCA cycle fluxes. For example, Figure 6.3 plots the predicted NADH-producing flux as a function of NAD and ADP concentrations. This analysis is one example of a *sensitivity analysis*, in which we have explored the sensitivity of flux to concentrations. In this section we lay out a theory of sensitivity analysis for biochemical reaction networks based on analyzing the response of linearized systems to various perturbations.

Metabolic control analysis (MCA) is a specialized theory that is concerned with particular sensitivity coefficients, *elasticity coefficients* and *control coefficients*. These coefficients tell us how a steady state of a biochemical system shifts in response to perturbation in enzyme activities or external (clamped) substrate concentration [53, 209].

6.3.1 Linear analysis near a steady state

Metabolic control analysis and related theories are based on examining how a system responds to infinitesimally small perturbations. Thus in general one can make use of a linearized set of governing kinetic equations. To be specific, an N-dimensional system governed by a non-linear system of differential equations

$$\frac{d\mathbf{x}}{dt} = \mathbf{f}(\mathbf{x}),\tag{6.67}$$

may be linearized about an asymptotically stable steady state $\mathbf{x}^*$, as follows:

$$\frac{d}{dt}(\delta x_i) = \sum_{j=1}^{N} \left.\frac{\partial f_i}{\partial x_j}\right|_{\mathbf{x}=\mathbf{x}^*} \cdot \delta x_j,\tag{6.68}$$

in which $\delta x_j = x_j - x_j^*$. Defining the linear stability matrix $\mathbb{A}$

$$A_{ij} = x_j^* \left.\frac{\partial f_i}{\partial x_j}\right|_{\mathbf{x}=\mathbf{x}^*}$$

we have

$$\frac{d}{dt}\delta x_i = \sum_{j=1}^{N} A_{ij}\left(\frac{\delta x_j}{x_j^*}\right).\tag{6.69}$$

For a reaction network of M biochemical reactions involving N biochemical species X_i ($i = 1, 2, \ldots, N$), the jth biochemical reaction is characterized by a set of stoichiometric coefficients $v = \{v_i^j\}$ and $\kappa = \{\kappa_i^j\}$:

$$v_1^j X_1 + v_2^j X_2 + \cdots + v_N^j X_N \rightleftharpoons \kappa_1^j X_1 + \kappa_2^j X_2 + \cdots + \kappa_N^j X_N,\tag{6.70}$$

($j = 1, 2, \ldots, M$), where integers v's and κ's can be zero. The $N \times M$ matrix $\mathbb{S}$ (with entries $S_{ij} = \kappa_i^j - v_i^j$) is known as the stoichiometric matrix.

Denoting the concentration of species X_i as x_i, we have $d\mathbf{x}/dt = \mathbf{f}(\mathbf{x}) = \mathbb{S}\mathbf{J} + \mathbf{J}_e$, where $\mathbf{x}$ and $\mathbf{J}_e$ are N-dimensional and $\mathbf{J}$ is an M-dimensional column vector. In a *closed* reaction system, $\mathbf{J}_e = \mathbf{0}$. In this case, thermodynamic equilibrium is the only positive stationary solution: the internal fluxes $\mathbf{J} = \mathbf{J}_+ - \mathbf{J}_- = \mathbf{0}$ for each and every reaction (see Chapter 9). Therefore, in a non-equilibrium steady state either one or more injection fluxes must be non-zero ($J_i^e \neq 0$), or certain concentrations x_i must be held at constant levels. We refer to the first case as *external flux injection*, the latter as *external concentration clamping*.

There is a close analog between an open biochemical system and an electrical circuit powered by a battery. Recall that there are two types of ideal batteries, those that provide constant current (current sources with zero internal conductance) and those that provide constant voltage (voltage sources with zero internal resistance). A real battery of course has a finite internal resistance and conductance. For a metabolic system in an ideal setting, one can either control the fluxes and let the concentrations change in response, or one can control the concentrations and let the fluxes change in response. Controlling fluxes can, but not necessarily, be accomplished by changing enzyme activities.

The linear stability matrix for chemical reaction systems is computed

$$A_{ij} = x_j^* \left. \frac{\partial f_i}{\partial x_j} \right|_{\mathbf{x}=\mathbf{x}^*} = x_j^* \sum_{k=1}^{M} S_{ik} \left. \frac{\partial J_k}{\partial x_j} \right|_{\mathbf{x}=\mathbf{x}^*} \tag{6.71}$$

and we define its inverse $\mathbb{A}^{-1} = \mathbb{B}$.[1] From Equation (6.71) we also define the matrix $\mathbb{R}$, where $\mathbb{A} = \mathbb{S}\mathbb{R}$ and

$$R_{ij} = x_j^* \left. \frac{\partial J_i}{\partial x_j} \right|_{\mathbf{x}=\mathbf{x}^*}. \tag{6.72}$$

We will see below that the matrix $\mathbb{R}$ is central to MCA.

6.3.2 Metabolic control analysis

6.3.2.1 Elasticity coefficients

Let us first consider the case where the concentration of a species, species n, is changed: $x_n^* \rightarrow x_n^* + \delta x_n$. Since the state $\mathbf{x}$ is asymptotically stable, the system will return to the original steady state if x_n is a dynamic variable. However, if x_n is a clamped concentration, then the system will achieve a new steady state. Locally

---

[1] The non-linear dynamics scheme $d\mathbf{x}/dt = \mathbb{S}\mathbf{J} + \mathbf{J}_e$ may conserve certain quantities. (See Section 9.4.3.) In such cases $\mathbb{A}$ is singular [159] and must be transformed into a non-singular matrix by removing the redundancies from the original dynamics scheme resulting in replacing the linearly dependent rows of $\mathbb{A}$ with vectors in its left null space. Here $\mathbb{A}$ is understood as the transformed non-singular matrix.

and immediately, before reaching the new steady state, the response is only in the flux of the reactions involving X_n. The local change is characterized by:

$$\epsilon_n^m \triangleq \left[\frac{x_n}{J_m}\frac{\partial J_m}{\partial x_n}\right]_{\mathbf{x}=\mathbf{x}^*} = \frac{R_{mn}}{J_m} \qquad (6.73)$$

which is called the local elasticity coefficient [54].

The coefficient ϵ_n^m is distinguished from the coefficient ε_n^m, which characterizes the steady state response [208]. When a new steady state is established following a change of $x_n^* \to x_n^* + \delta x_n$ in clamping the species n, the remaining $(N-1)$ concentrations satisfy the system of linear equations:

$$\sum_{\substack{j \neq n}}^{N} A_{ij}\frac{\delta x_j}{x_j^*} = -A_{in}\frac{\delta x_n}{x_n^*} \quad (1 \le i \le N, i \neq n), \qquad (6.74)$$

from setting Equation (6.69) to zero.

Solving Equation (6.74) for δx_i, we have $\delta x_i/x_i^* = B_{in}\delta x_n/B_{nn}x_n^*$ where B_{in} is the nth column vector of the matrix $\mathbb{A}^{-1}$. The new steady state established near $\{x_i^*\}$ is $\{x_i^* + \delta x_i\}$.[2]

The fluxes in the new steady state, $J_m + \delta J_m$, are given by:

$$\delta J_m = \sum_{\ell=1}^{N}\frac{\partial J_m}{\partial x_\ell}\cdot\delta x_\ell = \sum_{\ell=1}^{N}\epsilon_\ell^m\frac{B_{\ell n}\delta x_n}{B_{nn}x_n^*}J_m. \qquad (6.75)$$

Hence the steady state elasticity coefficient is

$$\varepsilon_n^m \triangleq \frac{x_n^*}{J_m}\left(\frac{\delta J_m}{\delta x_n}\right) = \sum_{\ell=1}^{N}\epsilon_\ell^m\frac{B_{\ell n}}{B_{nn}} = \frac{(\mathbb{R}\mathbb{B})_{mn}}{J_m B_{nn}}. \qquad (6.76)$$

The two quantities ϵ and ε have different properties as we shall see.

6.3.2.2 Control coefficients

We now consider the case where enzyme activity for the mth reaction E_m is changed: $E_m \to E_m + \delta E_m$. Assuming that the flux through the reaction is linearly proportional to the activity of the enzyme catalyzing the reaction, E_m, when $E_m \to E_m + \delta E_m$, the new steady state satisfies

$$\frac{d}{dt}(\delta x_i) = \sum_{j=1}^{N} A_{ij}\frac{\delta x_j}{x_j^*} + S_{im}J_m\frac{\delta E_m}{E_m} = 0, \qquad (6.77)$$

---

[2] See Exercise 3.

which arises from the linear expansion of Equation (6.67)

$$\frac{d}{dt}(\delta x_i) = \sum_j \frac{\partial f_i}{\partial x_j} \cdot (\delta x_j) + \frac{\partial f_i}{\partial E_m} \cdot (\delta E_m).$$

Solving Equation (6.77) for δx_j ($j = 1, 2, \ldots, N$), we have

$$\frac{\delta x_j}{x_j^*} = -\sum_{i=1}^{N} B_{ji} S_{im} J_m \frac{\delta E_m}{E_m}. \tag{6.78}$$

We define

$$\widehat{C}_j^m \triangleq \frac{E_m}{x_j^*}\left(\frac{\delta x_j}{\delta E_m}\right) = -(\mathbb{B}\mathbb{S})_{jm} J_m, \tag{6.79}$$

which is called the concentration control coefficient. Combining Equations (6.78) and (6.75), we have

$$C_n^m \triangleq \frac{E_n}{J_m}\left(\frac{\delta J_m}{\delta E_n}\right) = \delta_{mn} - \frac{1}{J_m}(\mathbb{R}\mathbb{B}\mathbb{S})_{mn} J_n, \tag{6.80}$$

where δ_{mn} is the Kronecker delta function: $\delta_{mn} = 0$ if $m \neq n$ and $= 1$ if $m = n$. The quantity C_n^m is called the flux control coefficient.

6.3.2.3 Summation theorems

There exist several important theorems related to sums of these control coefficients [209, 53]. From Equation (6.79) it is apparent that

$$\sum_{m=1}^{M} \widehat{C}_n^m = 0 \tag{6.81}$$

if there is no injection flux ($\mathbb{S}J = -J^e = 0$). Similarly, from Equation (6.80)

$$\sum_{j=1}^{M} C_j^n = 1 \tag{6.82}$$

if there is no injection flux. Thus both summation theorems depend on the injection flux being zero. In other words, the metabolic steady state is sustained by concentration clamping.[3]

---

[3] The significance of the injection fluxes can be best understood in terms of Euler's theorem of homogeneous functions [65]: if there are no injection fluxes in a system, it is sustained by clamped concentrations. Assuming every reaction in the system is catalyzed by an enzyme, then if each and every enzyme concentration is doubled, the flux in each reaction doubles. In other words, the flux is a homogeneous function of enzyme concentrations with order 1. This leads to the summation rule for C_n^m, Equation (6.82). At the other extreme, if the steady state is sustained by injection fluxes and there are no clamped concentrations, then the flux is a homogeneous function of enzyme concentrations with order 0. A similar argument exists for steady state concentrations as functions of enzyme concentrations leading to the summation rule for $\widehat{C}_j^n$.

One of the many important consequences of these summation theorems is that, in the case where all flux control coefficients are positive, all coefficients have values between 0 and 1. In this case, the reaction for which $C_i{}^j$ is greatest represents the reaction to which the flux J_j is most sensitive. In the limit that one flux control coefficient has a value close to 1 and all others have values close to 0, we can say that there exists a rate-limiting step; a change in activity of the rate-limiting enzyme would be expected to elicit a proportional change in the flux J_j.

Summation rules for elasticity coefficients can be obtained by noting $\mathbb{BSR} = \mathbb{I}$ $\Rightarrow (\mathbb{RBS})(\mathbb{RB}) = \mathbb{RB}$, which yields

$$\sum_{n=1}^{M} \epsilon_N^n C_n^m = \sum_{n=1}^{M} \varepsilon_N^n C_n^m = 0$$

$$\sum_{n=1}^{N} \epsilon_N^n \widehat{C}_m^n = -\delta_{mN}, \qquad \sum_{n=1}^{N} \varepsilon_N^n \widehat{C}_m^n = -\frac{B_{mN}}{B_{NN}}. \tag{6.83}$$

Finally, the steady state fluxes $\mathbf{J}$, stoichiometric matrix $\mathbb{S}$, and flux control coefficient matrix $\mathbb{C}$ satisfy

$$\sum_{m=1}^{M} S_{im} J_m C_j^m = 0. \tag{6.84}$$

Note that local slopes of the surface illustrated in Figure 6.3 at given concentrations of reactants provide a sensitivity measure that is equivalent to the steady state elasticity coefficients. Namely Figure 6.3 illustrates the sensitivity of flux to finite changes in clamped concentration values. One appeal of this analysis is that it illustrates the sensitivity of flux over a wide range of behavior in the system, rather than at a single specified steady state. Of course, analysis and visualization in multi-dimensional sensitivity analysis is a challenge. The control of one predicted variable by two parameters over a range of values can be illustrated by a surface or contour plot such as in Figure 6.3. Higher dimensional analyses require different strategies.

Concluding remarks

Here we have introduced a detailed formalism for building models of biochemical systems. This approach has the advantages that the influences of pH and metal ion concentrations on apparent thermodynamic properties are explicitly accounted for. This detailed accounting allows us to take advantage of the rich data available on dissociation constants and thermodynamic properties. Even so, the available data remain incomplete. While standard free energies of formation are available for many biochemical reference species, little information on enthalpies of formation is available. Thus it is not always possible to accurately adjust reaction free energies for temperature. When one looks for kinetic information, one finds that the available data are less complete, less consistent, and less well analyzed than the thermodynamic data. Thus, assigning values to kinetic parameters (such as in the kinetic model of the TCA cycle developed in this chapter) is typically not a straightforward or unambiguous task. To make continued progress in this field, strategies must be developed for obtaining and analyzing missing thermodynamic and kinetic data for enzyme-catalyzed biochemical reactions.

It is also important to note that the simplification that pH and metal ion concentrations remain constant is a reasonable approximation in many cases. For example, if a cell maintains a constant metabolic state, then we expect pH to remain constant under most circumstances. However, we are ultimately not interested in cases in which a constant normal state is maintained. Our ultimate goal is to understand how cell systems act and respond to stress and disease. Since metabolism provides the energy and raw materials for cellular function, including the signaling modules of Chapter 5, fully integrated analysis of cellular function in health and disease will hinge on the detailed modeling approach outlined here.

Exercises

6.1 Construct a computer program to reproduce the simulation outlined in Section 6.1.4. Compare the behavior of the conservation relations (H_o, M_o, and K_o) computed using different solvers and numerical settings.

6.2 Derive Equations (6.65) and (6.66).

6.3 Show that $\delta x_i / x_i^* = B_{in} \delta x_n / B_{nn} x_n^*$ solves Equation (6.74). (Recall that $\mathbb{B} = \mathbb{A}^{-1}$.)

7

Coupled biochemical systems and membrane transport

Overview

As we have seen in previous chapters, living systems require that material be transported in and out in order to maintain an operating state (or operating states) that is far from thermodynamic equilibrium. Material is transported into and out of cells via passive permeation and by a diverse set of channels, pumps, transporters, and exchangers. In this chapter we consider kinetic models of transport across membranes, with specific examples of coupled transport and reaction in metabolic and electrophysiological systems. In the final example a computational model of oxidative ATP synthesis (which occurs as a set of reactions transporting charged species across the mitochondrial inner membrane) is developed. This model may be integrated with the detailed kinetic model of the TCA cycle presented in Chapter 6, allowing us to simulate and explore how the coupled systems interact – the TCA cycle producing reduced cofactors and the oxidative phosphorylation systems transducing the free energy of oxidation of these cofactors to synthesize and transport ATP.

7.1 Transporters

In Section 3.2 we introduced the basic processes of advection, diffusion, and drift, by which material is transported in biophysical systems. In this chapter we focus on a specialized class of transport: transport across biological membranes. Transport of a substance across a membrane may be driven by passive permeation, as described by Equation (3.60), or it may be facilitated by a *carrier protein* or *transporter* that is embedded in the membrane. Thus transport of substances across membranes mediated by transporters is termed *carrier-mediated transport*. The most basic way to think about carrier proteins or transporters is as enzymes that catalyze reactions that involve transport.

7.1.1 Active versus passive transport

Transport reactions are usually categorized into two classes: active transport and passive transport. Active transport implies that some species is moving against its electrochemical gradient in the transport process. This movement against the gradient is accomplished through other chemical processes that move down the electrochemical gradient. A typical coupled reaction in active transport is the hydrolysis of ATP, as in the sodium/potassium ATPase pump (or Na-K pump) that hydrolyzes an ATP while transporting 3 Na^+ ions out of the cardiac muscle cell and 2 K^+ into the cell against their electrochemical gradients.

Regardless of whether a transport reaction is termed active or passive, the overall process spontaneously moves down the overall gradient in free energy, as we shall demonstrate in the following examples. Since all transport processes act in the direction of free energy dissipation, the distinction between active and passive transport is a question of whether or not a chemical transformation is coupled to the transport processes. Thus, perhaps the terminology of active and passive transport is not perfect. In fact, it would perhaps be more informative if the terms active and passive transport were replaced with the terms *reacting* and *non-reacting*, reflecting processes that do and do not involve chemical transformation.

7.1.2 Examples: a uniporter and an antiporter

As a first example, let us consider the transport of glucose across the cell membrane via a glucose transport (GLUT). There exist several different isoforms of this enzyme expressed in different mammalian cell types. Glucose transporters are called *uniporters* because they transport a single substance across the membrane. Here we analyze a generic model for GLUT-mediated transport of glucose across the cell membrane, illustrated in Figure 7.1. In this model, glucose binds to a binding site on the protein, which may be exposed to either side of the cell membrane, as illustrated in the figure.

The cartoon of Figure 7.1 is modeled by the molecular mechanism illustrated in the left panel of Figure 7.2. One glucose molecule is transported from the outside to the inside of the cell every time one GLUT undergoes a forward catalytic cycle through the states $1 \to 2 \to 3 \to 4 \to 1$. The overall reaction is

$$G_{out} \rightleftharpoons G_{in}.$$

Therefore there is no overall chemical reaction. Assuming that conditions such as pH and ionic strength are the same on both sides of the membrane, the equilibrium constant for the transport reaction is $K_{GLUT} = 1$ and the equilibrium Gibbs free energy is $\Delta G^o_{GLUT} = 0$.

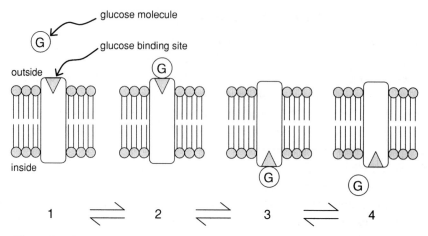

Figure 7.1 Transport of glucose across the cell membrane mediated by a glucose transporter. Four discrete states are illustrated. In state 1 the glucose binding site is exposed on the outside of the cell; in state 2 a glucose molecule is bound to the binding site on the outside of the cell; in state 3 the glucose-bound binding site is exposed to the inside of the cell; in state 4 the glucose is dissociated and the binding site is exposed to the inside of the cell.

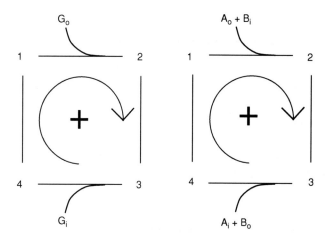

Figure 7.2 Left panel: reaction mechanism for glucose transporter illustrated in Figure 7.1. Right panel: reaction mechanism for antiport reaction $A_{out} + B_{in} \rightleftharpoons A_{in} + B_{out}$.

The principle of *detailed balance*, which states that at equilibrium the forward and reverse fluxes are equal for all component reactions in a mechanism [127], requires that in equilibrium each step in the mechanism shown in the left panel of Figure 7.2 is in equilibrium. Multiplying the equilibrium ratios for each reaction in the mechanism, we obtain:

$$\left(\frac{e_2}{[G_{out}]e_1}\right)_{eq} \cdot \left(\frac{e_3}{e_2}\right)_{eq} \cdot \left(\frac{[G_{in}]e_4}{e_4}\right)_{eq} \cdot \left(\frac{e_4}{e_1}\right)_{eq} = \left(\frac{[G_{in}]}{[G_{out}]}\right)_{eq} = K_{GLUT} = 1,$$

(7.1)

where we have denoted the concentrations of the enzyme in each state $e_1 = [E_1]$, $e_2 = [E_2]$, $e_3 = [E_3]$, and $e_4 = [E_4]$. Expressing the equilibrium ratios of Equation (7.1) in terms of mass-action rate constants for the steps we have

$$\frac{k_{12}}{k_{21}} \cdot \frac{k_{23}}{k_{32}} \cdot \frac{k_{34}}{k_{43}} \cdot \frac{k_{41}}{k_{14}} = 1.$$

(7.2)

To analyze this system we simplify the kinetic mechanism by assuming that the binding and unbinding of glucose from the transporter are rapid, with dissociation constant K_d on both sides of the membrane. This rapid-equilibrium assumption yields:

$$e_2 = [G_{out}]e_1/K_d$$
$$e_3 = [G_{in}]e_4/K_d,$$

(7.3)

where $K_d = k_{21}/k_{12} = k_{34}/k_{43}$. It follows that Equation (7.2) simplifies to

$$\frac{k_{23}}{k_{32}} \cdot \frac{k_{41}}{k_{14}} = 1.$$

(7.4)

Next we introduce the quasi-steady approximation, which yields:

$$J = k_{41}e_4 - k_{14}e_1 = k_{23}e_2 - k_{32}e_3$$

(7.5)

and a statement of conservation of total enzyme (or transporter):

$$E_o = e_1 + e_2 + e_3 + e_4.$$

(7.6)

Combining Equations (7.3), (7.5), and (7.6) to solve for e_1, we have

$$e_1 = \frac{E_o \left(k_{41} K_d + k_{32}[G_{in}]\right) K_d}{K_d^2 \left(k_{14} + k_{41}\right) + [G_{out}]K_d \left(k_{41} + k_{23}\right) + \cdots}$$
$$\times [G_{in}]K_d \left(k_{14} + k_{32}\right) + [G_{out}][G_{in}]\left(k_{23} + k_{32}\right).$$

(7.7)

Next we introduce the simplification that the conformational transformation of the protein that moves the binding site from one side of the membrane to the other is not affected by the presence of bound glucose. Therefore the kinetics governing transitions between states 2 and 3 is identical to that governing transitions between 1 and 4:

$$k_{23} = k_{14} = k_+$$

and

$$k_{32} = k_{41} = k_-.$$

Note that with the above simplification, Equation (7.4) is satisfied. Substituting these expressions into Equation (7.9) we obtain

$$e_1 = \frac{E_o K_d \left(\frac{k_-}{k_+ + k_-}\right)}{[G_{out}] + K_d}. \tag{7.8}$$

Using this expression for e_1, we obtain the quasi-steady flux expression

$$J = \frac{E_o K_d \left(\frac{k_+ k_-}{k_+ + k_-}\right)([G_{out}] - [G_{in}])}{([G_{out}] + K_d)([G_{in}] + K_d)}. \tag{7.9}$$

If we set $[G_{in}] = 0$, then

$$J = \frac{E_o \left(\frac{k_+ k_-}{k_+ + k_-}\right)[G_{out}]}{[G_{out}] + K_d} = \frac{V_{max}[G_{out}]}{[G_{out}] + K_d}, \tag{7.10}$$

which is the familiar Michaelis–Menten expression for a single-substrate irreversible enzyme flux.

To illustrate the general case with $[G_{in}] \neq 0$ we introduce the non-dimensional flux

$$v = \frac{J(k_- + k_+)}{E_o k_+ k_-}, \tag{7.11}$$

which is plotted in Figure 7.3 as a function of glucose concentration outside the cell for several different values of $[G_{in}]$.

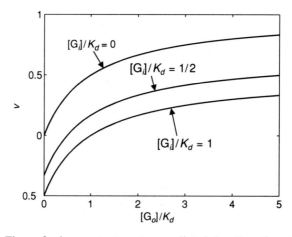

Figure 7.3 Flux of glucose transporter predicted by Equation (7.6). Non-dimensional flux v is defined in Equation (7.11).

A form for the flux more general than that of Equation (7.6) is obtained if we do not invoke the rapid pre-equilibrium assumption [108]:

$$J = \frac{E_o k_{12} k_{21} (k_{23} k_{41} [G_{out}] - k_{32} k_{14} [G_{in}])}{(K_{oi} [G_{out}][G_{in}] + K_o [G_{out}] + K_i [G_{in}] + K)}, \tag{7.12}$$

where the following assumptions have been made: $k_{12} = k_{43}$ and $k_{21} = k_{34}$. The rate constants in Equation (7.12) expressed in terms of the mass-action rate constants are: $K_{oi} = k_{12}^2/(k_{23} + k_{32})$, $K_o = k_{12}(k_{23}(k_{41} + k_{21}) + k_{41}(k_{32} + k_{21}))$, $K_i = k_{12}(k_{14}(k_{23} + k_{21}) + k_{32}(k_{14} + k_{21}))$, and $K = k_{21}(k_{14} + k_{41})(k_{21} + k_{23} + k_{32})$. Note that in this case Equation (7.2) reduces to Equation (7.4) because $k_{12} = k_{43}$ and $k_{21} = k_{34}$. Therefore the overall equilibrium

$$\left(\frac{[G_{in}]}{[G_{out}]} \right)_{eq} = 1$$

is obeyed by Equation (7.12).

Next we consider an *antiport* transport reaction, in which one substance on one side of the membrane is exchanged for another on the other side of the membrane. For example, an antiport reaction that exchanges A for B has the overall reaction

$$A_{out} + B_{in} \rightleftharpoons A_{in} + B_{out}.$$

As for the glucose transport reaction, there is no overall chemical reaction and the equilibrium constant for the process is 1.

A possible mechanism for this antiport reaction is illustrated in the right panel of Figure 7.2. Here the binding and unbinding of reactants to the antiporter protein are lumped into single reactions:

$$A_{out} + B_{in} + E_1 \rightleftharpoons E_2$$

and

$$A_{in} + B_{out} + E_4 \rightleftharpoons E_3,$$

which allows us to simplify the analysis and make the flux expression tractable. Assuming binding and unbinding are in rapid equilibrium, we have

$$\begin{aligned} e_2 &= [A_{out}][B_{in}]e_1/K_d \\ e_3 &= [A_{in}][B_{out}]e_4/K_d, \end{aligned} \tag{7.13}$$

where the dissociation constant K_d has units of M^2. Using the quasi-steady flux approximation (analogous to the above treatment of the glucose transporter), we obtain

$$J = \frac{E_o K_d \left(\frac{k_+ k_-}{k_+ + k_-} \right) ([\text{A}_{out}][\text{B}_{in}] - [\text{A}_{in}][\text{B}_{out}])}{([\text{A}_{out}][\text{B}_{in}] + K_d)([\text{A}_{in}][\text{B}_{out}] + K_d)} \tag{7.14}$$

for the flux expression for the antiporter.

A lesson to draw from the analysis of the two above examples is that transporters are treated as enzymes using techniques equivalent to those introduced in Chapter 4. The only difference is that transport reactions involve species on both sides of a membrane. Thus transport reaction may involve no overall chemical reaction, as is the case for the two examples in this section. Note that even when there is no overall chemical reaction, the equilibrium constant for a transport reaction can differ from unity. This is the case when charged species are transported across membranes, as described in the next section.

7.2 Transport of charged species across membranes

Many cells and subcellular organelles maintain an electrostatic potential difference across their membranes. This potential typically is important to the operation of the cell or organelle. For example, in nerve cells and other cells with excitable membranes such as muscle cells, the electrostatic potential is an important signal that governs cellular behavior. In these cells, some form of electrochemical signal that is sent to the cell can elicit an *action potential* – a transient change in the membrane potential that can trigger intracellular events, such as contraction of a muscle cell.

7.2.1 Thermodynamics of charged species transport

The thermodynamic potential that drives a chemical process involving the movement of charges across a membrane is given by:

$$\Delta\mu = \Delta\mu^o + \frac{\Delta\Psi}{N_C} \sum_{i \,\in\, inside} \nu_i z_i + k_B T \sum_i \nu_i \ln c_i \tag{7.15}$$

where $\Delta\Psi$ denotes the electrostatic potential difference between the two sides of the membrane. Associating the sides of the membrane with the inside and outside of a cell, $\Delta\Psi$ is the potential inside the cell relative to the potential outside – it is the inside potential minus the outside potential. The sign convention for Equation (7.15) is that the positive flux direction is defined to be from outside to inside. The constants k_B and N_C are Boltzmann's constant and the Coulomb constant, respectively. The notation $\sum_{i \,\in\, inside}$ denotes summation over all participating species on the inside of the membrane; z_i and ν_i represent the valence and stoichiometric number for the ith participating species, respectively; c_i denotes the concentration of the ith species.

The term $\sum_{i \,\in\, inside} \nu_i z_i$ in Equation (7.15) computes the total charge translocated across the membrane for the transport process. Note that since a chemical reaction does not create or destroy charge,

$$\sum_i \nu_i z_i = \sum_{i \,\in\, inside} \nu_i z_i + \sum_{i \,\in\, outside} \nu_i z_i = 0. \qquad (7.16)$$

Applying Equation (7.15) to isothermal isobaric transport, we have the following equation for the Gibbs free energy for the coupled chemical reaction and transport process:

$$\Delta G = \Delta G^o + F \Delta \Psi \sum_{i \,\in\, inside} \nu_i z_i + RT \sum_i \nu_i \ln c_i \qquad (7.17)$$

where F is Faraday's constant, which is equal to the Avogadro constant divided by the Coulomb constant. Note that Equation (7.17) is written in terms of chemical species, for which the charge is defined, and not in terms of chemical reactants. Thus ΔG in Equation (7.17) is the free energy for a chemical reaction, not the apparent free energy for a biochemical reaction.

As an example, let us consider the coupled transport of ADP and ATP through the adenine nucleotide translocase (ANT) exchanger located on the mitochondrial inner membrane. This transporter exchanges the species ATP^{4-} in the mitochondrial matrix for ADP^{3-} in intermembrane space between the outer and inner mitochondrial membranes, as illustrated in Figure 7.4. This exchange occurs against the concentration gradient, and is driven by the electrostatic potential across the membrane. The overall reference reaction for the transporter is:

$$ATP^{4-}_{in} + ADP^{3-}_{out} \rightleftharpoons ATP^{4-}_{out} + ADP^{3-}_{in} \qquad (7.18)$$

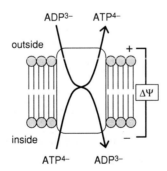

Figure 7.4 Cartoon of transport reaction catalyzed by adenine nucleotide translocase (ANT). ATP^{4-} in the mitochondrial matrix is exchanged for ADP^{3-} in intermembrane space between the outer and inner mitochondrial membranes.

where the subscripts "in" and "out" denote matrix and intermembrane space, respectively. Applying Equation (7.17), we have:

$$\Delta G_{ANT} = -F\Delta\Psi + RT\ln\left(\frac{[\text{ATP}^{4-}]_{\text{out}}[\text{ADP}^{3-}]_{\text{in}}}{[\text{ADP}^{3-}]_{\text{out}}[\text{ATP}^{4-}]_{\text{in}}}\right), \tag{7.19}$$

since the ΔG^o for this chemical reaction is zero. The potential difference $\Delta\Psi$ in Equation (7.19) is measured as the potential in the intermembrane space minus the potential in the matrix.

The ratio $([\text{ATP}^{4-}]/[\text{ADP}^{3-}])$ in the matrix is typically on the order of 1 and may be as low as 0.2, while in the cytoplasm and in the intermembrane space the ratio is approximately 50. The membrane potential in respiring mitochondria is approximately 160 mV. Under these conditions the free energy of the ANT exchanger is computed

$$\Delta G_{ANT} \approx -F\cdot(180\text{ mV}) + RT\ln(250)$$
$$\approx -5.98\ RT + 5.52\ RT = -0.46\ RT. \tag{7.20}$$

Thus under typical conditions the concentration free energy barrier of 5.52 RT is overcome by an electrostatic driving force of 5.98 RT.

7.2.2 Electrogenic transporters

Electrogenic transporters are membrane transport proteins that effect transport reactions that involve overall charge transport. Thus for electrogenic transport reactions the reaction is influenced by (and influences) the electrostatic potential across the membrane and the second term on the right-hand side of Equation (7.17) is non-zero.

Consider as an example the sodium–calcium (NC) exchanger, which exchanges 3 Na$^+$ ions for 1 Ca^{2+} across the cell membrane. The exchanger typically operates with calcium moving out of the cell and sodium ion moving in, as illustrated in Figure 7.5. The overall reaction for the NC exchanger is:

$$3\ \text{Na}^+_{out} + \text{Ca}^{2+}_{in} \rightleftharpoons 3\ \text{Na}^+_{in} + \text{Ca}^{2+}_{out}. \tag{7.21}$$

From Equation (7.17), the Gibbs free energy for the transport reaction is

$$\Delta G_{NAC} = F\Delta\Psi + RT\ln\left(\frac{[\text{Na}^+_{in}]^3[\text{Ca}^{2+}_{out}]}{[\text{Na}^+_{out}]^3[\text{Ca}^{2+}_{in}]}\right) \tag{7.22}$$

and the apparent equilibrium constant, accounting for the membrane potential, is $K_{NAC} = e^{-F\Delta\Psi/RT}$. In addition, if we assume that the transport proceeds according to the mechanism illustrated in Figure 7.5, the kinetic constants obey the relationship

$$\frac{k_{12}k_{23}k_{34}k_{41}}{k_{21}k_{32}k_{43}k_{14}} = e^{-F\Delta\Psi/RT}. \tag{7.23}$$

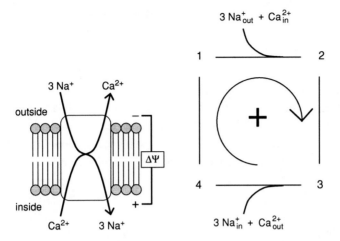

Figure 7.5 Cartoon (left) and mechanism (right) for sodium–calcium exchanger. This electrogenic transporter exchanges 3 Na^+ ions for 1 Ca^{2+} ion.

Analogous to the treatment of the antiporter in Section 7.1.2, the mechanism of Figure 7.5 assumes that the binding and unbinding steps are lumped into single reactions. This assumption is reasonable if the binding and unbinding reaction are maintained in rapid equilibrium: $e_2 = [Na^+_{out}]^3[Ca^{2+}_{in}]e_1/K_d$ and $e_3 = [Na^+_{in}]^3[Ca^{2+}_{out}]e_4/K_d$. These relationships allow us to simplify Equation (7.23) to

$$\frac{k_{23}k_{41}}{k_{32}k_{14}} = e^{-F\Delta\Psi/RT}. \qquad (7.24)$$

This equilibrium condition is satisfied by the following choice of kinetic constants

$$k_{23} = \hat{k}_+ = k_+e^{-\gamma F\Delta\Psi/RT}$$
$$k_{41} = \hat{k}_- = k_-e^{-(1-\gamma)F\Delta\Psi/RT}$$
$$k_{14} = k_+$$
$$k_{32} = k_- \qquad (7.25)$$

where k_+, k_-, and γ are arbitrary. Typically we expect $0 < \gamma < 1$.

The quasi-steady flux for this mechanism can be expressed

$$J = \frac{E_o(\hat{k}_+\hat{k}_-[Na^+_{out}]^3[Ca^{2+}_{in}] - k_+k_-[Na^+_{in}]^3[Ca^{2+}_{out}])}{K_0 + K_1[Na^+_{out}]^3[Ca^{2+}_{in}] + K_2[Na^+_{in}]^3[Ca^{2+}_{out}] + K_3[Na^+_{out}]^3[Ca^{2+}_{in}][Na^+_{in}]^3[Ca^{2+}_{out}]}, \qquad (7.26)$$

where $K_0 = (\hat{k}_- + k_+)K_d$, $K_1 = (\hat{k}_+ + \hat{k}_-)$, $K_2 = (k_+ + k_-)$, and $K_3 = (k_- + \hat{k}_+)/K_d$. Since $\hat{k}_+$ and $\hat{k}_-$ depend on the membrane potential, the flux depends on the membrane potential.

7.3 Electrophysiology modeling

Electrophysiology modeling – the modeling of the electrical properties of cells – was born with the Nobel Prize-winning work of Hodgkin and Huxley, reported in a series of papers in 1952 [94, 95, 96, 97, 98]. Hodgkin and Huxley demonstrated that the action potential (transient spike in membrane potential) in the squid giant axon (a convenient cell for study)[1] is primarily governed by sodium and potassium current across the membrane and introduced a computational model quantitatively describing the observed phenomena. Part of the legacy of Hodgkin and Huxley is that today the field of electrophysiology is one in which computational modeling is closely tied with experimental biology. Applications to a number of important cell types, including a deep literature on cardiac cell electrophysiology, have been developed.[2] In this section we introduce the basics of electrophysiology modeling using the original model of Hodgkin and Huxley as an example.

7.3.1 Ion channels

Ion channels are specialized transport proteins that facilitate the selective permeation of ions through cell membranes. Opening and closing of these channels (called *gating*) modulate the membrane potential in excitable cells, such as muscle and nerve cells. Modeling the voltage-dependent gating of ion channels, and the effects of ionic currents on membrane potential, is the basis of electrophysiology modeling.

Consider as an example the current of Na^+ ion through a sodium channel. As was illustrated in Section 1.7.2, there is an equilibrium membrane potential for which the passive sodium current is zero. This equilibrium potential is given by

$$V_{Na} = \frac{RT}{F} \ln \left(\frac{[Na^+_{out}]}{[Na^+_{in}]} \right).$$

When the membrane potential $\Delta \Psi$ equals V_{Na} there is no net flux of sodium through a passive sodium channel.

A useful and widely used model for the outward current through a sodium channel is:

$$I_{Na} = g_{Na}(\Delta \Psi - V_{Na}), \tag{7.27}$$

---

[1] In a 1929 lecture to the International Physiology congress, August Krogh said that "For a large number of problems there will be some animal of choice, or a few such animals, on which it can be most conveniently studied" [119]. Hans Krebs called this concept the "August Krogh Principle," which Hodgkin and Huxley brilliantly applied to arrive at the giant axon of the squid as a convenient model for their electrophysiology studies.

[2] An historical review of the field that tracks the extensions of Hodgkin and Huxley's pioneering work to the heart is given by Noble [152].

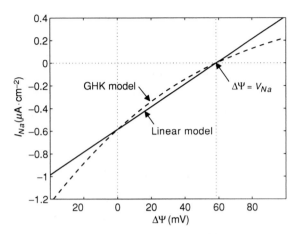

Figure 7.6 Current–voltage relationship for passive channel models of Equations (7.27) and (7.28). Sodium concentrations typical for the squid giant axon are used: $[\text{Na}^+_{out}] = 437\,\text{mM}$; $[\text{Na}^+_{in}] = 50\,\text{mM}$. The sodium equilibrium potential is $V_{Na} = 58.5\,\text{mV}$. Conductance g_{Na} is set to $0.01\,\text{mS·cm}^{-2}$. The permeability for the GHK model of Equation (7.28) is set so that both models predict the same current density at $\Delta\Psi = 0$. Figure adapted from [108].

where g_{Na} is the conductance of the channel. Typically current is measured as current density, amperes per unit cell membrane surface area, and conductance is measured in units of resistance per unit cell membrane surface area.

This linear current–voltage relationship can be compared to the Goldman–Hodgkin–Katz model of Equation (3.64), which in terms of current density is

$$I_{Na} = P_{Na} \frac{F^2 \Delta\Psi}{RT} \left(\frac{[\text{Na}^+_{in}] e^{F\Delta\Psi/RT} - [\text{Na}^+_{out}]}{e^{F\Delta\Psi/RT} - 1} \right). \tag{7.28}$$

Both Equations (7.27) and (7.28) predict zero current at $\Delta\Psi = V_{Na}$, as illustrated in Figure 7.6. Note that the current–voltage curve does not go through the point (0,0). The outward current is positive only when the membrane voltage exceeds the equilibrium voltage. For all membrane potential values below V_{Na}, the current is inward. Ionic current through membranes can be further understood in terms of the theory of electro-diffusion. For a mathematical treatment of the subject, see [102].

7.3.2 Differential equations for membrane potential

In electrophysiology modeling biological membranes are typically treated as capacitors with constant capacitance. The basic equation for a capacitor is:

$$C_m \frac{d\Delta\Psi}{dt} = - \sum I_{outward}, \tag{7.29}$$

where C_m is the capacitance of the membrane. The rate of change of membrane potential is proportional to the sum of currents across the membrane. The summation in Equation (7.29) is over all membrane currents. The sign of the right-hand side in Equation (7.29) arises from the convention that membrane potential is defined as the inside potential minus the outside potential. Therefore positive outward current tends to lower the potential across the membrane. The apparent capacitance of biological membranes is on the order of $1 \times 10^{-6}\,\mu\text{F}\cdot\text{cm}^{-2}$.

7.3.3 The Hodgkin–Huxley model

The Hodgkin–Huxley model involves three membrane currents due to potassium, sodium, and a leak current of charge through other pathways. The model assumes linear current–voltage relationships:

$$C_m \frac{dv}{dt} = -\bar{g}_K n^4 (v - v_K) - \bar{g}_{Na} m^3 h (v - v_{Na}) - \bar{g}_L (v - v_L) + I_{app}$$

$$\frac{dm}{dt} = \alpha_m (1 - m) - \beta_m m$$

$$\frac{dn}{dt} = \alpha_n (1 - n) - \beta_n n$$

$$\frac{dh}{dt} = \alpha_h (1 - h) - \beta_h h, \tag{7.30}$$

where $\bar{g}_K n^4$ is the potassium conductivity, $\bar{g}_{Na} m^3 h$ is the sodium conductivity, and $\bar{g}_L$ is the leak conductivity. The current I_{app} is the current applied across the membrane. In the experiments of Hodgkin and Huxley, I_{app} was imposed by a capillary electrode. In vivo, the applied current arises from an event (such as a synapse firing) that causes a transient increase in current or from the spread of an action potential along a nerve fiber.

The voltage v in the Hodgkin–Huxley model is the membrane potential measured relative to the equilibrium voltage V_{eq}: $v = \Delta\Psi - V_{eq}$, where V_{eq} is the potential when no current is applied. The experimentally determined equilibrium potentials (which depend on the ion gradients across the membrane) for the model are

$$v_K = -12\,\text{mV}, \quad v_{Na} = 115\,\text{mV}, \quad v_L = 10.6\,\text{mV}$$

and the constants $\bar{g}_K$, $\bar{g}_{Na}$, and $\bar{g}_L$ are

$$\bar{g}_K = 36\,\text{mS}\cdot\text{cm}^2, \quad \bar{g}_{Na} = 120\,\text{mS}\cdot\text{cm}^2, \quad \bar{g}_L = 0.3\,\text{mS}\cdot\text{cm}^2.$$

The capacitance has a value $C_m = 1 \times 10^{-6}\,\mu\text{F}\cdot\text{cm}^{-2}$.

The variables m, n, and h are phenomenological variables that describe the observed gating of the sodium and potassium channels in response to changes in the membrane potential. If the potential were held fixed, m, n, and h would obtain values:0

$$m_\infty = \frac{\alpha_m}{\alpha_m + \beta_m}$$

$$n_\infty = \frac{\alpha_n}{\alpha_n + \beta_n}$$

$$h_\infty = \frac{\alpha_h}{\alpha_h + \beta_h}, \tag{7.31}$$

where the αs and βs are empirical functions of the membrane potential v. The empirical functions for the αs and βs were developed to match experimental observation on the kinetics of sodium and potassium currents in the giant axon:

$$\alpha_m = 0.1 \, \frac{25 - v}{\exp\left(\frac{25-v}{10}\right) - 1}$$

$$\beta_m = 4 \, \exp\left(\frac{-v}{18}\right)$$

$$\alpha_n = 0.01 \, \frac{10 - v}{\exp\left(\frac{10-v}{10}\right) - 1}$$

$$\beta_n = 0.125 \, \exp\left(\frac{-v}{80}\right)$$

$$\alpha_h = 0.07 \, \exp\left(\frac{-v}{20}\right)$$

$$\beta_h = \frac{1}{\exp\left(\frac{30-v}{10}\right) + 1}, \tag{7.32}$$

where voltages are expressed in units of mV. The m_∞, n_∞, and h_∞ predicted by these functions are illustrated in Figure 7.7.

From this figure we can see that the conductivities given by $\bar{g}_K n^4$ and $\bar{g}_{Na} m^3 h$ will be relatively small when the potential is near zero. However, if the potential is increased to approximately $v = 10\,\text{mV}$, the "m-gate" will open. (The variable m obtains a value that increases with v.) As illustrated in Figure 7.7, the conductivity of the sodium channel, given by $\bar{g}_{Na} m^3 h$, will be zero in the limit that v becomes very high. However, the functions for the αs and βs are designed to capture the observed phenomenon that the sodium channels open on a timescale faster than that on which they close. Thus, the sodium channel opening at small positive v results in an inward sodium current because $(v - v_{Na}) < 0$, generating an increase in the membrane potential. This hyperpolarization will temporarily result in a positive feedback situation as increasing Na^+ current leads to increasing v and increasing Na^+ conductivity.

The temporary hyperpolarization caused by the sodium current is illustrated in Figure 7.8, which illustrates the predicted membrane potential transients for

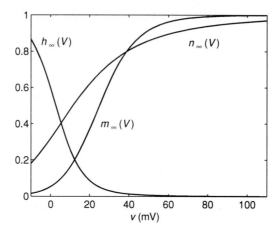

Figure 7.7 Functions m_∞, n_∞, and h_∞ predicted by Equations (7.31) and (7.32) for the Hodgkin–Huxley model. Figure adapted from [108].

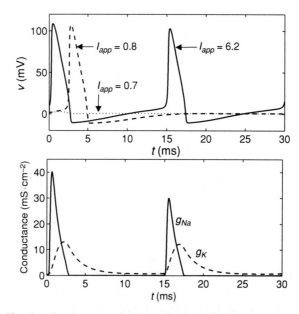

Figure 7.8 Simulated action potential from the Hodgkin–Huxley model. The upper panel plots action potential for three different values of applied current. The lower panel plots the predicted conductances of the sodium and potassium channels for the case of $I_{app} = 6.2 \, \mu\text{A} \cdot \text{cm}^{-2}$, for which sustained period firing of the nerve cell is predicted.

different values of the applied current. When the applied current is high enough to open the sodium channel to a substantial degree, the membrane hyperpolarizes due to the sodium current. Following hyperpolarization by the sodium current, the sodium channel closes and the potassium channel opens, both events happening on the timescale of a few milliseconds. The combined action of closing the sodium channel and opening the potassium channel causes the membrane to move back to its resting potential near v_K, which is -12 mV.

Figure 7.8 illustrates that applied currents too small ($I_{app} = 0.7$ µA·cm^{-2} in the figure) to open the sodium channel do not result in action potentials. Higher applied currents ($I_{app} = 0.8$ µA·cm^{-2} in the figure) result in a single action potential and return to a stable steady state near $v = 0$. Still higher currents ($I_{app} = 6.2$ µA·cm^{-2} in the figure) result in a periodic train of sustained action potentials.

The predicted channel conductivities for the case of sustained periodic action potentials are plotted in the lower panel of Figure 7.8, illustrating the predicted kinetics of channel opening and closing.

7.3.3.1 Computer code

A Matlab computer code for the Hodgkin–Huxley model is given below. The code to compute the time derivatives of the state variables (the right-hand side of Equation (7.30)) is:

```
function [f] = dXdT_HH(t,x,I_app);
% FUNCTION dXdT_HH
%   Inputs: t - time (milliseconds)
%           x - vector of state variables {v,m,n,h}
%           I_app - applied current (microA cm^{-2})
%
%   Outputs: f - vector of time derivatives
%                {dv/dt,dm/dt,dn/dt,dh/dt}
% Resting potentials, conductivities, and capacitance:
V_Na = 115;
V_K  = -12;
V_L  = 10.6;
g_Na = 120;
g_K  = 36;
g_L  = 0.3;
C_m  = 1e-6;
% State Variables:
v = x(1);
m = x(2);
n = x(3);
```

```
h = x(4);
% alphas and betas:
a_m = 0.1*(25-v)/(exp((25-v)/10)-1);
b_m = 4*exp(-v/18);
a_h = 0.07*exp(-v/20);
b_h = 1 ./ (exp((30-v)/10) + 1);
a_n = 0.01*(10-v)./(exp((10-v)/10)-1);
b_n = 0.125*exp(-v/80);
% Computing currents:
I_Na = (m^3)*h*g_Na*(v-V_Na);
I_K  = (n^4)*g_K*(v-V_K);
I_L  = g_L*(v-V_L);
% Computing derivatives:
f(1) = (-I_Na - I_K - I_L + I_app)/C_m;
f(2) = a_m*(1-m) - b_m*m;
f(3) = a_n*(1-n) - b_n*n;
f(4) = a_h*(1-h) - b_h*h;
f = f';
```

The model may be simulated and the predicted voltage transient plotted using the following commands:

```
% Initial equilibration with I_app = 0 to
% Generate initial condition xo for simulation:
I_app = 0;
[t,x] = ode15s(@dXdT_HH,[0 30],xo,[],I_app);
xo = x(end,:);
% Add nonzero applied current:
I_app = 6.2;
[t,x] = ode15s(@dXdT_HH,[0 30],xo,[],I_app);
% Plot computed action potential
plot(t,x(:,1));
```

See Section 3.1.4 for an outline of the use of computer solvers, such as the 'ode15s' function in Matlab, to integrate ordinary differential equations.

7.4 Large-scale example: model of oxidative ATP synthesis

The mitochondrion is the key cellular organelle responsible for transducing free energy from primary substrates into the ATP potential that drives the majority of energy-consuming processes in a cell. Thus the mitochondrion plays a central

Matrix. This large internal space contains a highly concentrated mixture of hundreds of enzymes, including those required for the oxidation of pyruvate and fatty acids for the citric acid cycle. The matrix also contains several identical copies of the mitochondrial DNA genome, special mitochondrial ribosomes, tRNAs, and various enzymes required for expression of mitochondrial genes.

Inner membrane. The inner membrane is folded into numerous cristae, greatly increasing its total surface area. It contains proteins with three types of functions: (1) those that carry out the oxidation reactions of the electron transport chain, (2) the ATP synthase that makes ATP in the matrix, and (3) transport proteins that allow the passage of metabolites into and out of the matrix. An electrochemical gradient of H+, which drives the ATP synthase, is established across this membrane, so the membrane must be impermeable to ions and most small charged molecules.

Outer membrane. Because it contains a large channel-forming protein (called porin), the outer membrane is permeable to all molecules 5000 daltons or less. Other proteins in this membrane include enzymes involved in mitochondrial lipid synthesis and enzymes that convert lipid substrates into forms that are subsequently metabolized in the matrix.

Intermembrane space. This space contains several enzymes that use the ATP passing out of the matrix to phosphorylate other nucleotides.

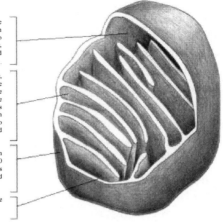

Figure 7.9 Structural organization of the mitochondrion. Figure reproduced from [1] with permission.

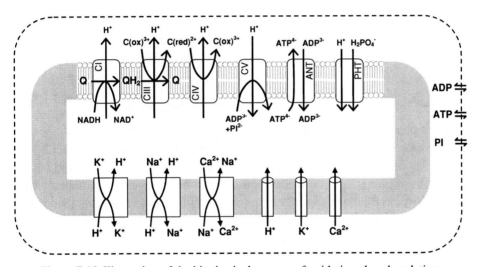

Figure 7.10 Illustration of the biophysical process of oxidative phosphorylation.

role in the majority of eukaryotic intracellular events. Its structural organization is illustrated in Figure 7.9.

The biophysical processes involved in oxidative ATP synthesis are illustrated in Figure 7.10. The reactions and stoichiometry, along with a kinetic model of the transport reactions of these processes, are described in the following section.

Briefly, a series of enzymes located on the inner membrane of the mitochondrion pump protons from the matrix into the inner membrane space, generating an electrostatic potential, $\Delta\Psi$, that is positive on the outside of the inner membrane. These proton pumps are denoted Complex I, III, and IV and are labeled CI, CIII, and CIV in the figure. The pumping of positively charged hydrogen ions against the electrostatic gradients is driven by the oxidation of NADH and QH_2 generated by the reactions of the TCA cycle, as discussed in Section 6.2.

The electrostatic gradient established by the proton pumps is consumed in synthesizing ATP from ADP and PI, and in transporting ATP out of the matrix. ATP synthesis is catalyzed by F_0F_1-ATPase, also called Complex V and labeled CV in the figure. The antiporter that exchanges ATP^{4-} inside the matrix with ADP^{3-} outside is called adenine nucleotide translocase, as described in the example in Section 7.2.2. Transport of inorganic phosphate is via an electrically neutral co-transport of H^+ and $H_2PO_4^-$ (or alternatively through antiport of OH^- and $H_2PO_4^-$). The phosphate–hydrogen cotransporter is labeled PHT.

A number of additional ion transporters exist on the mitochondrial inner membrane. Channels and exchangers responsible for Ca^{2+}, Na^+, and K^+ transport are illustrated in Figure 7.10.

The outer membrane is highly permeable to small molecules (see Figure 7.9). The model described below assumes passive permeability of ATP, ADP, and PI across this membrane; cations such as Ca^{2+} and Na^+ are assumed to rapidly equilibrate with no significant concentration gradient across the outer membrane.

7.4.1 Model of oxidative phosphorylation

Here we present a computational model of mitochondrial electrophysiology and oxidative phosphorylation is based on the models of one of the authors [13, 14] and Wu *et al.* [212]. The processes illustrated in Figure 7.9 are modeled based on the electrophysiology modeling approach outlined in Section 7.3. Thermodynamic constants for the transport reactions are computed from thermodynamic data tabulated in Table 6.1.

7.4.1.1 Electron transport chain reactions

The transport reaction of Complex I is

$$5\,H_x^+ + NADH^{2-} + CoQ \rightleftharpoons NAD^- + QH_2 + 4\,H_i^+, \tag{7.33}$$

where subscripts "x" and "i" denote H^+ ions in the matrix and intermembrane space (between inner and outer membranes), respectively. This reaction pumps four H^+ ions out of the matrix and across the inner membrane. The fifth H^+ ion on the left-hand side of the reaction participates in the chemical reaction $NADH^{2-} + CoQ + H^+ \rightleftharpoons NAD^- + QH_2$. Thus the H^+ ion pumping is driven by the redox reaction that transfers an electron pair from NADH to QH_2.

The apparent equilibrium constant for the reaction is computed

$$K'_{C1} = \left(\frac{[NAD][QH_2]}{[NADH][CoQ]} \right)_{eq} = \frac{[H_x^+]^5}{[H_i^+]^4} \exp \left[-(\Delta G^o_{C1} + 4F\Delta\Psi)/RT \right],$$
(7.34)

where $\Delta G^o_{C1} = -109.7 \, kJ \cdot mol^{-1}$ is the equilibrium free energy for the chemical reaction. The term $4F\Delta\Psi$ accounts for the charge transport across the membrane. In the absence of knowledge regarding the detailed kinetics of the transporter, we use simple mass action to describe the kinetics:

$$J_{C1} = x_{C1}(K'_{C1}[NADH][CoQ] - [NAD][QH_2]),$$
(7.35)

where $x_{C1} = 2.47 \times 10^4 \, mol \cdot s^{-1} \cdot M^{-2} \cdot (1 \, mito)^{-1}$ is the activity of the transporter. With the activity expressed in these units, the flux is computed in mass per unit time per unit volume of mitochondria.

In the Complex-III reaction, the reduced QH_2 is re-oxidized to CoQ and cytochrome C is reduced:

$$2 \, H_x^+ + QH_2 + 2 \, C(ox)^{3+} \rightleftharpoons CoQ + 2 \, C(red)^{2+} + 4 \, H_i^+,$$
(7.36)

where $C(ox)^{3+}$ and $C(red)^{2+}$ are the oxidized and reduced forms of cytochrome C, which is a protein that is present in the intermembrane space. This reaction involves a total charge transfer of two positive charges from the matrix to the intermembrane space – four H^+ ions appear in the intermembrane space, while two cytochrome molecules are reduced from a $+3$ charge to a $+2$ charge every time the reaction turns over.

The apparent equilibrium constant for the reaction is computed

$$K'_{C3} = \left(\frac{[CoQ][C(red)^{2+}]^2}{[QH_2][C(ox)^{3+}]^2} \right)_{eq} = \frac{[H_x^+]^2}{[H_i^+]^4} \exp \left[-(\Delta G^o_{C3} + 2F\Delta\Psi)/RT \right],$$
(7.37)

where $\Delta G^o_{C3} = +46.69 \, kJ \cdot mol^{-1}$ is the equilibrium free energy for the chemical reaction. The kinetics are modeled by the equation:

$$J_{C3} = x_{C3} \left(\frac{1 + [PI_x]/k_{PI1}}{1 + [PI_x]/k_{PI2}} \right) \left(\sqrt{K'_{C3}}[C(ox)^{3+}][QH_2]^{1/2} - [C(red)^{2+}][CoQ]^{1/2} \right),$$
(7.38)

where the term describes allosteric activation of Complex III by inorganic phosphate $(1 + [PI_x]/k_{PI1})/(1 + [PI_x]/k_{PI2})$. The activity of the transporter is $x_{C3} = 0.665 \, mol \cdot s^{-1} \cdot M^{-3/2} \cdot (1 \, mito)^{-1}$; the activation parameters are $k_{PI1} = 28.1 \, \mu M$ and $k_{PI2} = 3.14 \, mM$.

In the Complex-IV reaction, the reduced cytochrome C is re-oxidized and oxygen is reduced to form water:

$$4\,H_x^+ + 2\,C(\text{red})^{2+} + 1/2\,O_2 \rightleftharpoons 2\,C(\text{ox})^{3+} + H_2O + 2\,H_i^+. \tag{7.39}$$

In this reaction a total of four charges are transported across the inner membrane.

The apparent equilibrium constant for the reaction is computed

$$K'_{C4} = \left(\frac{[C(\text{ox})^{3+}]^2}{[C(\text{red})^{2+}]^2[O_2]^{1/2}} \right)_{eq} = \frac{[H_x^+]^4}{[H_i^+]^2} \exp\left[-(\Delta G^o_{C4} + 4F\Delta\Psi)/RT \right],$$
$$\tag{7.40}$$

where $\Delta G^o_{C4} = -202.2\,\text{kJ·mol}^{-1}$ is the equilibrium free energy for the chemical reaction. The kinetics are modeled by the equation:

$$J_{C4} = x_{C4} \left(\frac{[O_2]}{[O_2] + K_{O2}} \right) \left(\frac{[C(\text{red})^{2+}]}{C_o} \right)$$
$$\cdot \left(\sqrt{K'_{C4}}[C(\text{red})^{2+}][O_2]^{1/4} - [C(\text{ox})^{3+}](1M)^{1/4} \right), \tag{7.41}$$

where $C_o = [C(\text{red})^{2+}] + [C(\text{ox})^{3+}] = 2.7\,\text{mM}$ is the total concentration of cytochrome C. The activity of the transporter is $x_{C4} = 9.93 \times 10^{-5}\,\text{mol·s}^{-1}\text{·M}^{-1}\cdot$ (1 mito)$^{-1}$; the K_M for oxygen is $K_{O2} = 0.12\,\text{mM}$.

7.4.1.2 ATP synthesis

In the ATP synthesis reaction, catalyzed by the transporter complex F_0F_1-ATPase, n_A H^+ ions are transported from the intermembrane space to the matrix:

$$n_A\,H_i^+ + \text{ADP}_x^{3-} + \text{PI}_x^{2-} \rightleftharpoons \text{ATP}_x^{4-} + (n_A - 1)\,H_x^+ + H_2O, \tag{7.42}$$

where H_x^+ appears on the right-hand side of the reaction with the stoichiometric coefficient $n_A - 1$ because one H^+ ion is consumed by the chemical reaction. The value of n_A is assumed to be three in the model. The movement of H^+ ions down the electrostatic gradient allows the above overall transport reaction to proceed, forcing the chemical reaction $\text{ADP}_x^{3-} + \text{PI}_x^{2-} + H_x^+ \rightleftharpoons \text{ATP}_x^{4-} + H_2O$ to proceed against the free energy gradient.

The apparent equilibrium constant for the reaction is computed

$$K'_{F1} = \left(\frac{[\text{ATP}_x]}{[\text{ADP}_x][\text{PI}_x]} \right)_{eq} = \frac{[H_i^+]^{n_A} P_{ATPx}}{[H_x^+]^{n_A-1} P_{ADPx} P_{PIx}}$$
$$\cdot \exp\left[-(\Delta G^o_{ATPase} - n_A F\Delta\Psi)/RT \right], \tag{7.43}$$

where the equilibrium free energy for the reaction is $\Delta G^o_{ATPase} = -4.51\,\text{kJ·mol}^{-1}$.

Note that this equation differs from those for the reactions of Complex I, III, and IV in two important ways. First, the binding polynomials, P_{ATPx}, P_{ADPx}, and P_{PIx}, are included to account for cation binding to the biochemical reactants ATP, ADP, and PI in the matrix. (The reactants for the previous reactions do not significantly bind cations; see Table 6.1.) Second, the potential term $\Delta\Psi$ enters the exponent with a positive sign because in this case the overall charge transport for the forward reaction results in positive charges moving into the matrix.

The flux is modeled using mass-action kinetics

$$J_{F1} = x_{F1} \left(K'_{F1}[\text{ADP}_x][\text{PI}_x] - [\text{ATP}_x](1\text{M}) \right). \tag{7.44}$$

The activity of the transporter is $x_{F1} = 1000 \, \text{mol·s}^{-1}\cdot\text{M}^{-1}\cdot(1 \, \text{mito})^{-1}$.

7.4.1.3 Substrate transport

The overall transport reaction, in terms of references species, for the phosphate-hydrogen cotransporter is:

$$2 \, \text{H}_i^+ + \text{PI}_i^{2-} \rightleftharpoons 2 \, \text{H}_x^+ + \text{PI}_x^{2-}. \tag{7.45}$$

Thus the exchange is electroneutral and not dependent on the membrane potential. Assuming that the species $\text{HPO}_4{}^{2-}$ is cotransported with H^+ ion and that the transport flux does not saturate in H^+ concentration, we have from Equation (7.9) the equation for flux:

$$J_{PHT} = \frac{x_{PHT} \left([\text{H}_i^+][\text{H}_2\text{PO}_4^-] - [\text{H}_x^+][\text{H}_2\text{PO}_{4_x}^-]\right)}{\left([\text{H}_2\text{PO}_{4_i}^-] + K_{PHT}\right)\left([\text{H}_2\text{PO}_{4_x}^-] + K_{PHT}\right)}, \tag{7.46}$$

where the $\text{HPO}_4{}^{2-}$ species concentrations on either side of the membrane are computed the usual way:

$$[\text{H}_2\text{PO}_{4_x}^-] = \frac{[\text{HPO}_{4_x}^{2-}][\text{H}_x^+]}{K_{PI}^H \, P_{PIx}([\text{H}_x^+], [\text{K}_x^+], [\text{Mg}_x^{2+}])}$$

and

$$[\text{H}_2\text{PO}_{4_i}^-] = \frac{[\text{HPO}_{4_i}^{2-}][\text{H}_i^+]}{K_{PI}^H \, P_{PIi}([\text{H}_i^+], [\text{K}_i^+], [\text{Mg}_i^{2+}])}.$$

The binding polynomials for PI in the matrix and intermembrane space are denoted P_{PIx} and P_{PIi}, respectively. The hydrogen-ion dissociation constant K_{PI}^H is listed in Table 6.1. The activity of the transporter is $x_{PHT} = 2.0 \times 10^7 \, \text{mol·s}^{-1}\cdot\text{M}^{-1}\cdot(1 \, \text{mito})^{-1}$ and the Michaelis–Menten constant is $K_{PHT} = 1.0 \, \text{mM}$.

As discussed in Section 7.2.2, ATP is delivered to the intermembrane space via the adenine nucleotide translocase (ANT) enzyme. The reference transport reaction is

$$\text{ATP}_x^{4-} + \text{ADP}_i^{3-} \rightleftharpoons \text{ATP}_i^{4-} + \text{ADP}_x^{3-}. \tag{7.47}$$

The flux is modeled using an empirical expression based on the concentrations of ATP^{4-} and ADP^{3-} on both sides of the membrane, which are computed the usual way

$$[ATP_x^{4-}] = [ATP_x]/P_{ATPx}$$

$$[ADP_x^{3-}] = [ADP_x]/P_{ADPx}$$

$$[ATP_i^{4-}] = [ATP_i]/P_{ATPi}$$

$$[ADP_i^{3-}] = [ADP_i]/P_{ADPi}, \tag{7.48}$$

where P_{ATPx}, P_{ADPx}, P_{ATPi}, and P_{ADPi} are the binding polynomials for ATP and ADP in the matrix and intermembrane space.

$$J_{ANT} = x_{ANT} \left(\frac{[ADP_i^{3-}]}{([ADP_i^{4-}] + [ATP_i^{4-}]e^{-\psi_i})} - \frac{[ADP_x^{4-}]}{([ADP_x^{4-}] + [ATP_x^{4-}]e^{-\psi_x})} \right)$$

$$\cdot \frac{[ADP_i^{4-}]}{[ADP_i^{4-}] + K_{ADP}}, \tag{7.49}$$

where $\psi_i = \theta F \Delta \Psi / RT$, $\psi_x = (\theta - 1)F\Delta\Psi/RT$, and θ is an empirical parameter with value set to 0.60. The activity of the transporter is $x_{ANT} = 7.27 \times 10^{-3}$ mol·s^{-1}·(1 mito)$^{-1}$ and the Michaelis–Menten constant is $K_{ADP} = 3.6 \, \mu M$.

ATP, ADP, and PI are assumed to permeate across the outer membrane with biochemical transport reactions

$$ATP_c \rightleftharpoons ATP_i$$

$$ADP_c \rightleftharpoons ADP_i$$

$$PI_c \rightleftharpoons PI_i \tag{7.50}$$

in terms of biochemical species ATP, ADP, and PI on both sides of the membrane. The subscript "c" denotes concentrations in the cytoplasm compartment.

The fluxes, governed by passive permeation, are computed in terms of total reactant concentrations:

$$J_{ATPt} = \gamma p_A ([ATP_c] - [ATP_i])$$

$$J_{ADPt} = \gamma p_A ([ADP_c] - [ADP_i])$$

$$J_{PIt} = \gamma p_{PI} ([PI_c] - [PI_i]), \tag{7.51}$$

where $\gamma = 6 \, \mu m^{-1}$ is the outer membrane area per unit mitochondrial volume, $p_A = 85 \, \mu m \cdot s^{-1}$ is the permeability to adenine nucleotides, and $p_{PI} = 320 \, \mu m \cdot s^{-1}$ is the permeability to phosphate.

7.4.1.4 Cation transport

Additional cation transporters on the inner membrane are included in the model. The potassium–hydrogen exchanger, which has chemical transport equation

$$K_i^+ + H_x^+ \rightleftharpoons K_x^+ + H_i^+ \tag{7.52}$$

is important for regulation of mitochondrial pH and osmotic pressure. We model the flux using the mass-action expression:

$$x_{KH} \left([K_i^+][H_x^+] - [K_x^+][H_i^+] \right), \tag{7.53}$$

where $x_{KH} = 5.65 \times 10^6 \, \text{mol·s}^{-1}\text{·M}^{-2}\text{·(1 mito)}^{-1}$.

The integrity of the inner membrane is not perfect and an unproductive "leak" of H^+ ions into the matrix can occur:

$$H_i^+ \rightleftharpoons H_x^+. \tag{7.54}$$

The leak current of hydrogen ions is modeled using the Goldman–Hodgkin–Katz expression:

$$J_{Hle} = x_{Hle} \Delta \Psi \left(\frac{[H_i^+] e^{F \Delta \Psi / RT} - [H_x^+]}{e^{F \Delta \Psi / RT} - 1} \right), \tag{7.55}$$

where $x_{Hle} = 250 \, \text{mol·s}^{-1}\text{·M}^{-1}\text{·mV}^{-1}\text{·(1 mito)}^{-1}$.

7.4.1.5 Model differential equations

The differential equations for biochemical reactions and membrane potential arise directly from the stoichiometry of the reactions outlined above:

$$d\Delta\Psi/dt = (4J_{C1} + 2J_{C3} + 4J_{C4} - n_A J_{F1} - J_{ANT} - J_{Hle})/C_{IM}$$
$$d[QH_2]/dt = (J_{C1} - J_{C3})/W_x$$
$$d[C(\text{red})^{2+}]/dt = (2J_{C3} - 2J_{C4})/W_i$$
$$d[ATP_x]/dt = (J_{F1} - J_{ANT})/W_x$$
$$d[PI_x]/dt = (-J_{F1} + J_{PHT})/W_x$$
$$d[ATP_i]/dt = (J_{ATPt} + J_{ANT})/W_i$$
$$d[ADP_i]/dt = (J_{ADPt} - J_{ANT})/W_i$$
$$d[PI_i]/dt = (-J_{PHT} + J_{PIt})/W_i. \tag{7.56}$$

The constants W_x and W_i represent the matrix and intermembrane space water fractions and are estimated at 0.6514 and 0.0724 (1 water)·(1 mito)$^{-1}$, respectively.

Additional concentrations are computed based on conserved pools of metabolites:

$$[C(ox)^{3+}] = C_o - [C(red)^{2+}]$$
$$[CoQ] = Q_o - [QH_2]$$
$$[ADP_x] = A_o - [ATP_x] \tag{7.57}$$

where $C_o = 2.7\,\text{mM}$, $Q_o = 1.35\,\text{mM}$, and $A_o = 10\,\text{mM}$.

The concentrations of NAD and NADH are not modeled in the above equations. Since NADH-generating fluxes are not included, this model holds NAD and NADH concentrations at fixed concentrations. The behavior of the model as a function of NADH redox state is explored below.

Since the outer membrane is highly permeable to small molecules, it is assumed that it does not represent a significant barrier to H^+ and K^+ transport. Therefore the concentrations in the intermembrane space are set to reasonable concentrations for muscle cell cytoplasm:

$$[H_i^+] = [H_c^+] = 10^{-7.1}\,\text{M}$$

and

$$[K_i^+] = [K_c^+] = 0.150\,\text{M}.$$

In addition, cytoplasmic PI concentration is set at $[PI_c] = 1\,\text{mM}$. Cytoplasmic ATP and ADP are varied, as explained below.

The pH, K^+, and Mg^{2+} kinetics inside the mitochondrial matrix are governed by Equations (6.25), (6.26), and (6.27), derived in Chapter 6. In these equations, the hydrogen flux term is

$$\Phi^H = -\sum_{i=1}^{N_r} \frac{\partial[H_{bound}]}{\partial[L_i]} \frac{d[L_i]}{dt} - 4J_{C1} - 2J_{C3} - 4J_{C4}$$
$$+ (n_A - 1)J_{F1} + 2J_{PHT} + J_{Hle} - J_{KH}, \tag{7.58}$$

where the stoichiometric numbers associated with Complex I, III, and IV, F_0F_1-ATPase, phosphate–hydrogen cotransport, potassium–hydrogen exchange, and proton leak follow from the transport reactions. The stoichiometric coefficient for J_{C1} is 4 rather than 5, as appears in Equation (7.33), because it is assumed that for each NADH consumed, the reaction $NAD^+ + H_2O \rightleftharpoons NADH + H^+$ turns over.

Transport of magnesium ion across the inner membrane is not considered. Therefore the Mg^{2+} ion kinetics is governed by binding and unbinding to biochemical reactants:

$$\Phi^M = -\sum_{i=1}^{N_r} \frac{\partial[\text{Mg}_{bound}]}{\partial[\text{L}_i]} \frac{d[\text{L}_i]}{dt}.$$

Potassium ion is transported via the KH exchanger:

$$\Phi^K = -\sum_{i=1}^{N_r} \frac{\partial[\text{K}_{bound}]}{\partial[\text{L}_i]} \frac{d[\text{L}_i]}{dt} + J_{KH}.$$

The buffering terms in Equations (6.25), (6.26), and (6.27) are computed as described in Chapter 6:

$$\alpha_H = 1 + \frac{\partial[\text{H}_{bound}]}{\partial[\text{H}^+]} + \frac{B_x}{K_{Bx}(1 + [\text{H}_x^+]/K_{Bx})^2}$$

$$\alpha_{Mg} = 1 + \frac{\partial[\text{Mg}_{bound}]}{\partial[\text{Mg}^{2+}]}$$

$$\alpha_K = 1 + \frac{\partial[\text{K}_{bound}]}{\partial[\text{K}^+]}, \tag{7.59}$$

where we have added an additional H^+-buffering term to account for buffering by biochemical species not explicitly accounted for in the model. The constants B_x and K_{Bx} are set to 0.02 M and 10^{-7} M, respectively.

7.4.2 Model behavior

The predicted steady state behavior of the model is explored in Figures 7.11 and 7.12. Figure 7.11 plots the predicted membrane potential as a function of the NADH reduced fraction, $[\text{NADH}]/N_o$, where the total NADH is $N_o = [\text{NADH}] + [\text{NAD}] = 2.97$ mM. At $[\text{NADH}]/N_o = 0$, there is no driving force for the respiratory chain to maintain the membrane potential. Within the operating range illustrated, $\Delta\Psi$ is predicted to be between approximately 165 and 200 mV. For finite NADH, the membrane potential increases with increasing driving force $[\text{NADH}]/N_o$. As is illustrated in the figure, the relationship between NADH and $\Delta\Psi$ depends on the ADP concentration in the cytoplasm. As the ADP concentration is increased, the ANT flux increases and the load on the oxidative phosphorylation capacity increases. As this load is increased, the membrane potential decreases.

The relationship between the work rate (rate of delivery of ATP out of the mitochondrion) and $[\text{ADP}_c]$ is illustrated in Figure 7.12. Flux through the ANT transporter increases with $[\text{ADP}_c]$, with higher flux possible at higher NADH concentration, due to the effect of NADH on $\Delta\Psi$, which drives ANT.

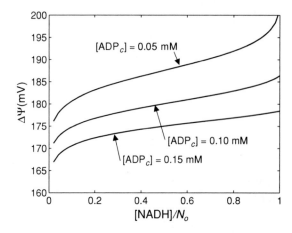

Figure 7.11 Relationship between NADH redox state and predicted mitochondrial membrane potential at different levels of cytoplasmic ADP. The cytoplasmic ADP concentration is fixed at the values indicated and the cytoplasmic ATP concentration is computed $[ATP_c] = 8.2\,mM - [ADP_c]$; cytoplasmic phosphate $[PI_c]$ is clamped at 1 mM.

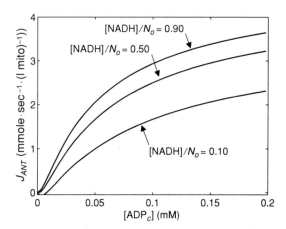

Figure 7.12 Predicted relationship between the rate of delivery of ATP out of the mitochondrion and the cytoplasmic ADP concentration. For these model predictions, the cytoplasmic ATP concentration is computed as a function of $[ADP_c]$ as described in the legend to Figure 7.11; cytoplasmic phosphate $[PI_c]$ is clamped at 1 mM.

7.4.3 Applications to in vivo systems

Of course, $[ADP_c]$ and $[NADH]/N_o$ do not vary independently in vivo, as in the preceding model analysis. Neither does cytoplasmic PI concentration stay fixed for varying work rates in cells. The integrated system behavior can only be captured by simultaneously simulating the generation of redox equivalents by the TCA cycle (as

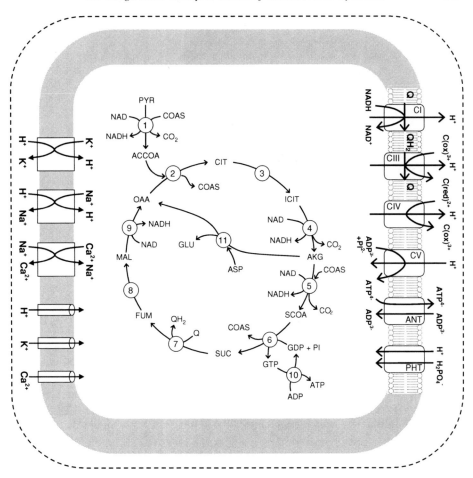

Figure 7.13 Illustration of integrated components of TCA cycle and oxidative phosphorylation model.

discussed in Chapter 6), oxidative phosphorylation by the mitochondrial respiratory system (as outlined above), and the reaction of cellular energetics, including ATP hydrolysis in the cytoplasm. A simulation study that puts all of these things together to understand muscle cell energetics has been published by Wu *et al.* [213].

Mitochondrial oxidative ADP phosphorylation is the primary source of ATP in skeletal muscle during aerobic exercise. Thus, to maintain the free-energy state of the cytoplasmic phosphoenergetic compounds ATP, ADP, and inorganic phosphate (PI), oxidative phosphorylation is modulated to match the rate of ATP utilization during exercise. The mitochondrial components of Wu *et al.*'s model are illustrated in Figure 7.13. Additional components include ATP hydrolysis, adenylate kinase,

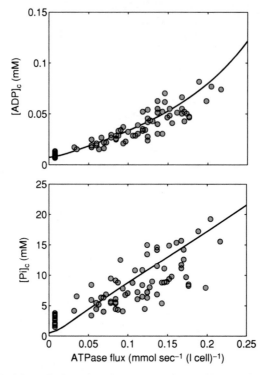

Figure 7.14 Model prediction of ADP concentration and inorganic phosphate concentration in cytoplasm as a function of ATP hydrolysis rate in the healthy subjects. Data are from [106]; solid lines are prediction of model of Wu *et al.* [213].

and creatine kinase reactions in the cytoplasm. The full model is described in detail in [213].

Among other predictions, the integrated model reveals that as work rate is varied, commensurate increases in the rate of mitochondrial ATP synthesis are effected by changes in concentrations of available ADP and inorganic phosphate. In other words, mitochondrial respiratory control is achieved in vivo by substrate feedback control. The predicted relationship between substrates and work rate is plotted in Figure 7.14. Model predictions are compared to data obtained from NMR spectroscopy of exercising flexor forearm muscle in healthy human subjects [106].

Concluding remarks

In this chapter we have shown how membrane transport processes are linked with in vivo biochemical function. The basic equations for and examples of electroneutral and electrogenic transport were introduced. The deep field of electrophysiology modeling was introduced based in large part on the example of Hodgkin and Huxley's classic model of the squid giant axon. For further reading in this area, we suggest Keener and Sneyd's text [108], particularly the first 12 chapters, which emphasize electrophysiology modeling in clearly explained detail.

The ultimate example studied in this chapter is the mitochondrial respiratory system and oxidative ATP synthesis. This system, in which biochemical network function is tightly coupled with membrane transport, is essential to the function of nearly all eukaryotic cell types. As an example of a critically important system and an analysis that makes use of a wide range of concepts from electrophysiology to detailed network thermodynamics, this model represents a milestone in our study of living biochemical systems. To continue to build our ability to realistically simulate living systems, the following chapter covers the treatment of spatially distributed systems, such as advective transport of substances in the microcirculation and exchange of substances between the blood and tissue.

Exercises

7.1 Express the equilibrium potential in terms of ion conductances for a membrane permeable to both Na^+ and K^+ ions with relative conductivities g_{Na} and g_K. [Hint: assume that the flux of each ion is proportional to conductivity multiplied by the driving force, which can be expressed as the difference between the membrane potential and the Nernst equilibrium potential for a given ion.]

7.2 Use computer simulation to determine if the solution to the Hodgkin–Huxley model of Section 7.3.3 is periodic at $I_{app} = 6.2\,\mu A \cdot cm^{-2}$. What happens when the applied current is lowered to $6.0\,\mu A \cdot cm^{-2}$?

Part III

Special topics

8

Spatially distributed systems and reaction–diffusion modeling

Overview

Previous chapters have treated cells – or cellular organelles such as mitochondria – as well-mixed bags of reacting biochemical components, without accounting for spatial heterogeneities in the concentrations of reacting species. This well-mixed-compartment approximation is indeed an appropriate assumption for certain applications. However, it is clear that intracellular concentration gradients exist. For example, the intracellular transport of oxygen, like that of many important species, is driven by diffusion. Diffusion is driven by concentration gradients. Thus, for oxygen to be transported from the outside of a cell to sites of oxidative phosphorylation, significant intracellular gradients must exist.

This chapter introduces the concept of reaction–diffusion modeling – that is, modeling of coupled diffusion and reaction of chemical species. Following a brief overview of the mathematics of diffusion, classical models of oxygen transport to tissue are explored. Studying these analytically tractable models will provide the reader with the basic tools for modeling and analysis of spatially distributed transport of other species. Although a major focus is placed on oxygen transport, general models for transport and exchange between blood and tissue regions of solutes are developed. For more complex models, the focus here is on establishing appropriate governing equations without going into the details of numerical methods for simulations. The final section of this chapter provides a survey of applications in three-dimensional modeling of transport in biological systems.

8.1 Diffusion-driven transport of solutes in cells and tissue

Section 3.2 introduced the governing equations for three physical processes responsible for transporting material in living systems: advection, drift, and diffusion. Advection refers to the process by which solutes are transported with the bulk movement of fluid, as in oxygen carried in the blood; drift refers to movement of material directed by some force field, such as the movement of a charged particle in an electric field; diffusion is the process by which the unbiased thermal motion of particles results in the net transport down concentration gradients. Diffusion is the most important of these three phenomena in transporting small molecules, including oxygen and metabolic intermediates, within cells.

8.1.1 The diffusion equation: assumptions and applications

Fick's first law of diffusion [56], which arises as a consequence of random unbiased movement of particles, states that the mass flux of particles in a continuous concentration field is proportional to the negative of the spatial concentration gradient. Fick's first law in a general form is [39]

$$\vec{\Gamma} = -\mathbb{D}\nabla c = -\begin{pmatrix} D_{11} & D_{12} & D_{13} \\ D_{21} & D_{22} & D_{23} \\ D_{31} & D_{32} & D_{33} \end{pmatrix} \nabla c. \tag{8.1}$$

Here $\vec{\Gamma}$ is the mass flux density and $c(\vec{x}, t)$ is the concentration of a solute, continuously distributed in the spatial field $\vec{x}$. For this general anisotropic case $\mathbb{D}$ is the positive definite *diffusion matrix*.[1]

The conservation statement $\partial c/\partial t = -\nabla \cdot \vec{\Gamma}$ yields the diffusion equation:[2]

$$\frac{\partial c}{\partial t} = \nabla \cdot \mathbb{D} \nabla c, \tag{8.2}$$

or expressed using Einstein notation,

$$\frac{\partial c}{\partial t} = \frac{\partial}{\partial x_i} D_{ij} \frac{\partial}{\partial x_j} c.$$

(This shorthand implies summation over all repeated indices. In the case of this expression the repeated indices are i and j.)

Off-diagonal entries of $\mathbb{D}$ arise from the fact that the coordinate system in which Equations (8.1) and (8.2) are expressed does not necessarily coincide with the principle directions of anisotropic diffusion defined by $\mathbb{D}$. The principle diffusion directions correspond to the eigenvectors of $\mathbb{D}$, with the highest rate of diffusion occurring in the direction associated with the largest eigenvalue of $\mathbb{D}$. To see this we introduce the coordinate transformation $\vec{\zeta} = \mathbb{R}\vec{x}$ (or $\zeta_i = R_{ij}x_j$). Application

---

[1] Compare Equation (8.1) to Equation (3.49), which applies to the isotropic case. In the case of Equation (8.1), diffusive transport proceeds at different rates in different directions.

[2] Equation (8.2) is also called the heat equation and was introduced by Fourier to simulate the distribution of temperature in solids. In mathematics it is characterized as a parabolic differential equation.

of the chain rule to Equation (8.2) yields

$$\frac{\partial c}{\partial t} = \frac{\partial}{\partial \zeta_k} R_{ki} D_{ij} R_{lj} \frac{\partial}{\partial \zeta_l} c$$

$$= \frac{\partial}{\partial \zeta_k} \left(\mathbb{R} \mathbb{D} \mathbb{R}^T \right)_{kl} \frac{\partial}{\partial \zeta_l} c$$

$$= \nabla_\zeta \cdot \mathbb{R} \mathbb{D} \mathbb{R}^T \nabla_\zeta c, \tag{8.3}$$

where $\nabla_\zeta \cdot$ and ∇_ζ are the divergence and gradient operators in the ζ coordinate system.

If the coordinate transformation $\mathbb{R}$ is chosen so that $\mathbb{D} = \mathbb{R}^T \Lambda \mathbb{R}$, where Λ is a diagonal matrix, then (recalling the fact that the eigenvectors of a positive definite matrix are orthogonal) $\Lambda = \mathbb{R} \mathbb{D} \mathbb{R}^T$ and the diffusion equation is

$$\frac{\partial c}{\partial t} = \Lambda \nabla_\zeta^2 c, \tag{8.4}$$

for the homogeneous case where Λ is constant in space. In other words, the eigen decomposition of $\mathbb{D}$ yields the diffusion equation

$$\frac{\partial c}{\partial t} = \lambda_1 \frac{\partial^2 c}{\partial \zeta_1^2} + \lambda_2 \frac{\partial^2 c}{\partial \zeta_2^2} + \lambda_3 \frac{\partial^2 c}{\partial \zeta_3^2}, \tag{8.5}$$

where the coordinates ζ_i are the directions defined by the eigenmodes of the diffusion matrix and the constants λ_i are the diffusion coefficients associated with the eigenmodes.

Equation (8.2) can be shown to apply equivalently to either a continuous concentration field or the position probability density of a single particle undergoing Brownian motion [174]. This equation is used to model transport processes in a wide range of natural phenomena from population distribution in ecology [146] to pollutant distribution in groundwater [30]. One of the earliest (and still important) applications to transport within cells and tissues is to describe the transport of oxygen from microvessels to the sites of oxidative metabolism in cells.

8.1.2 Oxygen transport to tissue and the Krogh–Erlang model

Typical estimates of the molecular diffusivity of oxygen in tissue are in the range of $2 \times 10^{-5} \, \text{cm}^2 \cdot \text{sec}^{-1}$. The rate of oxygen consumption in the heart during exercise may be as high as 2×10^{-4} moles per liter per second. Based on these numbers we can estimate how far oxygen can passively diffuse in tissue before it is consumed. If we consider a slab of tissue with oxygen supplied on one side at a fixed concentration, we can analyze this system using the one-dimensional homogeneous version

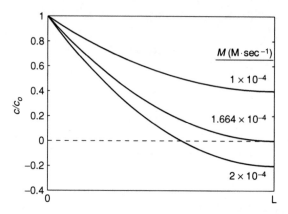

Figure 8.1 Concentration profiles predicted by Equation (8.8) with $D = 2 \times 10^{-5}\,\mathrm{cm^2 \cdot sec^{-1}}$ and $L = 25\,\mu\mathrm{m}$ and values of M indicated in the figure.

of Equation (8.2):

$$\frac{\partial c}{\partial t} = D\frac{\partial^2 c}{\partial x^2} - M, \tag{8.6}$$

where the term $-M$ has been added to account for an assumed constant rate of consumption. Steady state solutions to this equation can be obtained:

$$\frac{\partial^2 c}{\partial x^2} = \frac{M}{D}$$

$$\frac{\partial c}{\partial x} = \frac{M}{D}x + A$$

$$c = \frac{M}{D}x^2 + Ax + B, \tag{8.7}$$

where A and B are constants. Assuming the tissue slab has fixed width L, oxygen is supplied at $x = 0$ at concentration c_o, and there is no flux into or out of the slab at $x = L$, we apply the boundary conditions $c(x = 0) = c_o$ and $dc/dx|_{x=L} = 0$, and obtain the concentration profile

$$c(x) = \frac{M}{D}\left(\frac{x^2}{2} - Lx\right) + c_o. \tag{8.8}$$

Concentration profiles predicted by Equation (8.8) for $c_o = 2.6 \times 10^{-5}\,\mathrm{M}$ are plotted in Figure 8.1. (Given an oxygen solubility of $\alpha = 1.3 \times 10^{-6}\,\mathrm{M \cdot mmHg^{-1}}$, $c_o = 2.6 \times 10^{-5}\,\mathrm{M}$ corresponds to an oxygen partial pressure of $c_o/\alpha = 20\,\mathrm{mmHg}$.)

Predicted oxygen concentration decreases as the rate of oxygen consumption increases. In fact, Equation (8.8) predicts that oxygen concentration becomes negative when the rate of oxygen consumption is greater than $2Dc_o/L^2$. For the values of c_o, D, and L used in Figure 8.1, this maximal consumption value is

$M = 1.664 \times 10^{-4}\,\mathrm{M \cdot sec^{-1}}$. Given values of c_o, D, and M, the maximal width of the slab that maintains positive oxygenation is computed $L_{max} = (2Dc_o/M)^{1/2}$. At $D = 2 \times 10^{-5}\,\mathrm{cm^2 \cdot sec^{-1}}$, $M = 2 \times 10^{-4}\,\mathrm{M \cdot sec^{-1}}$, and $c_o = 2.6 \times 10^{-5}\,\mathrm{M}$, the maximal diffusion distance is $L_{max} \approx 22.8\,\mu\mathrm{m}$. This value provides an approximation of the maximal distance over which diffusion can effectively supply oxygen to tissue at this rate of oxygen consumption. Clearly this length is much shorter than the typical dimensions of most multicellular organisms. Certainly the distance between most cells in the human body and the atmosphere is much greater than $23\,\mu\mathrm{m}$.

Higher organisms use a circulatory system to deal with such relatively short diffusion distances. In vertebrates oxygenated blood flows from the left side of the heart through a branched network of successively smaller arterial vessels until it reaches the microcirculation, a collections of vessels of diameter of the order of a few micrometers. Most of the exchange of oxygen, nutrients, and wastes with the tissues occurs at the level of the microcirculation, which (for tissues other than the lung) drains into a collection of successively larger vessels leading to the right side of the heart.[3]

In tissues such as skeletal muscle and the heart, the primary sites of exchange of solutes between the blood and tissue are the capillaries – the smallest vessels with walls composed of a single endothelial cell. Capillaries in striated muscle tend to run parallel to the muscle fibers, which are approximately cylindrical cells with diameters ranging from approximately 15 to 20 μm (typical cardiac muscle cells) to over 50 μm (skeletal muscle cells). Based on the observation that capillaries tend to be distributed regularly in the plane perpendicular to the axis of their orientation (and orientation of the muscle cells), Krogh (1919) introduced the approximation that each capillary supplies "oxygen independently of all others to a cylinder of tissue surrounding it" [118]. This approximation, called the Krogh cylinder, has been the basis of much of the field of analysis of oxygen transport to tissue ever since.

The Krogh–Erlang model of oxygen transport to tissue is illustrated in Figure 8.2. Two concentric cylinders, corresponding to the central capillary vessel and the surrounding tissue, are illustrated in the lower panel of the figure. Blood flows through the capillary, carrying oxygen with it. Oxygen passively diffuses and is consumed in the tissue. While Krogh analyzed the predicted concentration profile in the radial direction and did not treat oxygen transport in the axial (z) direction, it is a natural extension of the model to determine the oxygen profile as a function of both the axial and radial variables in this model.

---

[3] In the early 1600s William Harvey deduced the existence of the microcirculation. Until then it was thought that venous blood and arterial blood made up independent pools. Capillary vessels were first observed later in the century by Marcello Malpighi, verifying the existence of a microcirculation connecting the arterial and venous networks.

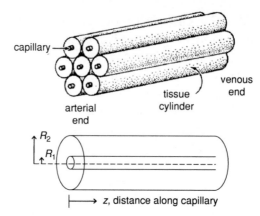

Figure 8.2 Cylindrical geometry of the Krogh–Erlang model of blood–tissue exchange. The upper panel, from Middleman [141], illustrates the assumed parallel arrangement of capillaries with each vessel independently supplying a surrounding cylinder of tissue. A diagram of the model geometry is provided in the lower panel. Figure in upper panel is reprinted with the permission of John Wiley & Sons, Inc.

Before analyzing the model, it is valuable to state explicitly the additional assumptions and approximations that are a part of this model. First, we will assume that oxygen consumption is homogeneous and constant in the tissue. Second, we will ignore diffusion of oxygen in the axial direction. Both of these assumptions are reasonable – oxygen consumption is expected to be fairly homogeneous unless a portion of the tissue becomes hypoxic and diffusion along the z direction is not important compared to advective transport in the capillary. These assumptions, which allow us to determine closed form model solutions, are easily relaxed in numerical simulations of oxygen transport to tissue.

Under the simplifying assumptions of the Krogh–Erlang model, the steady state oxygen distribution in the tissue at any position z is governed by the steady state diffusion equation in radial coordinates

$$\frac{1}{r}\frac{d}{dr}r\frac{d}{dr}c = \frac{M_t}{D}, \tag{8.9}$$

where M_t is the rate of oxygen consumption in the tissue. This equation has general solution

$$c(r) = \frac{M_t}{4D}r^2 + a\ln r + b, \tag{8.10}$$

where a and b are constants. Assuming no flux ($dc/dr = 0$) at $r = R_2$ and $c(r = R_1) = c_1$, the radial concentration profile is

$$c(r) = \frac{M_t}{2D}\left[\left(\frac{r^2 - R_1^2}{2}\right) + R_2^2\ln(R_1/r)\right] + c_1. \tag{8.11}$$

Krogh used this solution to express the concentration difference from the capillary ($r = R_1$) to the outside of the tissue cylinder:

$$c_1 - c(R_2) = \frac{M_t}{2D} \left[-\left(\frac{R^2 - R_1^2}{2} \right) + R_2^2 \ln (R_2/R_1) \right]. \tag{8.12}$$

Because Krogh credited the help of "the mathematician Mr. Erlang" [118] in deriving this formula, this model is appropriately referred to as the Krogh–Erlang model.

Reasonable parameter values for the Krogh–Erlang model can be estimated for a specific tissue. For example, the measured density of capillaries in the heart is approximately $\rho = 2500 \, \text{mm}^{-2}$. From this the outer radius of the effective tissue cylinder can be computed $R_2 = (\pi\rho)^{-1/2} = 11.3 \, \mu\text{m}$; capillary radii ($R_1$) are approximately $2 \, \mu\text{m}$; the diffusion coefficient of molecular oxygen in tissue is approximately $2 \times 10^{-5} \, \text{cm}^2 \cdot \text{sec}^{-1} = 2000 \, \mu\text{m}^2 \cdot \text{sec}^{-1}$. Given $M_t = 1.74 \times 10^{-4} \, \text{M} \cdot \text{sec}^{-1}$, the difference $c_1 - c(R_2)$ is approximately $7 \times 10^{-6} \, \text{M}$, or 5 mmHg of partial pressure.

In the capillary oxygen binds to hemoglobin in red blood cells. In fact, the majority of oxygen in the blood is bound to hemoglobin with only a few percent at most freely dissolved. One hemoglobin molecule has four binding sites for O_2 and the oxygen–hemoglobin binding is highly competitive with a Hill coefficient of approximately 2.5. Over the physiological range of oxygen concentration, the oxygen–hemoglobin binding is effectively approximated by the simple Hill equation

$$S_{Hb} = \frac{P^n}{P^n + P_{50}^n} \tag{8.13}$$

where P is the oxygen partial pressure, $P_{50} \approx 25 \, \text{mmHg}$ is the 50% saturation pressure, and $n \approx 2.5$ is the Hill coefficient. (Partial pressure is related to free concentration by the expression $c = \alpha P$, where $\alpha \approx 1.3 \times 10^{-6} \, \text{M} \cdot \text{mmHg}^{-1}$ is the oxygen solubility.)

The concentration of oxygen bound to hemoglobin in blood is given by

$$c_T = H_c C_{Hb} S_{Hb} \tag{8.14}$$

where H_c is the *hematocrit* – the volume fraction of the blood composed of red blood cells, approximately 0.45 – and $C_{Hb} \approx 0.0231 \, \text{M}$ is the concentration of oxygen binding sites in a red cell. If we assume that Equation (8.14) defines the total oxygen concentration in blood (ignoring the free dissolved oxygen), then the transport equation for oxygen in blood is given by the advection equation:

$$\frac{\partial c_T}{\partial t} = -v \frac{\partial c_T}{\partial z} - \frac{2\pi D}{R_1} \left. \frac{\partial c}{\partial r} \right|_{r=R_1}, \tag{8.15}$$

where v is the velocity of blood in the capillary. Applying this one-dimensional advection equation assumes that no significant concentration gradient exists in the radial direction inside the capillary. The final term in Equation (8.15) is the diffusive flux out of the capillary into the tissue.

In the steady state, mass flux out of the capillary is equal to the rate of consumption in the tissue. The mass consumption rate in the tissue (mass per unit time) is equal to the volume times M_t, the mass consumed per unit volume per unit time. Similarly, if M_c is the rate of oxygen loss from the capillary expressed as mass per unit volume per unit time, then $V_c M_c$ is the mass flux out of the capillary, where V_c is the volume of the capillary. Equating $V_t M_t$ and $V_c M_c$, where V_t is the volume of tissue, M_c is equal to $V_t M_t / V_c$. Thus in the steady state, Equation (8.15) becomes

$$-v\frac{\partial c_T}{\partial z} - \frac{V_t}{V_c} M_t = 0, \tag{8.16}$$

with solution

$$c_T(z) = c_o - \left(\frac{V_t}{V_c}\frac{M_t}{v}\right) z, \tag{8.17}$$

where c_o is the total oxygen concentration at the entrance to the capillary at $z = 0$. For arterial blood with oxygen partial pressure of 100 mmHg, the total oxygen concentration is $c_o = 9.3$ mM.

Blood flow to tissue is usually measured as volume per unit time per unit volume or mass of tissue. Typical blood flow to heart tissue is of the order of 1–4 ml per minute per ml of tissue. If we denote tissue flow F, then flow and velocity in the Krogh–Erlang model are related

$$F = \frac{V_c}{V_t}\frac{v}{L} = \frac{\pi R_1^2 v}{\pi R_2^2 L}, \tag{8.18}$$

where L is the length of the cylinder. Substituting this expression into Equation (8.17), we have

$$c_T(z) = c_o - \left(\frac{M_t}{F}\right)\frac{z}{L}. \tag{8.19}$$

With $M_t = 2 \times 10^{-4} \text{ M} \cdot \text{sec}^{-1}$ and $F = 2 \text{ ml} \cdot \text{min}^{-1} \cdot \text{ml}^{-1}$, the predicted outlet concentration is $c_T(L) = c_o - 6$ mM, or 3.3 mM, when $c_o = 9.3$ mM. Thus given these conditions, approximately 65 percent of the arterial input oxygen is extracted from the blood as it passes from inlet to outlet.

The predicted partial pressure in the capillary can be computed by inverting Equation (8.14) to convert from total oxygen to partial pressure. The oxygen profile in the tissue as a function of r and z is given by combining Equation (8.19), which gives the capillary oxygen concentration as a function of z, with Equation (8.11),

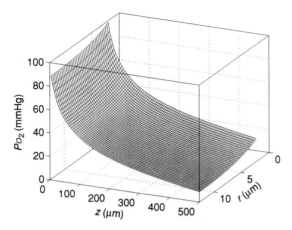

Figure 8.3 Oxygen profile predicted by the Krogh–Erlang model of oxygen transport to tissue, based on the geometry of Figure 8.2. Parameter values used are $D = 1.5 \times 10^{-5} \, cm^2 \cdot sec^{-1}$, $M = 2 \times 10^{-4} \, M \cdot sec^{-1}$, $F = 2 \, ml \cdot min^{-1} \cdot ml^{-1}$, $c_o = 9.3 \, mM$, $L = 500 \, \mu m$, and $\rho = 2500 \, mm^{-2}$, $R_1 = 2 \, \mu m$. The computed value of R_2 is $(\pi\rho)^{-1/2} = 12.62 \, \mu m$ for this geometry.

which gives the radial oxygen profile at each z position. The combined solution is illustrated in Figure 8.3, for parameter values specified in the legend.

8.1.3 Facilitated diffusion

The preceding analysis of oxygen transport based on the Krogh–Erlang model assumed that oxygen diffuses and is consumed homogeneously with no other processes affecting the transport of oxygen in the tissue space. This assumption of course is a simplification used to allow us to focus on a model that represents the key transport phenomena while remaining accessible to analysis. One potentially important phenomenon ignored in this analysis is oxygen binding to myoglobin, a heme-containing protein present in varying concentrations in muscle cells. In fact, the role of myoglobin in affecting oxygen transport in tissue has yet to be unambiguously established. Here we explore the potential role of myoglobin in facilitating passive diffusion of oxygen based on a mathematical analysis introduced by James Murray [145].

Myoglobin is a single-chain protein, homologous to one of the four chains of the hemoglobin protein, with a single oxygen binding site. It is present in typical mammalian cardiomyocytes in concentrations of the order of 100–400 μM. This binding affinity of oxygen to myoglobin is of the order of 2 mmHg; therefore at typically intracellular P_{O_2} of 20 mmHg, which corresponds to approximately 26 μM, the concentration of oxygen bound to myoglobin can be an order of magnitude

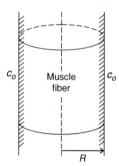

Figure 8.4 Cylindrical model for the muscle fiber. Figure redrawn from Murray [145].

higher than that freely dissolved. The tissues of sea mammals contain even greater concentrations of myoglobin. For example, myoglobin concentration in skeletal muscle is as high as 4 mM in some dolphin species [45], pointing to the adaptive role of myoglobin as an oxygen storage buffer in these species. However, the potential of myoglobin as a useful buffer in highly oxidative tissue such as the heart is weak. Assuming a relatively low consumption rate of $M = 1 \times 10^{-4}\,\mathrm{M \cdot sec^{-1}}$, then at most a few seconds worth of oxygen is stored at myoglobin concentrations in the range of 100–400 μM.

In the early 1950s investigators began to explore the role of myoglobin in the passive diffusion of oxygen in cells (for a review see [211]), and theoretical models soon followed [214, 144, 145]. While the molecular diffusion coefficient of myoglobin in cells may be two orders of magnitude lower than that of molecular oxygen, it is possible that, due to the non-linear oxygen–myoglobin binding curve, passive diffusion of oxygen-bound myoglobin contributes significantly to overall oxygen transport in certain concentration regimes.

Murray [145] analyzed this problem based on the cylindrical geometry illustrated in Figure 8.4. Here, the muscle fiber is assumed to be a homogeneous cylinder, with oxygen supplied via the capillary network to the outer boundary of the fiber.

The oxygen–myoglobin association is assumed to be a first-order reaction

$$O_2 + Mb \underset{k_-}{\overset{k_+}{\rightleftharpoons}} MbO_2, \tag{8.20}$$

which is assumed to proceed by simple mass-action kinetics. The net rate of binding (forward flux minus reverse flux) is given by:

$$b = k_+ C_{Mb}(1 - S_{Mb})c - k_- C_{Mb}S_{Mb}, \tag{8.21}$$

where c is the concentration of freely dissolved oxygen, $C_{Mb} = [Mb] + [MbO_2]$ is the total myoglobin saturation, and S_{Mb} is the fractional oxygen–myoglobin saturation

$$S_{Mb} = \frac{[MbO_2]}{C_{Mb}}. \tag{8.22}$$

Given these definitions, the steady state reaction diffusion equations for oxygen and oxy-myoglobin in the muscle fiber in radial coordinates are

$$D_{O_2}\frac{1}{r}\frac{d}{dr}r\frac{d}{dr}c = b + M$$

$$C_{Mb}D_{Mb}\frac{1}{r}\frac{d}{dr}r\frac{d}{dr}S_{Mb} = -b, \tag{8.23}$$

where r is the radial coordinate and M is the rate of oxygen consumption in the fiber. Dissolved oxygen and myoglobin-bound oxygen diffuse with two different molecular diffusion coefficients, D_{O_2} and D_{Mb}, respectively. Summing these two equations and multiplying by r we obtain

$$D_{O_2}\frac{d}{dr}r\frac{d}{dr}c + C_{Mb}D_{Mb}\frac{d}{dr}r\frac{d}{dr}S_{Mb} = Mr. \tag{8.24}$$

Integrating this expression once we obtain

$$D_{O_2}\frac{d}{dr}c + C_{Mb}D_{Mb}\frac{d}{dr}S_{Mb} = \frac{Mr}{2} + \frac{A}{r}, \tag{8.25}$$

where A is an integration constant. The condition at $r = 0$ ($dc/dr|_{r=0} = dS_{Mb}/dr|_{r=0} = 0$), which arises from the axisymmetry of the model, yields $A = 0$. (Alternatively, the physical condition that the solution remains finite at $r = 0$ yields the same result.) Integrating one more time, we have

$$D_{O_2}c + C_{Mb}D_{Mb}S_{Mb} = \frac{Mr^2}{4} + B. \tag{8.26}$$

Next we apply the boundary condition that the concentration at the exterior of the muscle fiber is prescribed: $c(r = R) = c_o$ and $S_{Mb}(r = R) = S^o$. Thus we have

$$D_{O_2}c_o + C_{Mb}D_{Mb}S^o = \frac{MR^2}{4} + B. \tag{8.27}$$

Subtracting Equation (8.27) from Equation (8.26) yields

$$D_{O_2}(c - c_o) + C_{Mb}D_{Mb}(S_{Mb} - S^o) = \frac{M}{4}(r^2 - R^2). \tag{8.28}$$

Solving this equation for S_{Mb} we obtain

$$S_{Mb} = S^o + \frac{1}{C_{Mb}D_{Mb}}\left[\frac{M}{4}(r^2 - R^2) + D_{O_2}c_o\right] - \frac{D_{O_2}c}{D_{Mb}C_{Mb}} \tag{8.29}$$

for the spatial dependence of myoglobin saturation. Substituting this expression into Equation (8.23) yields a single differential equation for c, the free oxygen concentration:

$$D_{O_2} \frac{1}{r} \frac{d}{dr} r \frac{dc}{dr} = M - k_- C_{Mb} \left[S^o + \frac{1}{C_{Mb} D_{Mb}} \left(\frac{M}{4} (r^2 - R^2) + D_{O_2} c_o \right) \right]$$
$$+ k_+ C_{Mb} \left[1 - S^o + \frac{1}{C_{Mb} D_{Mb}} \left(\frac{M}{4} (r^2 - R^2) + D_{O_2} c_o \right) \right] c$$
$$+ k_- \frac{D_{O_2}}{D_{Mb}} c + k_+ \frac{D_{O_2}}{D_{Mb}} c_o^2. \tag{8.30}$$

Introducing the non-dimensional (unitless) variables $c_1 = c/c_o$, $r_1 = r/R$, this expression is drastically simplified:

$$\frac{1}{r_1} \frac{d}{dr_1} r_1 \frac{d}{dr_1} c_1 = (\alpha + \gamma r_1^2) + (\beta + \lambda r_1^2) c_1 + \delta c_1^2, \tag{8.31}$$

where the non-dimensional parameters α, γ, β, λ, and δ have the following definitions

$$\alpha = \frac{MR^2}{c_o D_{O_2}} - \frac{k_- R^2 C_{Mb} S^o}{c_o D_{O_2}} + \frac{k_- MR^4}{4c_o D_{Mb} D_{O_2}} - \frac{k_- R^2}{D_{Mb}}$$

$$\beta = \frac{C_{Mb} k_+ R^2}{D_{O_2}} \left[1 - S^o + \frac{MR^2}{4C_{Mb} D_{Mb}} - \frac{c_o D_{O_2}}{D_{Mb} D_{Mb}} \right] + \frac{k_- R^2}{D_{Mb}}$$

$$\gamma = -\frac{k_- MR^4}{4c_o D_{Mb} D_{O_2}}, \quad \delta = \frac{k_+ R^2 c_o}{D_{Mb}}, \quad \lambda = -\frac{k_+ MR^4}{4D_{Mb} D_{O_2}}. \tag{8.32}$$

A set of parameter values that represents reasonable choices for skeletal muscle is:

$$R = 25 \, \mu\text{m}$$
$$k_+ = 2.4 \times 10^7 \, \text{M}^{-1} \cdot \text{sec}^{-1}$$
$$k_- = 65 \, \text{sec}^{-1}$$
$$c_o = 1.3 \times 10^{-5} \, \text{M}$$
$$C_{Mb} = 2.8 \times 10^{-4} \, \text{M}$$
$$M = 1.75 \times 10^{-4} \, \text{M} \cdot \text{sec}^{-1}$$
$$D_{O_2} = 2 \times 10^{-5} \, \text{cm} \cdot \text{sec}^{-1}$$
$$D_{Mb} = 2 \times 10^{-7} \, \text{cm} \cdot \text{sec}^{-1}, \tag{8.33}$$

where, with certain exceptions noted as follows, these values are the same as those used in the study of Murray [145]. Here the assumed value of D_{Mb} is ten times smaller than that assumed by Murray. The value chosen by Murray corresponds to the estimated molecular diffusion coefficient for myoglobin in dilute aqueous solution. In cells, the molecular diffusivity of globular proteins such as myoglobin is severely restricted and the value used here represents a much more realistic value than that used by Murray and others in early theoretical studies of myoglobin-facilitated oxygen transport. Because of this lower value for the myoglobin diffusivity, facilitated diffusion is apparent only at oxygen concentration lower than the concentration range studied by Murray. Therefore, we use a value of c_o that corresponds to approximately 10 mmHg, which is approximately three times smaller than that used by Murray.

If the myoglobin–oxygen binding reaction occurs rapidly enough that it is maintained near equilibrium, then S^o can be computed

$$S^o = \frac{c_o}{c_o + k_-/k_+} \approx 0.83,$$

for the given parameter values.

Substituting these parameter values into Equation (8.32) yields the following estimates for the non-dimensional parameters:

$$\alpha = -1781.8$$
$$\beta = -4427$$
$$\gamma = -610.35$$
$$\delta = 9750,$$
$$\lambda = -2929.7. \tag{8.34}$$

To analyze the non-linear system governed by Equation (8.31) we note that all of the non-dimensional parameters have magnitude much greater than 1. Taking advantage of this fact, we introduce a non-dimensional parameter $\epsilon \ll 1$ and re-scale Equation (8.31):

$$\epsilon \frac{1}{r_1} \frac{d}{dr_1} r_1 \frac{d}{dr_1} c_1 = (a + gr_1^2) + (b + lr_1^2)c_1 + dc_1^2, \tag{8.35}$$

where

$$a = \epsilon\alpha, \; b = \epsilon\beta, \; g = \epsilon\gamma$$
$$d = \epsilon\delta, \; l = \epsilon\lambda. \tag{8.36}$$

Since the value of ϵ is arbitrary, we are free to choose any value that makes our analysis convenient. Given the non-dimensional parameter values in Equation (8.34), we choose $\epsilon = 10^{-3}$, so that (away from any boundary layer) the left-hand side of this equation is of the order of ϵ while the right-hand side is of the order

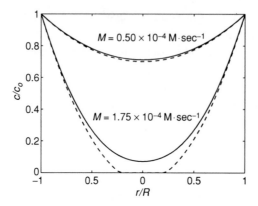

Figure 8.5 Oxygen profiles predicted by Equations (8.38) and (8.41) for the model illustrated in Figure 8.4. Solutions are illustrated for two different consumption values for the case with (solid lines; Equation (8.38)) and without (dashed lines; Equation (8.41)) myoglobin present.

of 1. Therefore, the first-order solution can be obtained from

$$(a + gr_1^2) + (b + lr_1^2)c_1 + dc_1^2 = 0, \text{ or}$$
$$(\alpha + \gamma r_1^2) + (\beta + \lambda r_1^2)c_1 + \delta c_1^2 = 0. \tag{8.37}$$

Thus, the concentration profile is approximated

$$c_1 = -\left(\frac{\beta + \lambda r_1^2}{2\delta}\right) + \frac{1}{2\delta}\left[(\beta + \lambda r_1^2)^2 - 4\delta(\alpha + \gamma r_1^2)\right]^{1/2}, \tag{8.38}$$

which satisfies the boundary conditions $c_1(1) = 1$, or $c(R) = c_o$.

With no myoglobin present, the governing equation for this model reduces to

$$D_{O_2}\frac{1}{r}\frac{d}{dr}r\frac{d}{dr}c = M, \tag{8.39}$$

or in non-dimensional form,

$$\frac{1}{r_1}\frac{d}{dr_1}r_1\frac{d}{dr_1}c_1 = \frac{MR^2}{D_{O_2}c_o}, \tag{8.40}$$

which, given the imposed boundary conditions, has solution

$$c_1 = \frac{MR^2(r_1^2 - 1)}{4c_o D_{O_2}} + 1. \tag{8.41}$$

Oxygen concentration profiles predicted by Equations (8.38) and (8.41), for the cases with and without myoglobin respectively, are illustrated in Figure 8.5. Only when the oxygen concentration falls near zero does the presence of myoglobin (and myoglobin-facilitated diffusion) have a significant impact on the concentration profile.[4]

In numerical simulations of oxygen transport accounting for myoglobin binding and diffusion, both species $[O_2]$ and $[MbO_2]$ (or S_{Mb}) are usually not explicitly accounted for as state variables. Instead, typically either free oxygen or partial pressure is treated as a state variable and the oxygen–myoglobin saturation is assumed to be maintained in equilibrium everywhere in the tissue, with myoglobin saturation given by

$$S_{Mb}(c) = \frac{c}{c + k_- / k_+}. \tag{8.42}$$

In this case the total diffusive oxygen flux is computed from the sum of diffusion terms for free and bound oxygen [161]:

$$\begin{aligned}
\vec{\Gamma} &= -\left[D_{O_2} \nabla c + D_{Mb} C_{Mb} \nabla S_{Mb} \right] \\
&= -\left[D_{O_2} + D_{Mb} C_{Mb} \frac{\partial S_{Mb}}{\partial c} \right] \nabla c.
\end{aligned} \tag{8.43}$$

The conservation equation for total (free plus bound) oxygen is

$$\begin{aligned}
\frac{\partial}{\partial t}(c + C_{Mb} S_{Mb}) &= -\nabla \cdot \vec{\Gamma} - M \\
\frac{\partial c}{\partial t} + C_{Mb} \frac{\partial S_{Mb}}{\partial c} \frac{\partial c}{\partial t} &= \nabla \cdot \left[D_{O_2} + D_{Mb} C_{Mb} \frac{\partial S_{Mb}}{\partial c} \right] \nabla c - M
\end{aligned} \tag{8.44}$$

or

$$\frac{\partial c}{\partial t} = \frac{\nabla \cdot \left[D_{O_2} + D_{Mb} C_{Mb} \frac{\partial S_{Mb}}{\partial c} \right] \nabla c - M}{1 + C_{Mb} \frac{\partial S_{Mb}}{\partial c}}, \tag{8.45}$$

which is a partial differential equation for c that does not require invoking a separate differential equation for bound oxygen.

8.2 Advection–diffusion modeling of solute transport in tissues

From Figure 8.3 it is apparent that the arterial–venous oxygen concentration difference is much greater than concentration differences in the radial direction occurring at a specific axial position. This is generally the case for solutes that are highly extracted from the blood in their passage through the microcirculation – the drop in concentration from the input to the output of a capillary is much larger than

---

[4] Since recent measurements using NMR methods yield an estimate of $D_{Mb} = 7.85 \times 10^{-7}\,\mathrm{cm \cdot sec^{-1}}$ at 35 C in the myocardium [131], which is significantly higher than the values used here, the role of facilitated diffusion of oxygen by myoglobin remains unclear.

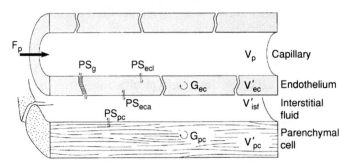

Figure 8.6 Diagram of a transport model for simulating exchange of solutes between blood and tissue. Four distinct regions are included: parenchymal cells (functional cells of a given organ), interstitial fluid (ISF) that bathes the tissue, endothelial cells that make up the capillary wall, and intra-capillary space, in which blood flows. In general, exchange of solutes occurs directly between the ISF and the parenchymal cells, between the ISF and the endothelial cells, between the endothelial cells and the blood, and directly between the ISF and the blood through clefts in the capillary wall. Associated with each of these transport processes is a permeability-surface area (PS) product. Also each region has associated with it a volume of distribution (the V's in the figure) associated with a given solute. The symbols G_{ec} and G_{pc} represent biochemical reaction processes, here assumed to be confined to endothelial cells and parenchymal cells. Figure from Schwartz *et al.* [182], used with permission.

the drop in concentration from the capillary wall to the periphery. In addition, there exist structural features of the tissue that are not explicitly accounted for in the Krogh–Erlang model (endothelial cell and parenchymal cell membranes) that represent resistance barriers to the transport of certain solutes. The result is that stepwise changes in concentration occurring across membranes and across the capillary wall may be more significant than continuous concentration gradients within the volumes bounded by those barriers [11]. Even the transport of oxygen (a solute to which biological membranes do not pose significant transport resistance) may be effectively simulated for certain applications by models that are continuously distributed in the axial direction only, with relatively few concentric regions (representing the blood, the capillary wall, interstitial fluid, and parenchymal cells) separated by apparent permeability barriers [43].

Figure 8.6 shows a diagram of a general model of blood–tissue solute transport, used to analyze data on the transport of labeled solutes introduced in the blood or perfusate flow supplied to individual organs. The development and analysis of models of this sort to analyze solute transport in physiological systems is a field pioneered by Sangren and Sheppard [178], Renkin [172], and Crone [40]. Optically detectable probes (such as Evans Blue dye bound to albumin) can be used in conjunction with model analysis to probe the intravascular transport of solutes governed by the distribution of flow in the vascular network. Radioactively labeled bio-active molecules (such as ^{125}I-albumin, ^{14}C-sucrose, and ^{15}O$_2$) can be used to probe how the non-labeled analogs of the molecules are transported and metabolized in tissue.[5]

8.2.1 Axially distributed models of blood–tissue exchange

Before examining how a model of the level of detail of the four-region model illustrated in Figure 8.6 is constructed, we first examine the two-region model analyzed in 1953 by Sangren and Sheppard [178]. This model will allow us to explore the kinetics of blood–tissue exchange based on an analytically tractable set of governing equations.

To develop the two-region Sangren–Sheppard model, consider a substance that traverses the endothelial cell clefts but does not enter endothelial cells (such as L-glucose, which is not taken up by cells.) This solute is assumed to exchange passively between the capillary and interstitial fluid (ISF) spaces. Applying a one-dimensional approximation, the governing equation for solute concentration in the blood is the advection equation:

$$\frac{\partial c_B}{\partial t} = -v\frac{\partial c_B}{\partial z} - \frac{PS}{V_B}(c_B - c_{ISF}), \qquad (8.46)$$

where $c_B(z, t)$ and $c_{ISF}(z, t)$ are the concentrations in the capillary blood and ISF spaces, v is the blood velocity in the capillary, PS is the permeability-surface area product for the exchange between the capillary and ISF (typically measured in units of volume per unit tissue mass), and V_B is the blood volume (typically measured in units of volume per unit tissue mass.) The second term on the right-hand side of Equation (8.46) accounts for passive permeation between the capillary and ISF spaces.

The governing equation for the concentration in the ISF is similar to Equation (8.46). However, there is no advection term in the ISF equation since we assume the ISF to be stagnant.

$$\frac{\partial c_{ISF}}{\partial t} = +\frac{PS}{V_{ISF}}(c_B - c_{ISF}). \qquad (8.47)$$

Note that the exchange term on the right-hand side of Equation (8.47) has the opposite sign to that of Equation (8.46). Also, the volume in the denominator is

---

[5] Incidentally, August Krogh was one of the pioneers of the use of radioactive tracers to investigate biological transport. Krogh's mentor and colleague at the University of Copenhagen was Christian Bohr. Krogh's collaborators in the application of radioisotopes were George de Hevesy and Christian's son, the physicist Niels Bohr [125].

the ISF space volume, V_{ISF} – the volume of distribution of the given solute in the interstitial fluid.

By multiplying Equation (8.46) by V_B and Equation (8.47) by V_{ISF} and summing the resulting equations, we can show that the system governed by these equations conserves mass:

$$\frac{\partial}{\partial t} [V_B c_B + V_{ISF} c_{ISF}] = -V_B v \frac{\partial c_b}{\partial z}. \tag{8.48}$$

Since this model does not account for chemical transformation or diffusion in the axial direction, the rate of change of mass of solute at any location along the capillary is driven by advection alone. If the blood velocity were zero, then the total mass density at any location z would remain constant.

Equations (8.46) and (8.47) describe the Sangren–Sheppard model. While the equations are straightforward and can be thought of as the minimal model that captures the important biophysical phenomena of solute exchange along a capillary, this model represents nearly the maximal level of complexity that can be effectively analyzed without invoking numerical approximations to simulate it.[6]

Given an initial condition specifying concentration in the capillary and ISF, and a boundary condition at the capillary inlet, solutions to Equations (8.46) and (8.47) may be obtained. For the sake of exploring the kinetic properties of this system, and studying how a solute injected into the microcirculation washes out of a tissue, we determine the *impulse response function* that arises from the following conditions.

We specify that there is no solute present in either the capillary or ISF space at the beginning of the experiment:

$$c_B(z, t = 0) = 0, \ 0 < z \le L$$
$$c_{ISF}(z, t = 0) = 0, \ 0 \le z \le L, \tag{8.49}$$

where L is the length of the capillary. To represent the situation where an impulse of solute is injected into the capillary inflow at time $t = 0$, the boundary condition

$$c(z = 0, t) = \frac{q_o}{F} \delta(t) \tag{8.50}$$

is applied, where q_o is the finite mass injected into the capillary and F is the blood flow to the tissue ($F = v V_B / L$).

The function $\delta(t)$ is the Dirac delta function, which has properties

$$\delta(s) = 0, \ s \ne 0$$

---

[6] Actually, the distinction between analytically and numerically obtained model solutions is rarely clear. Analytical solutions to governing differential equations are often expressed in terms of special functions such as exponentials, which must be approximated numerically. Here we will see that the solutions to the Sangren and Sheppard model are conveniently expressed in terms of a class of special functions called modified Bessel functions.

and

$$\int_{-\infty}^{+\infty} \delta(s)ds = 1. \tag{8.51}$$

The function $q_o \delta(t)/F$ can be thought of here as a spike of finite mass q_o injected into the capillary at position $z = 0$ at time $t = 0$. If flow (F) and volumes (V_B and V_{ISF}) are expressed relative to total mass of tissue (for example, in units of $ml \cdot min^{-1} \cdot g^{-1}$ and $ml \cdot g^{-1}$, respectively) then the injected mass q_o is expressed in units of moles per mass of tissue. The finite mass injected at $z = 0$ results in an infinitely high concentration in an infinitely small volume. Thus, this function is a mathematical abstraction used to represent the limiting case of instantaneous injection of mass at time $t = 0$.

Given these initial and boundary conditions, the concentration in the capillary as a function of t and z is given by

$$c_B(z, t) = \frac{q_o}{F} e^{-\frac{PSz}{vV_B}} \delta\left(t - \frac{z}{v}\right)$$

$$+ \frac{q_o}{F} e^{-\frac{PS}{V_{ISF}}\left[t-\left(1-\frac{V_{ISF}}{V_B}\right)\right]} \frac{PS}{\sqrt{V_B V_{ISF}\left(\frac{vt}{z} - 1\right)}} I_1(\beta), \ t \geq z/v,$$

$$c_B(z, t) = 0, \ t < z/v, \tag{8.52}$$

where $I_1(\beta)$ is the first-order modified Bessel function of the first kind [153]; and the argument β is given by

$$\beta = 2PS\left[\frac{(t - z/v)z}{v V_B V_{ISF}}\right]^{1/2}.$$

This equation for the concentration profile in the capillary has two terms. The first term is a delta function $\delta(t - z/v)$ that travels through the capillary at velocity v and decays in strength exponentially as it travels from the inlet to the outlet of the capillary. The decay is due to permeation out of the capillary; thus the decay constant is proportional to the permeability-surface area product of the capillary wall. If $PS = 0$, then the impulse travels along the capillary and reaches the outlet at undiminished strength at time $t = L/v$. The second term in Equation (8.47) represents the concentration profile in the capillary that trails behind the impulse.

The concentration profile in the ISF is given by

$$c_{ISF}(z, t) = \frac{q_o}{F} \frac{PS}{V_{ISF}} e^{-\frac{PS}{V_{ISF}}\left[t-\left(1-\frac{V_{ISF}}{V_B}\right)\right]} I_0(\beta), \ t \geq z/v$$

$$c_{ISF}(z, t) = 0, \ t < z/v, \tag{8.53}$$

where I_0 is the zeroth-order modified Bessel function of the first kind [153].

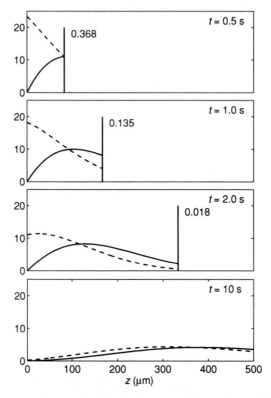

Figure 8.7 Concentration profiles for the Sangren–Sheppard model following impulse injection of solute into the capillary. The solution predicted by Equations (8.52) and (8.53) is plotted at four different times, for parameter values indicated below. Capillary blood concentrations are plotted as solid lines and ISF concentrations are plotted as dashed lines. The impulse at $z = vt$ is indicated by a vertical line and the relative strength $e^{-PSz/(vV_B)}$ is indicated in the plots. Parameter values are $V_B = 0.05\,\mathrm{ml \cdot g^{-1}}$, $V_{ISF} = 0.20\,\mathrm{ml \cdot g^{-1}}$, $PS = 6.0\,\mathrm{ml \cdot min^{-1} \cdot g^{-1}}$, $F = 1.0\,\mathrm{ml \cdot min^{-1} \cdot g^{-1}}$, $L = 500\,\mathrm{\mu m}$, $q_o = 10^{-3}\,\mathrm{mol \cdot g^{-1}}$. The velocity is $v = FL/V_B = 166.7\,\mathrm{\mu m \cdot s^{-1}}$. Concentrations are plotted in Molar units.

Concentration profiles predicted by the Sangren–Sheppard model, following impulse injection of solution into the capillary, are illustrated in Figure 8.7. The figure plots capillary and ISF solute concentrations as a function of distance along the capillary at different times following the initial impulse.

8.2.2 Analysis of solute transport in organs

While it may be elegant to obtain analytic closed-form model solutions, such as Equations (8.52) and (8.53) (introduced by Sangren and Sheppard as solutions to their model governing equations [178]), modeling of transport in biological systems nearly always requires simulation based on numerical (computer-generated) solutions to governing equations. For example, simulating the transport of labeled solutes that move between many regions (such as illustrated in Figure 8.6) is practically possible only by solving the governing equations using computers. Here, we examine how to construct the governing equations for this class of transport models (axially distributed models) without going into the details of how to simulate these models using computers.

The governing equations for the multiple-region models such as that of Figure 8.6 are constructed analogously to those of the two-region Sangren–Sheppard model. For example, for this model the solute concentration in the blood is governed by the equation:

$$\frac{\partial c_p}{\partial t} = -\frac{FL}{V_p}\frac{\partial c_p}{\partial z} - \frac{PS_g}{V_p}(c_p - c_{isf}) - \frac{PS_{ecl}}{V_p}(c_p - c_{ec}), \qquad (8.54)$$

which is analogous to Equation (8.46) with the addition of a term accounting for passive permeation between the blood and the endothelial cell space. Here PS_g is the permeability-surface area product for gaps in the capillary wall and PS_{ecl} is the permeability-surface area product for transport between the blood and capillary wall endothelial cells. The concentrations c_p, c_{isf}, and c_{ec} are the solute concentrations in the plasma, interstitial fluid, and in the endothelial cell regions, respectively.

This equation may be further modified to account for molecular diffusion and/or flow-mediated dispersion in the axial direction:

$$\frac{\partial c_p}{\partial t} = -\frac{FL}{V_p}\frac{\partial c_p}{\partial z} - \frac{PS_g}{V_p}(c_p - c_{isf}) - \frac{PS_{ecl}}{V_p}(c_p - c_{ec}) + D_p\frac{\partial^2 c_B}{\partial z^2}, \qquad (8.55)$$

where D_p is the axial dispersion coefficient for the plasma region.

Similarly a general governing equation in the endothelial cells is

$$\frac{\partial c_{ec}}{\partial t} = +\frac{PS_{ecl}}{V'_{ec}}(c_p - c_{ec}) - \frac{PS_{eca}}{V'_{ec}}(c_{ec} - c_{isf}) + D_{ec}\frac{\partial^2 c_{ec}}{\partial z^2} - G_{ec}, \qquad (8.56)$$

where PS_{eca} is the permeability-surface area product for transport between endothelial cells and the ISF and D_{ec} is the dispersion coefficient for the endothelial cell region. The term G_{ec} represents biochemical consumption of the solute in endothelial cells. Here the volume V'_{ec} represents the apparent volume of distribution of the given solute, which due to binding may be different from the physical volume of the region [182].

In the interstitial fluid, the general governing equation is

$$\frac{\partial c_{isf}}{\partial t} = +\frac{PS_{eca}}{V'_{isf}}(c_{ec} - c_{isf}) - \frac{PS_{pc}}{V'_{isf}}(c_{isf} - c_{pc}) + D_{isf}\frac{\partial^2 c_{isf}}{\partial z^2} - G_{isf}, \qquad (8.57)$$

where the subscript pc denotes parenchymal cell, D_{isf} is the ISF space dispersion coefficient, and G_{isf} is a consumption term.

In the parenchymal cell,

$$\frac{\partial c_{pc}}{\partial t} = +\frac{PS_{pc}}{V'_{pc}}(c_{isf} - c_{pc}) + D_{pc}\frac{\partial^2 c_{pc}}{\partial z^2} - G_{pc}, \tag{8.58}$$

where G_{pc} is the consumption term for the parenchymal cell space. Simulation and analysis of data on transport of labeled solutes in the heart based on Equations (8.55)–(8.58) are presented, for example, in Gorman *et al.* [72] and Schwartz *et al.* [182].

8.2.3 *Whole-organ metabolic modeling*

Equations of the sort developed in the previous section are useful in simulating the transport of solutes in tissues and organs. Typical applications make use of simulating multiple parallel pathways (made up of multiple axially distributed models) to account for heterogeneities in flow and other variables observed in tissues. In addition, this basic model formulation can serve as the basis of integrated models of transport and metabolism. For example, by accounting for oxygen transport using a model of this sort, and accounting for oxygen-dependent energy metabolism using the model introduced in the previous chapter, we can simulate the impact of ischemia (reduction in flow) on cardiac energy metabolite concentrations [14], as is illustrated in Figure 8.8.

8.3 Three-dimensional modeling

Modern computing resources make it possible to simulate transport and reaction in three space dimensions based on model geometries constructed to realistically represent anatomies of cells and tissues. Such applications allow researchers to simulate the intricate concentration fields in tissues generated by the non-idealized microvascular geometries, as illustrated in Figure 8.9. Three-dimensionally distributed modeling of reaction and transport in non-idealized geometry requires specialized numerical tools for computational modeling. A number of software packages, including the biologically focused Virtual Cell package [187], are available for modeling reaction–diffusion systems.

Concluding remarks

This chapter has developed the basic concepts of modeling diffusive transport and coupled diffusion, advection, and reaction in physiological systems. The emphasis

Parallel-pathway transport model:

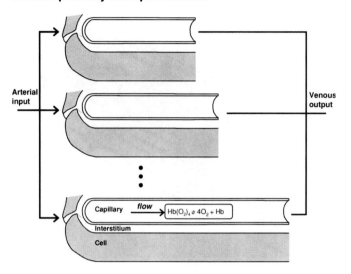

Simulation results:

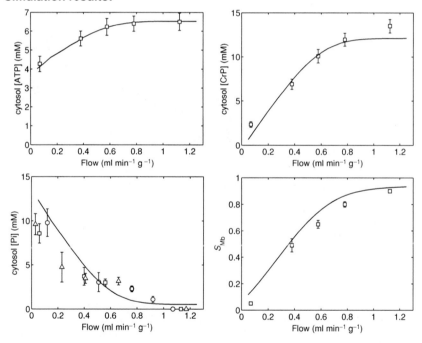

Figure 8.8 Integrated modeling of transport and metabolism. An axially distributed model of oxygen transport and metabolism [14], with parallel pathways used to simulate heterogeneity in path length and flow is used to simulate an experiment in which coronary flow is reduced and myoglobin saturation and concentrations of phosphate metabolites are measured. Data are from Zhang *et al.* [218].

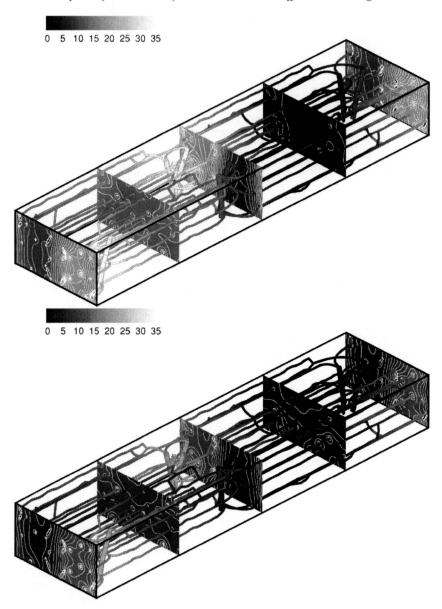

Figure 8.9 Predicted three-dimensional distributions of oxygen partial pressure in tissue. Model simulations from Tsoukias *et al.* [196] are based on a realistic model of the microvascular network associated with skeletal muscle. Predicted partial pressure distributions are illustrated for a control simulation (upper panel) and for a simulation of hemodilution – reduced hematocrit – plus addition of blood substitute (lower panel.) Predicted oxygen tension in mmHg is indicated by grayscale. Figure provided courtesy of Nicolaos Tsoukias.

here was on developing the appropriate equations to represent these processes for models of oxygen transport in tissue and the exchange of solutes between the blood and tissue. In analyzing the foundational Krogh–Erlang model, we saw how advection and hemoglobin binding of oxygen in blood, passive diffusion of oxygen into tissue, and metabolic consumption are coupled in tissue. The non-linear phenomena of facilitated diffusion were explored, also based on an idealized geometrical model. Time-dependent solute transport in tissue was studied based on the Sangren–Sheppard model.

All of these simple models have in common the fact that they are accessible to mathematical analysis, while more complex models are not. Yet whether one is dealing with idealized (analyzable) models or complex three-dimensional models, it is essential that the governing equations appropriately represent the underlying physical phenomena. To serve as a resource for this purpose, examples involving time-dependent and steady state transport, simple and facilitated diffusion, and passive permeations between regions were studied.

However, regardless of the range of phenomena investigated here, the scope of this field is much bigger than can be contained in a single chapter. For further study, a number of resources are available. The book by Middleman [141] is a classic that is still relevant more than 30 years after its original publication. For the analysis of solute transport in tissues, the handbook chapter by J. B. Bassingthwaighte and C. A. Goresky (who, along with a number of colleagues, introduced key techniques and technologies in this area) gives a deep review of the field through the mid 1980s, with an emphasis on analyzing transient washout of tracers from organs to probe the biochemical fates of the labeled solutes [11].

Exercises

8.1 Examine the model of passive flux through a membrane introduced in Section 3.2.4. How does the flux expression change if it is assumed that the transported solute (for example, oxygen) is consumed in the membrane?

8.2 Develop a numerical simulation to verify that the results of the asymptotic analysis in Section 8.1.3 are valid. Compare simulation results to the predictions illustrated in Figure 8.5.

8.3 Explore the behavior of the Sangren–Sheppard model for different values of the unitless parameter PS/F. Replot the simulations illustrated in Figure 8.7 for the limiting case as $PS/F \to \infty$. Explain why the model behaves in the way it does.

9

Constraint-based analysis of biochemical systems

Overview

When mechanistic information is available or obtainable for the components of a system, it is possible to develop detailed analyses and simulations of that system. Such analyses and simulations may be deterministic or stochastic in nature. (Stochastic systems are the subject of Chapter 11.) In either case, the overriding philosophy is to apply mechanistic rules to predict behavior. Often, however, the information required to develop mechanistic models accounting for details such as enzyme and transporter kinetics and precisely predicting biochemical states is not available. Instead, all that may be known reliably about certain large-scale systems is the stoichiometry of the participating reactions. As we shall see in this chapter, this stoichiometric information is sometimes enough to make certain concrete determinations about the feasible operation of biochemical networks.

Analysis of biochemical systems, with their behaviors constrained by the known system stoichiometry, falls under the broad heading *constraint-based analysis*, a methodology that allows us to explore computationally metabolic fluxes and concentrations constrained by the physical chemical laws of mass conservation and thermodynamics. This chapter introduces the mathematical formulation of the constraints on reaction fluxes and reactant concentrations that arise from the stoichiometry of an integrated network and are the basis of constraint-based analysis.

As we shall see, linear algebraic constraints arising from steady state mass balance form the basis of metabolic flux analysis (MFA) and flux balance analysis (FBA). Thermodynamic laws, while introducing inherent non-linearities into the mathematical description of the feasible flux space, allow determination of feasible reaction directions and facilitate the introduction of reactant concentrations to the constraint-based framework.

9.1 Motivation for constraint-based modeling and analysis

The traditional modeling approach in biochemistry is differential equation-based enzyme kinetics. Consequently, the majority of this book so far has been devoted to kinetic modeling. Many examples demonstrate the power and feasibility of kinetic modeling applied to a few enzymatic reactions at a time. It remains to be demonstrated, however, that that approach can be effectively scaled up to an in vivo system of hundreds of reactions and species with thousands of parameters [194]. More importantly, it is clear that the kinetic approach is not yet feasible for many large systems simply because the necessary kinetic information is not yet available.

The constraint-based approach facilitates prediction of system function based on limited information, such as only the stoichiometry of the reactions in a system. Doing so, this approach circumvents several difficulties currently at the center of the analysis of biological networks. It facilitates integration of experimental data of disparate types and from disparate sources, while increasing the accuracy in its prediction [107]. It does not require a-priori knowledge (or assumptions) regarding all of the mechanisms and parameters for a given system. However, when a-priori knowledge exists, such as data on enzyme kinetics or measurements of in vivo concentrations and fluxes, this knowledge may be introduced in the form of constraints, on equal footing with molecular genetic observations on the topology and information flow in a biochemical network. While traditional differential equation-based modeling provides predictions with relatively high information content, its predictions tend to be highly sensitive to assumptions and parameter values that may rest on shaky ground. The constraint-based approach, by design, provides a low level of false prediction, but can not often provide precise predictions.

At this most simple modeling level (considering only reaction stoichiometry), whole-genome metabolic models of several single-celled organisms have been developed [162, 171, 195, 202]. Thus, this approach to modeling and analysis holds great promise as a tool for large systems.

9.2 Mass-balance constraints

9.2.1 Mathematical representation for flux balance analysis

Metabolic fluxes are responsible for maintaining the homeostatic state of the cell. This condition may be translated into the assumption that the metabolic network functions in or near a non-equilibrium steady state (NESS). That is, all of the concentrations are treated as constant in time. Under this assumption, the biochemical fluxes are balanced to maintain constant concentrations of all internal metabolic species. If the stoichiometry of a system made up of M species and N fluxes is known, then the stoichiometric numbers can be systematically tabulated in a

$M \times N$ matrix, known as the stoichiometric matrix $\mathbb{S}$ [31, 32]. The S_{ij} entries of the stoichiometric matrix are determined by the stoichiometric numbers appearing in the reactions in the network. For example, if the jth reaction has the form:

$$v_1^j X_1 + \ldots + v_M^j X_M \rightleftharpoons \kappa_1^j X_1 + \ldots + \kappa_M^j X_M, \tag{9.1}$$

where X_i represents the ith species, then the stoichiometric matrix has the form $S_{ij} = \kappa_i^j - v_i^j$.

The fundamental law of conservation of mass dictates that the vector of steady state fluxes, $\mathbf{J}$, satisfies

$$\mathbb{S}\mathbf{J} = \mathbf{b}, \tag{9.2}$$

where $\mathbf{b}$ is the vector of boundary fluxes that transport material into and out of the system.[1] As an example, consider a simple network of three unimolecular reactions:

$$A \overset{J_1}{\rightleftharpoons} B, \quad B \overset{J_2}{\rightleftharpoons} C, \quad C \overset{J_3}{\rightleftharpoons} A \tag{9.3}$$

where all reactions are treated as reversible; left-to-right flux is the direction defined as positive. The network of Equation (9.3) is represented by the stoichiometric matrix:

$$\mathbb{S} = \begin{matrix} A \\ B \\ C \end{matrix} \begin{bmatrix} -1 & 0 & +1 \\ +1 & -1 & 0 \\ 0 & +1 & -1 \end{bmatrix}. \tag{9.4}$$

Next, consider that species A is transported into the system at rate b_A and species B is transported out at rate b_B. Then the mass-balance equations $\mathbb{S}\mathbf{J} = \mathbf{b}$ can be expressed:

$$\begin{bmatrix} -1 & 0 & +1 \\ +1 & -1 & 0 \\ 0 & +1 & -1 \end{bmatrix} \begin{bmatrix} J_1 \\ J_2 \\ J_3 \end{bmatrix} = \begin{bmatrix} -b_A \\ +b_B \\ 0 \end{bmatrix}. \tag{9.5}$$

Algebraic analysis of this equation reveals that mass-balanced solutions exist if and only if $b_A = b_B$. Equation (9.5) can be simplified to $J_2 = J_3 = J_1 - b_A$. Thus, mass balance does not provide unique values for the internal reaction fluxes. In fact, for this example, solutions exist for

$$J_1 \in (-\infty, +\infty), \quad J_2 \in (-\infty, +\infty), \quad J_3 \in (-\infty, +\infty). \tag{9.6}$$

---

[1] If the values of the boundary fluxes are not known, Equation (9.2) can be written as $\mathbb{S}\mathbf{J} = \mathbf{0}$ in which the boundary fluxes have been incorporated into the $M \times (N + N')$ matrix $\mathbb{S}$ where N' is the number of boundary fluxes [168].

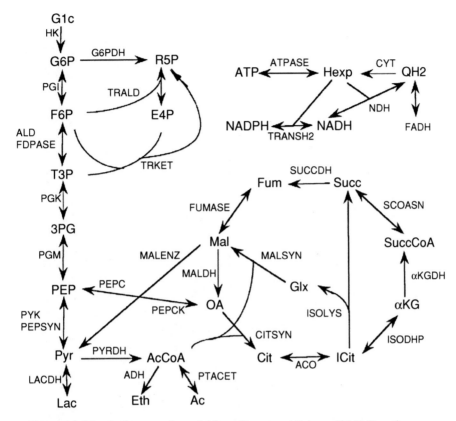

Figure 9.1 Metabolic network model from Varma and Palsson [204]. Reactions are indicated by enzyme abbreviations indicated; arrows represent reactions. Cofactors, such as ATP and NADH, are not shown for all reactions. Figure is from [204].

Equation (9.6) illustrates the fact that often the mass-balance constraint poses an underdetermined problem; typically it is necessary to identify additional constraints and/or to formulate a model objective function to arrive at meaningful estimates for biochemical fluxes.

9.2.2 Energy metabolism in E. coli

The research group of Bernhard Palsson and colleagues has developed a number of applications in constraint-based analysis, many using *Escherichia coli* as model organism. Some of their foundational work is reported in a 1993 publication on analysis of a portion of this organism's intermediary metabolism [204]. The reaction network studied is illustrated in Figure 9.1 and the associated stoichiometric matrix is given in Figure 9.2.

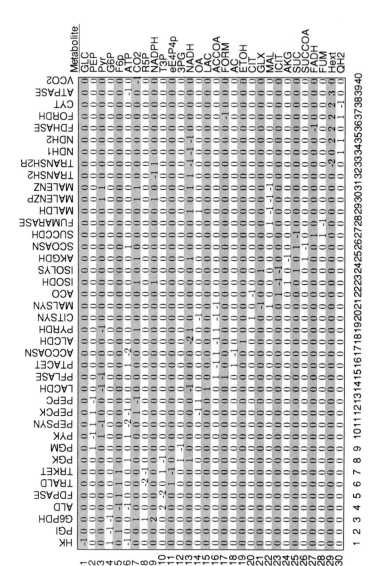

Figure 9.2 Stoichiometric matrix for model of Varma and Palsson [204]. Enzyme abbreviations, corresponding to reactions illustrated in Figure 9.1, are indicated for each column, corresponding to a reaction. Metabolite abbreviations for metabolite associated with rows of the matrix are indicated as well. VCO2 is a transport flux, transporting CO_2 out of the system. Other transport fluxes are not included here.

The metabolic network described by this matrix is simplified in the following ways. First, in some cases several reactions are lumped together when the network is so constrained that the fluxes through the combined reactions are the same in a steady state. For example the reaction of the third column in the matrix (labeled "G6PDH") is actually the combination of two serial reactions

$$G6P + NADP \rightleftharpoons 6PGL + NADPH \tag{9.7}$$

and

$$6PGL + NADP \rightleftharpoons CO_2 + R5P + NADPH, \tag{9.8}$$

where the following abbreviations are used: G6P: glucose-6-phosphate; 6PGL: 6-phospho-D-glucono-1,5-lactone; NADP: nicotinamide adenine dinucleotide phosphate (oxidized); NADPH: nicotinamide adenine dinucleotide phosphate (reduced). Since there are no other reactions considered in this model that involve 6PGL, this intermediate need not be considered in the stoichiometric matrix and both of the above reactions must have the same flux and are conveniently lumped together.

Second, some reactants that make up a free inter-converting pool are combined into a single metabolite [204]. For example the two glycolytic intermediates dihydroxyacetone phosphate and glyceraldehyde 3-phosphate are represented by the lumped species T3P (for triose-3-phosphate).

Third, for convenience some reactants are left out of the network. For example, rows are not included in the stoichiometric matrix for cofactor molecules NAD, NADP, and coenzyme A [204]. Leaving these cofactors out does not affect the overall flux balance of the reaction network.

The model also assumes that physiologically reversible reactions operate in either direction. Here the reactions PGI (reaction 2), TRALD (reaction 6), TRKET (reaction 7), PGK (reaction 8), PGM (reaction 9), LACDH (reaction 14), PFLASE (reaction 15), PTACET (reaction 16), ACO (reaction 22), ISODH (reaction 23), SCOASN (reaction 26), FUMARASE (reaction 28), TRANSH2 (reaction 32), and ATPASE (reaction 39) are considered reversible. The fluxes of all other reactions are constrained to operate in the direction defined as positive by the stoichiometric matrix.

Given the basic network structure we can ask a series of questions regarding the limits and flexibility of its operation. For example, we can ask the question illustrated in Figure 9.3: how much ATP may be generated from oxidizing one glucose molecule?

To address this question, we need to account for glucose input and ATP output in our stoichiometric model. If we wish to treat both of these fluxes as unknowns it is convenient to include them explicitly into the stoichiometric matrix. Therefore

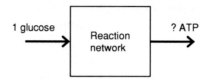

Figure 9.3 Illustration of the kind of question that can be addressed with stoichiometric constraint-based network analysis.

the matrix in Figure 9.2 is appended with columns corresponding to these transport fluxes. These columns have the following non-zero entries: $S_{1,41} = +1, S_{6,42} = -1$. Thus the 41st entry of the flux vector corresponds to glucose input and the 42nd entry corresponds to ATP output.

To compute the maximal ATP production per glucose consumed, we pose the following linear optimization problem: maximize J_{42} under the constraints:

$$\mathbb{S}\mathbf{J} = 0$$
$$J_{41} \leq 1$$
$$J_i > 0, \quad i \in I_{IR}, \tag{9.9}$$

where I_{IR} corresponds to the set of irreversible reactions. The constraint $J_{41} \leq 1$ sets the maximal glucose uptake to 1, in arbitrary units. So the optimal value of J_{42} will correspond to the maximal number of ATP molecules generated for each glucose consumed.

The problem described above is a *linear programming* problem – that is, an optimization problem with a linear objective function and linear constraints. Here the linear object is quite simple (maximize J_{42}). The linear constraints include both linear equalities ($\mathbb{S}\mathbf{J} = 0$) and inequalities ($J_i > 0$); yet both sets of constraints are linear in the sense that they involve no non-linear operations on the unknowns ($\mathbf{J}$).

Carrying out this optimization, we find that 18.667 ATP may be synthesized in this system for each glucose molecule consumed. The flux distribution at this optimal solution is illustrated in Figure 9.4. Here, 14.667 ATP/glucose are synthesized by the ATPase reaction of the oxidative phosphorylation system, 2 by glycolysis and 2 by the TCA cycle.

Analogous to the question of maximal ATP production, we can use linear programming to compute the maximum possible production of other cofactors. For example, this reaction network may be used in the cell to generate NADPH to be used in other metabolic pathways. To compute the maximal NADPH yield, we have to add a transport flux for NADPH to $\mathbb{S}$ by adding a new column with non-zero entry $S_{9,43} = -1$.

To compute the maximal NADPH production per glucose consumed, we pose the following linear optimization problem: maximize J_{43} (the NADPH output flux)

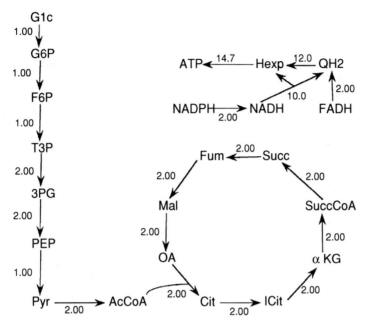

Figure 9.4 Flux values in model of Varma and Palsson [204] corresponding to maximal ATP production per glucose consumed. Figure is from [204].

under the constraints:

$$\mathbb{S}\mathbf{J} = 0$$
$$J_{41} \leq 1$$
$$J_{42} \geq 0$$
$$J_i > 0, \quad i \in I_{IR}. \tag{9.10}$$

Here we have added the additional constraint $J_{42} \geq 0$ to the list given in Equation (9.10). Since we have assigned the ATP output flux to the 42nd column, this constraint ensures that the ATP production is zero or positive, with no energy source other than glucose.

The computed optimal NADPH production is 12 molecules produced for every glucose molecule consumed. The flux distribution at this optimal solution is illustrated in Figure 9.5.

9.3 Thermodynamic constraints

In addition to the stoichiometric mass-balance constraint, constraints on reaction fluxes and species concentration arise from non-equilibrium steady state biochemical thermodynamics [91]. Some constraints on reaction directions are

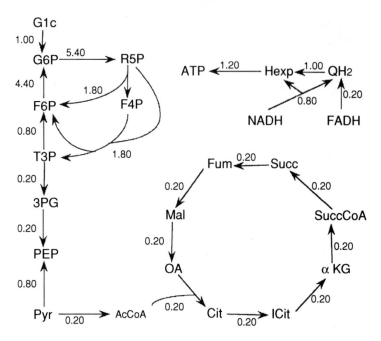

Figure 9.5 Flux values in model of Varma and Palsson [204] corresponding to maximal NADPH production per glucose consumed. Figure is from [204].

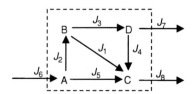

Figure 9.6 Small network example to illustrate the application of thermodynamic constraints in constraint-based analysis.

necessary to determine physically feasible and biologically reasonable flux patterns in constraint-based analysis. In a series of recent papers, we have demonstrated that application of thermodynamic constraints can be used to determine feasible flux directions from networks structure [17, 16, 83, 168, 167, 216].

9.3.1 The basic idea

The basic idea can be demonstrated based on the simple small-scale network illustrated in Figure 9.6.

The dashed line encloses the internal reactions, labeled J_1 through J_5; the processes J_6, J_7, and J_8 are transport fluxes that transport material into or out of the

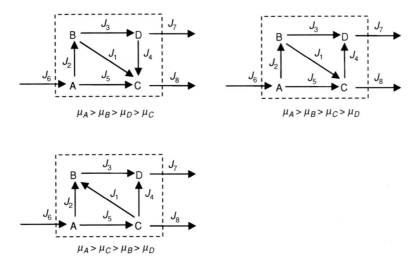

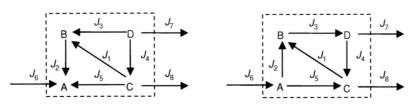

Figure 9.7 Enumeration of all feasible flux sign patterns for example from Figure 9.6. Figure adapted from [216].

Figure 9.8 Example of infeasible sign patterns. Figure adapted from [216].

system. If we assume that the directions of the transport fluxes are known or imposed on the system, then thermodynamics requires that the internal reaction direction be associated with feasible thermodynamic potentials for the reactions in the system. For this small system we can enumerate all possible flux direction patterns, which are illustrated in Figure 9.7.

Recall that there is a chemical potential associated with each reactant in the network. The first illustrated flux pattern is associated with chemical potentials obeying the inequalities $\mu_A > \mu_B > \mu_D > \mu_c$, where $\mu_A, \mu_B, \mu_C,$ and μ_D are the potentials associated with the reactants A, B, C, and D. The chemical potential inequalities associated with the other two flux patterns in Figure 9.7 are indicated in the figure.

Before digging further into the mathematical details, the reader may wish to try sketching alternative flux direction patterns. Doing so, one will find that no other pattern agrees with the transport flux directions and has a feasible thermodynamic potential. For example consider the direction patterns illustrated in Figure 9.8.

The first pattern in Figure 9.8 is disqualified because there is no way to achieve steady state balance of mass of A or D given this sign pattern. The second pattern is disqualified because there is no feasible chemical potential pattern associated with this flux pattern.

9.3.2 Mathematical details

Recall that for a biochemical reaction at a constant temperature T, such as

$$A + B \rightleftharpoons C + D, \tag{9.11}$$

there exists a forward flux J_+ and a backward flux J_-, with net flux $J = J_+ - J_-$. The concentrations of the reactants and products in this reaction are related to the chemical potential of the reaction via

$$\Delta\mu = \mu_C + \mu_D - \mu_A - \mu_B = k_B T \ln(J_-/J_+), \tag{9.12}$$

where k_B is the Boltzmann constant. Equivalently, for isothermal isobaric systems, we have [19][2]

$$\Delta G = RT \ln(J_-/J_+). \tag{9.13}$$

We see from these equations that J and $\Delta\mu$ (or ΔG) always have opposite signs, and are both zero only when a reaction is in equilibrium.

For a system of reactions, let $\boldsymbol{\mu}$ and $\boldsymbol{\Delta\mu}$ be the column vectors that contain the potentials for all the species and the potential differences for all the reactions, respectively. The chemical potential differences are computed

$$\boldsymbol{\Delta\mu} = \mathbb{S}^T \boldsymbol{\mu} \tag{9.14}$$

or $\Delta\mu_j = \mu_i S_{ij}$ using the Einstein notation introduced in the previous chapter. Next we recall the fundamental mass balance equation (Equation (9.2)):

$$\mathbb{S}\mathbf{J} = \mathbf{b}, \tag{9.15}$$

where here the transport fluxes are listed on the right-hand side in the vector $\mathbf{b}$ and the matrix $\mathbb{S}$ contains the stoichiometry of only the internal reactions. We combine Equations (9.14) and (9.15) as follows:

$$\boldsymbol{\mu}^T \mathbb{S} = \boldsymbol{\Delta\mu}^T$$
$$\boldsymbol{\mu}^T \mathbb{S}\mathbf{J} = \boldsymbol{\Delta\mu}^T \mathbf{J} = \boldsymbol{\mu}^T \mathbf{b}, \tag{9.16}$$

yielding the equality between heat dissipated per unit time by the internal reactions $(-\boldsymbol{\Delta\mu}^T \mathbf{J})$ and rate at which chemical energy is supplied by the boundary fluxes

---

[2] Recall, this relationship was derived in Section 3.1.2.

$(-\boldsymbol{\mu}^T \mathbf{b})$. This is the statement of energy conservation (first law of thermodynamics) for this system [168]:

$$-\boldsymbol{\Delta\mu}^T \mathbf{J} = -\boldsymbol{\mu}^T \mathbf{b}$$
$$-\sum_J \Delta\mu_j J_j = -\sum_j \mu_j b_j. \tag{9.17}$$

If $\mathbf{r}$ is a vector that has the property $\mathbb{S}\mathbf{r} = 0$, then

$$\mathbf{r}^T \boldsymbol{\Delta\mu} = \mathbf{r}^T \mathbb{S}^T \boldsymbol{\mu} = 0. \tag{9.18}$$

More generally, if $\mathbb{R}$ is a matrix that contains a basis for the right null space of $\mathbb{S}$ (i.e., $\mathbb{S}\mathbb{R} = 0$), then

$$\mathbb{R}^T \boldsymbol{\Delta\mu} = \mathbb{R}^T \mathbb{S}^T \boldsymbol{\mu} = 0. \tag{9.19}$$

Using Equation (9.19), the thermodynamic constraint that there must exist a feasible thermodynamic driving force for any flux pattern can be expressed as follows:

There must exist a vector $\boldsymbol{\Delta\mu}$ for which
$$\mathbb{R}^T \boldsymbol{\Delta\mu} = 0 \tag{9.20}$$
where $\mathbb{R}$ contains a basis for the right null space of $\mathbb{S}$ and
$$\Delta\mu_j J_j < 0 \tag{9.21}$$
for every reaction in the system [17].

This theorem is a generalization of Kirchhoff's loop law and Tellegen's theorem in electrical circuit analysis[3].

For the example of Equation (9.3), Equation (9.20) is expressed as

$$[1\ 1\ 1]\begin{bmatrix} \Delta\mu_1 \\ \Delta\mu_2 \\ \Delta\mu_3 \end{bmatrix} = 0. \tag{9.22}$$

Here, the matrix $\mathbb{S}$ of Equation (9.4) has a one-dimensional right null space, for which the vector $[1\ 1\ 1]^T$ is a basis. Equation (9.22) corresponds to summing the reaction potentials about the closed loop formed by the reactions in Equation (9.3).

Under this constraint of Equation (9.21), the bounds on the fluxes of Equation (9.5) are narrowed from those of Equation (9.6) to:

$$J_1 \in (0, b_A), \quad J_2 \in (-b_A, 0), \quad J_3 \in (-b_A, 0). \tag{9.23}$$

---

[3] It can also be seen as a consequence of the first and second laws of thermodynamics.

Thus, in general, the thermodynamic constraint narrows the feasible flux space, but not necessarily to a unique solution. Knowledge of the boundary fluxes translates into constraints on the reaction directions. Thus, the feasible reaction directions are a function of an open system's (i.e., a cell's) interaction with its environment.

9.3.3 Feasible sign patterns

Identifying constraints on reaction directions is essential for applications of metabolic flux analysis. However, in many applications the procedure used for determining reaction directions is not concretely defined. Typically, a subset of the reactions in a model is assigned as irreversible and the feasible directions are assigned based on information in pathway databases [59]. In these applications, by treating certain reactions as implicitly unidirectional, biologically reasonable results can often be obtained without considering the system thermodynamics as outlined above.

As an alternative to ad hoc procedures for assigning reaction directions, it is possible to determine reactions directions from first principles based on the thermodynamic constraint defined in the preceding section and knowledge of the direction of the transport flux directions. In fact, it is possible to mathematically state the thermodynamic constraint of Equations (9.20) and (9.21) in an alternative form in terms of the sign pattern of the vector $\mathbf{J}$ [16].

To define the thermodynamic constraint in terms of sign patterns, it is first necessary to define the concept of a *sign vector*. A sign vector is a vector with possible entries 0, $+$, and $-$. The operation sign $(\cdot)$ is defined to return the sign vector associated with a vector of real numbers. For example sign$\{-0.1, +5, 0, -2.1\} = \{-, +, 0, -\}$.

Next, it is necessary to define the concept of *orthogonality* of sign vectors. Two sign vectors a and b are said to be orthogonal ($a \perp b$) if either (1) the supports of a and b have no indices in common, or (2) there is an index i for which a_i and b_i have the same signs and there is another index j ($j \neq i$) for which a_j and b_j have opposite signs. Given these definitions, the thermodynamic constraint may be stated as:

For a stoichiometric matrix $\mathbb{S}$ associated with the internal reactions of a given system, with right null space $\mathbb{R}$, the vector of internal fluxes $\mathbf{J}$ is thermodynamically feasible if and only if

$$\text{sign}(\mathbf{J}) \perp \text{sign}(\mathbf{r}) \qquad (9.24)$$

for every $\mathbf{r} \in \mathbb{R}$.

Internal reactions:

Reaction cycles:

Figure 9.9 All of the possible sign patterns of the internal reactions of the network of Figure 9.6 captured in the cycles C defined in Equation (9.25).

This alternative statement of the thermodynamic constraint is shown to be equivalent to Equations (9.20) and (9.21) by Beard *et al.* [16]. Here we illustrate the concept by returning to the example of Figure 9.6. The feasible sign patterns of Figure 9.6 are feasible because they are orthogonal to the sign vectors of all vectors belonging to the right null space $\mathbb{R}$.

All possible sign patterns in the right null space are captured in the set of three cycles:[4]

$$
C = \begin{array}{c} J_1 \\ J_2 \\ J_3 \\ J_4 \\ J_5 \end{array}
\begin{bmatrix}
- & - & 0 \\
0 & - & + \\
+ & 0 & + \\
+ & 0 & + \\
0 & + & -
\end{bmatrix}, \tag{9.25}
$$

where each row corresponds to a cycle, with signed entries corresponding to directions of reactions indicated. These cycles are illustrated in Figure 9.9.

Based on these sign patterns, which enumerate all possible sign patterns of vectors in $\mathbb{R}$, the sign pattern on the right-hand side in Figure 9.8 is judged infeasible because it is not orthogonal to the first sign pattern in Figure 9.9. Here the definition of sign orthogonality is apparent. The vectors are not orthogonal because the signs of all non-zero entries in the pattern

---

[4] It is not necessary to explicitly enumerate the cycles with sign exactly opposite to those listed in Equation (9.25).

are the same as the signs of those entries of the following pattern from Figure 9.8.

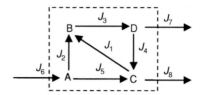

Therefore the above sign pattern is not thermodynamically feasible because it is not orthogonal to the sign pattern of a vector from the right null space of the stoichiometric matrix of internal reactions.

The above example is convenient for illustration because it is relatively small and involves only uni-unimolecular reactions. For large systems with arbitrary reaction stoichiometry, it turns out that the assignment of feasible reactions is an NP-complete computational problem [216]. Therefore, application of thermodynamic constraints in genome-scale problems is an area of ongoing research.

9.4 Further concepts in constraint-based analysis

9.4.1 Feasible concentrations from potentials

Introducing the chemical potential (or free energy) and the thermodynamic constraint provides a solid physical chemistry foundation for the constraint-based analysis approach to metabolic systems analysis. Treatment of the network thermodynamics not only improves the accuracy of the predictions on the steady state fluxes, but can also be used to make predictions on the steady state concentrations of metabolites. To see this, we substitute the relation between reaction Gibbs free energy $(\Delta_r G')_j$ of the jth reaction and the concentrations of biochemical reactants

$$(\Delta_r G')_j = (\Delta_r G'^o)_j + RT \sum_i S_{ij} \ln c_i \qquad (9.26)$$

into the requirement that $(\Delta_r G')_j J_j \leq 0$, resulting in the inequality

$$(\Delta_r G'^o)_j J_j + RT \sum_i S_{ij} \ln c_i J_j \leq 0. \qquad (9.27)$$

Here $(\Delta_r G'^o)_j$ is the standard transformed equilibrium Gibbs free energy for reaction j, which may be obtained from a standard chemical reference source.

If a flux vector $\mathbf{J}$ is thermodynamically feasible, then there exist concentrations c_i that satisfy the above inequality. In fact, Equation (9.27) defines a feasible space for the metabolites concentrations as a convex cone in the log-concentration space.

If the set of feasible concentrations is empty, then the vector $\mathbf{J} = \{J_j\}$ is thermodynamically infeasible.

9.4.2 Biochemical conductance and enzyme activity

From the traditional biochemical kinetics standpoint, both steady state biochemical concentrations and reaction fluxes are predictable from known enzyme reactions with appropriate rate constants and initial conditions. In steady state, the fluxes are computable from the concentrations of the reactants and products. However, a realistic challenge we confront is that our current understanding of the reaction mechanisms and measurements of rate constants are significantly deficient. From the standpoint of constraint-based analysis, the ratio between the J and $\Delta_r G'$ of a particular reaction is analogous to the conductance, which can be shown to be proportional to the enzyme activity of the corresponding reaction. We emphasize that the magnitudes of both J and $\Delta_r G'$ are functions of the reaction networks topology. Therefore, each one alone will not be sufficiently informative of the level of enzyme activity (i.e., the level of activity due to gene expression or post-translational modification).

9.4.3 Conserved metabolite pools

In addition to the constraint on concentrations imposed by Equation (9.27), a reaction network's stoichiometry imposes a set of constraints on certain conserved concentration pools [2]. These constraints follow from the equation for the kinetic evolution of the metabolite concentration vector:

$$\mathbb{P}d\mathbf{c}/dt = \mathbb{S}\mathbf{J}, \tag{9.28}$$

where $\mathbb{P}$ is a diagonal matrix, with diagonal entries corresponding to the partition coefficients, or fractional intracellular spaces, associated with each metabolite in the system. In Equation (9.28), columns corresponding to the boundary fluxes have been grouped into the matrix $\mathbb{S}$. Here, the vector $\mathbf{J}$ includes both internal reaction fluxes and boundary fluxes. The left null space of the matrix $\mathbb{P}^{-1}\mathbb{S}$ may be computed and a basis for this space stored in a matrix $\mathbb{L}$, such that:

$$\mathbb{L}d\mathbf{c}/dt = \mathbb{L}\mathbb{P}^{-1}\mathbb{S}\mathbf{J} = \mathbf{0}. \tag{9.29}$$

It follows from Equation (9.29) that the product $\mathbb{L}\mathbf{c}$ remains constant and defines a number of conserved pools of metabolic concentrations. For example, if we were to consider the glycolytic series as an isolated system, with no net flux of phosphate-containing metabolites into or out of the system, then as phosphate is shuttled among the various metabolites, the total amount of phosphate in the system is conserved.

9.4.4 Biological objective functions and optimization

The simple examples introduced above make use of computers to optimize mathematical objective functions, given defined mathematical constraints. Such an optimization problem is defined by maximizing a flux under the constraints of Equation (9.9). While we do not go into the mathematical/computational details of how to find solutions to optimization problems here, readers may find an accessible introduction to optimization theory, which represents a mature field of modern applied mathematics, elsewhere. (See for example, Strang's *Introduction to Applied Mathematics* [190].) We also note that optimization theory has been a major mathematical engine behind bioinformatics and genomic analysis. Constraint-based approaches have also been very successful in biological structural modeling ranging from distance geometry calculations for protein structure prediction from NMR to structural determination of large macromolecular complexes.

Mathematical optimization deals with determining values for a set of unknown variables $x_1, x_2, \ldots, x_n$, which best satisfy (optimize) some mathematical objective quantified by a scalar function of the unknown variables, $F(x_1, x_2, \ldots, x_n)$. The function F is termed the *objective function*; bounds on the variables, along with mathematical dependencies between them, are termed *constraints*. Constraint-based analysis of metabolic systems requires definition of the constraints acting on biochemical variables (fluxes, concentrations, enzyme activities) and determining appropriate objective functions useful in determining the behavior of metabolic systems.

Therefore, the objective functions used play a crucial role in constraint-based analysis. A given objective function can be thought of as a mathematical formulation of a working hypothesis for the function of a particular cell or cellular system. These objective functions should not be considered to be as theoretically sound as the physicochemical constraints; but they may be informative and biologically relevant. They can serve as concrete statements about biological functions and powerful tools for quantitative predictions, which must be checked against experimental measurements. One of the surprising discoveries in constraint-based modeling is how well certain simple objective functions have described biological function.

That a cell functions precisely following some rule of optimality is, of course, highly suspect. There may be an evolutionary argument in favor of certain objective functions, but the ultimate justification lies in the correctness of its predictions.

In this sense, the constraint-based optimization approach provides a convenient way to efficiently generate quantitative predictions of biological hypotheses formulated in terms of objective functions.[5] The value of this approach is in facilitating the systematic prediction-experimental verification-hypothesis modification cycle, ideally leading to new discoveries.

For bacterial cells, growth rate (rate of biomass production) has been a widely used objective function. This objective is constructed as a net flux out of the cell of the components of biomass (amino acids, nucleotides, etc.) in their proper stoichiometric ratios, which translates into a linear function of the reaction fluxes. Based on this elegant paradigm, predictions from flux-balance analysis (FBA) of the fate of the *E. coli* MG1655 cell following deletions of specific genes for central metabolic enzymes have been remarkably accurate [48]. When combined with thermodynamic analysis (or *energy balance analysis*), it has been shown [17] that cells with non-essential genes deleted can redirect the metabolic fluxes under relatively constant enzyme activity levels, with few changes due to gene expression regulation and/or post-translational regulation. The FBA/EBA combined approach predicts which enzyme must be up regulated, which must be down regulated, and which reactions must be reversed, given perturbations to the genotype and/or cellular environment. Using this combined approach, a clear relation is established between the enzyme regulation and constraint-based analysis of metabolism.

Different objective functions can be used in studying other biological systems and problems. When addressing cellular metabolic pathway regulation, robustness has proven to be a useful objective. Robustness can be defined as a measure of the amount that metabolic fluxes, steady state concentrations, or enzymes activities change following perturbations. For example, minimization of flux adjustment has been used to model the metabolic response of *E. coli* JM101 with pyruvate kinase knockout [184]. In this study, it is assumed that the cell acts to maintain its wild-type flux pattern in response to the challenge imposed by a gene knockout. However, a minimal change in the flux pattern may require an unrealistic level of metabolic control. We have shown that the objective of minimal changes in enzyme activities predicts the key regulatory sites in switching between glycogenic and gluconeogenic operating modes in hepatocytes [18]. This approach facilitates inverse analyses, where the regulatory system is treated as a black box and control mechanisms are identified from measurements of the inputs and outputs to the system.

---

[5] Objectives that have proved useful include optimal energetic efficiency or optimal cell growth rate [48, 204].

9.4.5 Metabolic engineering

Metabolic engineering is "the purposeful design of metabolic networks" [189] in microbial organisms for specific tasks. Thus the aim of metabolic engineering is different from that of biological research aimed at understanding existing organisms [8, 189]. In metabolic engineering, one is more interested in the capacity and optimal behavior of biological hardware rather than its natural function per se. For this reason the metabolic engineering community has been a major developer and proponent of the constraint-based analysis approach.

9.4.6 Incorporating metabolic control analysis

As briefly outlined in Section 6.3, one of the theoretical frameworks in quantitative analysis of metabolic networks is metabolic control analysis. In metabolic control analysis, the enzyme elasticity coefficients provide empirical constraints between the metabolites' concentrations and the reaction fluxes. These constraints can be considered in concert with the interdependencies in the $\mathbf{J}$ and $\mathbf{c}$ spaces that are imposed by the network stoichiometry. If the coefficients $\epsilon_k^i = (c_k/J_i)\partial J_i/\partial c_k$ are known, then these values bind the fluxes and concentrations to a hyperplane in the $(\mathbf{J}, \mathbf{c})$ space.

Concluding remarks

One can view biochemical systems as represented at the most basic level as networks of given stoichiometry. Whether the steady state or the kinetic behavior is explored, the stoichiometry constrains the feasible behavior according to mass balance and the laws of thermodynamics. As we have seen in this chapter, some analysis is possible based solely on the stoichiometric structure of a given system. Mass balance provides linear constraints on reaction fluxes; non-linear thermodynamic constraints provide information about feasible flux directions and reactant concentrations.

Applying mass-balance and thermodynamic constraints typically leaves one without a precisely defined (unique) solution for reaction fluxes and reactant concentration, but instead with a mathematically constrained feasible space for these variables. Exploration of this feasible space is the purview of constraint-based analysis. It has so far been left unstated that any application in this area starts with the determination of the reactions in a system, from which the stoichiometric matrix arises. This first step, network reconstruction, integrates genomic and proteomic data to determine carefully the enzymes present in an organism, cell, or subcellular compartment. The network reconstruction process is described elsewhere [107].

Exercises

9.1 Consider the following uni-unimolecular reaction network.[6]

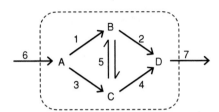

Construct the stoichiometric matrix for this system, given the reaction numbering defined in the figure. Assume that internal reactions 1 through 4 are irreversible with feasible directions indicated in the figure. Reactions 6 and 7 are transport reactions, also irreversible with directions indicated. If the maximum uptake rate of A is 1 (in arbitrary units), what is the maximal output of D? Given that production of D is optimal, is the internal flux distribution unique?

9.2 Set up the previous exercise as a linear programming problem and solve using a computer. (The built-in Matlab utility *linprog* is one possible package to use.) Investigate deleting reactions and applying additional linear (inequality) constraints to the reaction fluxes in this example. Is the production of D in this example robust to changes in the network structure?

9.3 Use the flux balance constraint and the thermodynamic feasibility to show that for a closed chemical reaction system, i.e., $\mathbf{b} = \mathbf{0}$ in Equation (9.2), the only possible steady state is $\mathbf{J} = \Delta\mu = \mathbf{0}$. That is, the steady state of a closed chemical reaction system is necessarily a chemical equilibrium.

9.4 Use linear programming to compute the optimal solutions illustrated in Figures 9.4 and 9.5 for the reaction network defined by the stoichiometric matrix given in Figure 9.2. (Note that to obtain the reported results, the model must include the transport fluxes and the irreversibility constraints defined in Section 9.2.2.)

9.5 Is the flux sign pattern in exercise 9.1 thermodynamically feasible? Would it be thermodynamically feasible for reaction 4 to operate in the reverse direction? Why or why not?

---

[6] This reaction network is motivated from an example in Joyce and Palsson [107].

10

Biomacromolecular structure and molecular association

Overview

The theories of chemical kinetics and thermodynamics – largely the theoretical foundation of this book – are concerned primarily with describing how distinct molecular states are distributed temporally and among ensembles of molecules. Based on these theories, previous chapters have developed mathematical models of biochemical systems that account for molecular state changes that include chemical reactions, conformational changes, and non-covalent binding interactions, with little attention paid to the underlying structural details. Yet there is much to be learned from examining the structural basis of biomolecular states and functions. After all, structure and function are inseparable.

In fact, a tremendous amount of information is available on the structures of biological macromolecules; descriptions of structures of proteins and nucleic acids make up major portions of modern textbooks in biochemistry and molecular biology. The Protein Data Bank and the Nucleic Acid Database are online archives that contain sequence and structural data on thousands of specific molecules and complexes of molecules. This structural information comes from in vitro experiments, with structures inferred from the x-ray diffraction patterns of crystallized molecules, spectroscopic measurements using multi-dimensional nuclear magnetic resonance, and a host of other methodologies.

Even though in vitro experiments necessarily remove biomolecules from the cellular environment, the structures and dynamics of individual macromolecules provide insights to their biological functions. For example, structural studies have revealed that the protein hemoglobin is made up of four interacting subunits, two α subunits and two β subunits. Furthermore, each subunit has two distinct conformational states, called the R state and the T state, and the energy of interaction between two neighboring subunits in different states is different from that of two subunits in the same state. This phenomenon is the structural basis of the observed allosteric cooperativity in oxygen binding to hemoglobin [21]. In sickle-cell anemia, a single mutation in the β subunit of hemoglobin causes the globular protein to form fibers, leading to the sickle shape of red blood cells [21]. Thus the structural dysfunction at the molecular scale can underlie disease.

In some cases, the structure essentially is the function. The most celebrated example is the structure of the DNA double helix, which "immediately suggests a possible copying mechanism for the genetic material" [207]. More often than not, however, a definitive understanding of biological and physiological function cannot be reached in terms of isolated biomolecular structures and dynamics alone. Many important biological processes involve many interacting molecules. Hence, the serious attention paid in the current era to networks of macromolecular interaction.

Theoretical approaches to structural biophysics, like the theories of transport and reaction kinetics explored in other chapters of this book, are grounded in physical chemistry concepts. Here we explore a few problems in molecular structural dynamics using those concepts. The first two systems presented, helix-coil transitions and actin polymerization, introduce classic theories. The material in the remainder of the chapter arises from the study of macromolecular interactions and is motivated by current research aimed at uncovering and understanding how large numbers of proteins (hundreds to thousands) interact in cells [7].

10.1 Protein structures and α-helices

The structure of a biological macromolecule – a large protein or nucleic acid, for example – determines its static and dynamic properties in aqueous solution and thus in large part its biological function(s).[1] The term "structure" should be understood broadly: the nucleotide and peptide sequences of DNA and protein molecules are components of their chemical structures. Beyond the primary sequences, biological macromolecules take on the complex three-dimensional structures of the sort that one will find illustrated in any biology textbook. These structures give us a static picture of biological macromolecules. Yet it is usually the dynamic character of a macromolecule that is the key to understanding the link between its structures and its functions.

Both the structures and the dynamics of macromolecules are studied in terms of statistical thermodynamics. In the following section, we introduce the helix-coil transition theory that accounts for formation of the ubiquitous α-helical structure of peptide chains in aqueous solution. To a large extent, current research on protein

---

[1] In the era of systems biology, great attention is paid to the structures of networks of reactions and interacting molecules (i.e., the topological connectivities). In some ways network structures have replaced molecular structures as the central object of biological attention.

folding, combining simple ideas from statistical thermodynamics with powerful computation, is built upon insight developed from helix-coil transition theory.

10.1.1 The theory of helix-coil transition

Images of three-dimensional protein structures are often rendered using cartoon-like ribbons to indicate common features of the carbon backbone, such as α-helices and β-sheets. Casually browsing the structures illustrated in a biology textbook or an online database, one is struck by the frequent appearance of the right-handed helical structures known as α-helices. The structural features of α-helices are illustrated in Figures 10.1 and 10.2.

As is apparent from Figure 10.1, an α-helical structure imposes fairly rigid constraints on the relative positions of successive residues in a peptide chain. Thus there is a loss of entropy that must be overcome energetically in order for an α-helix to form. To explain the underlying biophysics of this system, John Schellman introduced a theory of helix-coil transitions that is motivated by the Ising model for one-dimensional spin system in physics [180, 170].

We can characterize the primary structural degrees of freedom in a polypeptide chain (ignoring side-chain conformations) by the two dihedral angles (ϕ and ψ) associated with the C_α of each residue. As illustrated in Figure 10.1, these angles determine the relative positions of two successive peptide bonds. A chain of M residues has $M - 1$ sets of (ϕ, ψ) angles.[2] When three consecutive (ϕ, ψ) combinations fall within the relatively narrow region of α-helical structure indicated in Figure 10.1, a hydrogen bond may form between the carbonyl group of the ith peptide and the amide group of the $(i + 4)$th peptide, as illustrated in Figure 10.2. Thus the formation of a hydrogen bond requires three consecutive helical states for dihedral angles associated with α carbons.

Let us consider a polypeptide chain with M residues, characterized by $N = M - 1$ (ϕ, ψ) pairs. The helix-coil theory characterizes each dihedral pair as either helical (within the α-helical region) or coil (not helical).[3] Thus there are three possible energetic states: coil, non-hydrogen-bonded helical, and hydrogen-bonded helical. We denote the energies of these states as E_c, E_{nh}, and E_{hh}, respectively. Since we do not want to count each hydrogen bond twice, we use the convention that only the residue nearer the C-terminus (i.e., the residue on the right within a hydrogen bond in Figure 10.2B) is counted as hydrogen bonded.

---

[2] There is no (ϕ, ψ) set associated with the N-terminal α carbon. The (ϕ, ψ) pair associated with the C-terminal α carbon defines the relative orientation of the C-terminal peptide bond and the terminal COOH group.

[3] The term *coil* refers to a random unstructured chain in polymer science. The term may be confusing because in common use the word *coil* does not necessarily invoke a picture of a random structure.

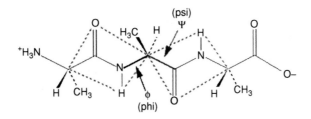

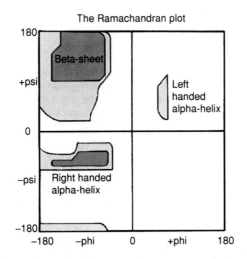

Figure 10.1 Basic polypeptide geometry. The upper panel shows a short peptide sequence of three amino acids joined by two peptide bonds. A relatively rigid planar structure, indicated by dashed lines, is formed by each peptide bond. The relative positions of two adjacent peptide bond planes is determined by the rotational dihedral angles ϕ and ψ associated with the C_α of each peptide. The relative frequency of ϕ and ψ angles occurring in proteins observed in a database of structures obtained from crystallography is illustrated in the lower panel. In this plot, called a Ramachandran plot, the shaded regions denote (ϕ, ψ) pairs that occur with some frequency in the database. The white region corresponds to (ϕ, ψ) values not observed in crystal structures of proteins due to steric hindrance. The most commonly occurring (ϕ, ψ) values correspond to β-sheets and right-handed α-helices. Left-handed α-helical conformations occur with lower frequency.

Given the energies associated with the possible states, the Boltzmann probability law tells us that the respective probabilities of a residue being in a coil state, a non-hydrogen-bonded helical state, and hydrogen-bonded helical state are proportional to $e^{-E_c/k_BT} = 1$, $v = e^{-E_{nh}/k_BT}$, and $w = e^{-E_{hh}/k_BT}$. Thus $w > 1$ implies that helix formation is energetically favorable. The formation of hydrogen bonds is energetically favorable and can stabilize the formation of helices [129, 170].

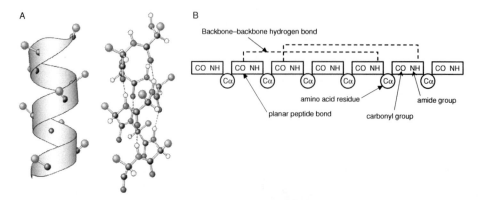

Figure 10.2 A polypeptide can form a helical structure in which the carbonyl group of the ith peptide bond is aligned with the amide group of the $(i + 4)$th peptide bond and forms a hydrogen bond (A) if all the three amino acid residues, i.e., C_α, are in their helical state (B). Image in (A) reprinted from Berg *et al.* [21] with permission.

We call the relative probability weights $(1, v,$ and $w)$ the Boltzmann weights – they are the non-normalized probabilities of possible states. For a polypeptide chain, its conformation may be described by a sequence of N states as shown in the table below. The corresponding total energy of the conformation is simply the sum of the energies for each residue. Then the corresponding Boltzmann weight will be computed as the product of the weights corresponding to the states.

Sequence of states	Boltzmann weight
$cchhhhcc\ldots$	$1 \cdot 1 \cdot v \cdot v \cdot w \cdot w \cdot 1 \cdot 1 \cdots$
$hhhchhhh\ldots$	$v \cdot v \cdot w \cdot 1 \cdot v \cdot v \cdot w \cdot w \cdots$

Here states c and h correspond to coil and helical, respectively. Determination of whether a helical state is hydrogen bonded or not, and assignment of the Boltzmann weight for a given state, requires knowing the states of the two preceding residues. The state hhh has energy $v \cdot v \cdot w$ because the sequence of three helical (ϕ, ψ) pairs allows for the formation of one hydrogen bond.

Computing the partition function for an N-state chain requires enumerating all possible states. The clever trick associated with the helix-coil transition theory is to generalize this calculation using the statistical-weight matrix:

$$
\mathbb{M} = \begin{array}{c|cccc}
 & c\bar{c} & h\bar{c} & c\bar{h} & h\bar{h} \\
\hline
cc & 1 & 0 & v & 0 \\
hc & 1 & 0 & v & 0 \\
ch & 0 & 1 & 0 & v \\
hh & 0 & 1 & 0 & w
\end{array}
\qquad (10.1)
$$

in which each row corresponds to the first pair of states in a three-state sequence, and each column corresponds to the second two states in the sequence. The entries in the matrix are the Boltzmann weights given to the third state. For example, the entry in the row hc and column $c\overline{h}$ corresponds to the sequence $hc\overline{h}$. Hence, the Boltzmann weight for the $\overline{h}$ is v. The zeros in the matrix represent incompatibility between a given column and a given row.

For an N-state system, the first two states can take the form cc, hc, ch, or hh, with corresponding Boltzmann weights 1, v, v, and v^2. Putting these weights into a vector in the respective order that corresponds to the columns of $\mathbb{M}$, and multiplying by $\mathbb{M}$ yields a vector that lists the weights of all possible three-state sequences:

$$\begin{bmatrix} 1, & v, & v, & v^2 \end{bmatrix} \mathbb{M} = \begin{bmatrix} 1+v, & v+v^2, & v+v^2, & v^2+v^2w \end{bmatrix}. \tag{10.2}$$

These entries can be summed by taking the product

$$\begin{bmatrix} 1, & v, & v, & v^2 \end{bmatrix} \mathbb{M} \begin{bmatrix} 1 \\ 1 \\ 1 \\ 1 \end{bmatrix} = \begin{bmatrix} 1+v, & v+v^2, & v+v^2, & v^2+v^2w \end{bmatrix} \begin{bmatrix} 1 \\ 1 \\ 1 \\ 1 \end{bmatrix}$$

$$= 1+v+v+v^2+v+v^2+v^2+v^2w, \tag{10.3}$$

where these eight terms correspond to the 2^3 possible three-state sequences ccc, hcc, chc, hhc, cch, hch, chh, and hhh.

This result is generalized for an N-state polypeptide ($N+1$ residues):

$$Z(v, w) = \begin{bmatrix} 1, & v, & v, & v^2 \end{bmatrix} \mathbb{M}^{N-2} \begin{bmatrix} 1 \\ 1 \\ 1 \\ 1 \end{bmatrix}, \tag{10.4}$$

which gives us the partition function – the weighted sum of all possible states. From Equation (1.21), we can compute $-k_B T \ln Z$, which is the free energy of the entire polypeptide chain, with the all-coil conformation $cccc \cdots c$ as the reference state. In other words, the probability of the all-coil state is Z^{-1}.

The partition function of Equation (10.4) is the source of much information on the thermodynamics of helix-coil conformations. We note that $Z(v, w)$ is a polynomial of w of order $N-2$:

$$Z(v, w) = \sum_{k=0}^{N-2} a_k(v)w^k, \tag{10.5}$$

in which $a_k(v)w^k$ is the Boltzmann weight for the peptide with exactly k hydrogen-bonded helical residues. Thus, $a_k w^k$ is proportional to the probability of the

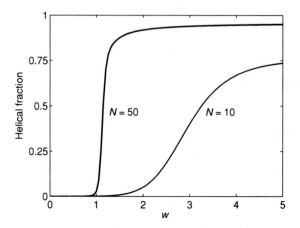

Figure 10.3 The helix-coil transition curves according to Equation (10.6), with $v = 0.01$. In realistic laboratory experiments, one can change the temperature or the solvent conditions to change w, hence observing the transition.

polypeptide having k hydrogen bonds. To obtain the average number of hydrogen bonds, we have

$$\langle n_h \rangle = \frac{\sum_{k=0}^{N-2} k a_k w^k}{\sum_{k=0}^{N-2} a_k w^k} = \left(\frac{\partial \ln Z}{\partial \ln w} \right)_v. \tag{10.6}$$

Figure 10.3 shows the number of hydrogen bonds in a peptide, normalized by the length of the polypeptide $\langle n_h \rangle / N$. We see that with increasing length of a peptide, the coil to helix transition becomes sharper, more cooperative.

Matlab computer codes for the helix-coil model are given below. First, we introduce a function that computes and returns the vector $[1, v, v, v^2] \, \mathbb{M}^k$, given the inputs v, w, and k:

```
function LM = lm(v,w,k)

leftv = [1, v, v, v^2];
M = [1, 0, v, 0; 1, 0, v, 0; 0, 1, 0, v; 0, 1, 0, w];
LM = leftv*M^k;
```

Next, a function that computes and returns the column vector

$$\mathbb{M}^k \begin{bmatrix} 1 \\ 1 \\ 1 \\ 1 \end{bmatrix}$$

```
function RM = rm(v,w,k)

rightv = [1, 1, 1, 1]';
M = [1, 0, v, 0; 1, 0, v, 0; 0, 1, 0, v; 0, 1, 0, w];
RM = M^k*rightv;
```

The following function calls the above two functions and computes the helical fraction of a chain of length N given input parameters v and w.

```
% This function computes the mean number of
% h-bonds according to the helix-coil theory

function HCONTENT = hcontent(v,w,N)

% Computing the partition function Z
rightv = [1, 1, 1, 1]';
Z = lm(v,w,N-2)*rightv;

% The derivative of matrix M with respect to w
% is the matrix D given below:
D = [0, 0, 0, 0; 0, 0, 0, 0; 0, 0, 0, 0; 0, 0, 0, 1];

% The dZ/dw can be carried out by the chain rule
dZdw = 0;
for i = 0:N-3
    dZdw = dZdw + lm(v,w,i)*D*rm(v,w,N-3-i);
end

% The mean number of h-bonds
HCONTENT = w*dZdw/Z;
```

This function can be called in Matlab using the following syntax to compute the number of hydrogen bonds as a function of the w parameter, with $v = 0.01$ and $N = 50$.

```
function Hvsw = hvsw()

v = 0.01;
N = 50;
Hvsw = [];
for w = 0:0.05:5
```

```
    Hvsw = [Hvsw,HCONTENT(v,w,N)];
end

w = [0:0.05:5];
plot(w,Hvsw);
```

We note that the α-helical structure of a polypeptide chain is a component of secondary structures in proteins. An α-helix is determined by the local orientations of neighboring peptide bonds. Thus the theory of helix-coil transitions deals with only one way in which hydrogen bonds along a peptide chain can contribute to secondary structure. The theory completely neglects any interaction between amino acids more than four residues apart. It also neglects interactions involving the side chains of the amino acids. Taking these tertiary interactions into consideration in a theory is a task that is outside of the scope of this chapter. Still, the helix-coil theory has found wide applications including local secondary structures of proteins and soluble, bio-active small peptides. It has also been applied to study the structure of DNA [160].

10.2 Protein filaments and actin polymerization

The previous section dealt with the secondary structure of a peptide chain and demonstrated how the inclination to form α-helices, as a function of chain length and hydrogen-bonding energy, is explained on the basis of the thermodynamic theory of Chapter 1. In this section we examine how a higher level of structural organization – the association of certain globular proteins to form filaments – is described by the kinetic theory of Chapter 2. For example globular actin monomers (G-actin units) can string together to form a filament called F-actin, which is made up of two parallel strings of G-actin monomers twisted about one another as shown in Figure 10.4A. F-actin makes up an important component of the cytoskeleton and is involved in cell motility. Another important protein filament in cells is the microtubule that is made of tubulin monomers. Microtubules play key roles in intracellular transport of organelles and in cell division.

The processes of polymerization of protein monomers into filaments, such as actin and tubulin, can be investigated based on a simple kinetic model. To develop a model, we let A_1 represent the monomer and A_ℓ represent the filaments made up of ℓ subunits. Our simple kinetic model has two distinct steps:

$$A_1 + A_1 \underset{k_{-1}^*}{\overset{k_{+1}^*}{\rightleftharpoons}} A_2, \quad A_1 + A_\ell \underset{k_{-1}}{\overset{k_{+1}}{\rightleftharpoons}} A_{\ell+1}, \tag{10.7}$$

where the first reaction involves the association of two monomers into a dimer and the second reaction represents elongation of a polymer of length $l \geq 2$. The rate

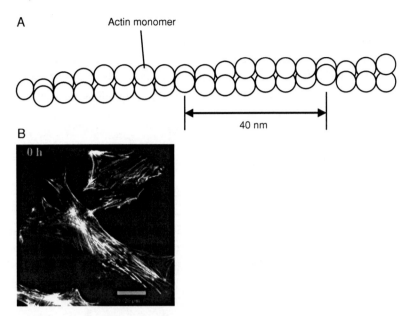

Figure 10.4 (A) A schematic of globular actin monomer forming a protein filament, called F-actin. This filament is one of the important components of muscle cells, as well as the cytoskeleton of other cells. (B) Oriented actin filaments inside a fibroblast cell, called stress fibers, seen through a fluorescence microscope. Image obtained from Nguyen *et al.* [148] and reprinted with permission.

constants k_{+1} and k_{-1} are associated with the polymerization and depolymerization rates for the elongation of the polymer. The first reaction, called nucleation, has rate constants k_{+1}^* and k_{-1}^* different from all other reaction steps. Generally the nucleation step is less favorable than elongation; this phenomenon is reflected in the rate constants by setting $\frac{k_{+1}^*}{k_{-1}^*} \ll \frac{k_{+1}}{k_{-1}}$. As is the case for the helix-coil system, the low likelihood of nucleation can be compensated for in long polymers.[4]

10.2.1 Nucleation and critical monomer concentration

If the ratio

$$\frac{k_{+1}^*/k_{-1}^*}{k_{+1}/k_{-1}}$$

---

[4] In the twisting double-helical actin chain, the nucleation step involves only one monomer–monomer interaction while each monomer placed on the chain following nucleation involves two monomer–monomer bonds. In the helix-coil theory, the formation of two consecutive helical states represents the nucleation step that allows for a third helical state and hydrogen bond formation.

is very small, then the formation of a nucleus is an extremely unlikely event. Therefore, when there is a nucleus, the dynamics of the reaction system in Equation (10.7) is dominated by filament growth rather than forming additional nuclei. In this scenario, if we use A_p to denote all the polymers with length greater than 1 and A_1 for monomer, the system simplifies to a single chemical reaction:

$$A_1 + A_p \underset{k_{-1}}{\overset{k_{+1}}{\rightleftharpoons}} A_p. \tag{10.8}$$

The chemical equilibrium of Equation (10.8) is

$$[A_1]_{eq} = \frac{[A_1]_{eq}[A_p]_{eq}}{[A_p]_{eq}} = \frac{k_{-1}}{k_{+1}}. \tag{10.9}$$

This is a very interesting result. It suggests that, no matter what is the total initial monomer concentration, the polymer will continue to grow until the monomer concentration reaches the equilibrium value $\frac{k_{-1}}{k_{+1}}$. The constant $\frac{k_{-1}}{k_{+1}}$ is called the *critical concentration* for the polymerization. According to this simple model, if the initial monomer concentration is lower than the critical concentration, there will be no polymerization.

10.2.2 Theory of nucleation-elongation of actin polymerization

The above concept of critical concentration can be further understood through the more realistic reaction system of Equation (10.7). The multiple chemical equilibria of Equation (10.7) are given by

$$\frac{[A_2]_{eq}}{\left([A_1]_{eq}\right)^2} = \frac{k_{+1}^*}{k_{-1}^*} = K^*, \quad \frac{[A_{\ell+1}]_{eq}}{[A_\ell]_{eq}[A_1]_{eq}} = \frac{k_{+1}}{k_{-1}} = K \quad (\ell \geq 2). \tag{10.10}$$

Mass conservation yields the conservation relationship

$$[A_1]_{eq} + 2[A_2]_{eq} + \cdots + \ell[A_\ell]_{eq} + \cdots = A_0 \tag{10.11}$$

where A_0 is the total monomer concentration in the system. Expressing each $[A_\ell]_{eq}$ in terms of $[A_1]_{eq}$ and substituting back into Equation (10.11), the system of equations can be collapsed into one:

$$[A_1]_{eq} + 2K^* \left([A_1]_{eq}\right)^2 + K^* \sum_{\ell=3}^{\infty} \ell K^{\ell-2} \left([A_1]_{eq}\right)^\ell = A_0. \tag{10.12}$$

Equation (10.12) is conveniently written

$$x + 2(1-\sigma)x^2 + 3(1-\sigma)x^3 + 4(1-\sigma)x^4 + \cdots = x_t,$$

where $x = K[A_1]_{eq}$, $x_t = KA_0$, and $\sigma = 1 - \frac{K^*}{K}$.

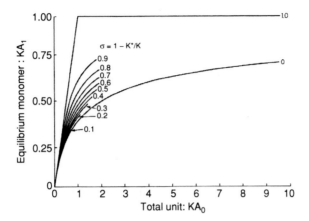

Figure 10.5 The equilibrium monomer concentration as a function of the total monomers for a nucleation-elongation protein polymerization process described by Equations (10.7) and (10.13). The constant $K = k_{+1}/K_{-1}$ is the association constant of a monomer to the polymer; $[A_1]_{eq}$ is the equilibrium free monomer concentration; and A_0 is the total monomer concentration. The variable $\sigma = 1 - K^*/K$ depends on the difference in equilibrium constants between elongation and nucleation steps. At $\sigma = 0$ there is no difference in equilibrium constants; at $\sigma = 1$ nucleation is highly unfavorable compared to elongation and there is infinite cooperativity in the polymerization process.

The above equation can be rearranged to give an equation for free unbound monomer concentration (represented by the variable $x = K[A_1]_{eq}$) and the total monomer concentration (represented by the variable $x_t = K A_0$):

$$\frac{x - 2\sigma x^2 + \sigma x^3}{(1 - x)^2} = x_t \quad (x \leq x_t). \qquad (10.13)$$

Figure 10.5 plots the relationship between $K[A_1]_{eq}$ and $K A_0$ predicted by Equation (10.13). When $K A_0 \ll 1$, all the monomers tend to stay in the monomeric form, regardless of the nucleation kinetics. Therefore $K[A_1]_{eq} \approx K A_0$ in the limit of $K A_0 \to 0$.

When nucleation is highly unfavorable (i.e., $\sigma \approx 1$) the polymer system exhibits a biphasic behavior depending on the total monomer concentration A_0. In this case there is a sharp phase transition between the all-monomer state for $A_0 < 1/K$, where $1/K$ is the critical monomer concentration. When A_0 exceeds $1/K$ the free monomer concentration stays fixed at $[A_1]_{eq} = 1/K$. This type of non-smooth behavior at $x = 1$ for $\sigma = 1$ is called a transcritical bifurcation in non-linear dynamics [191]. It is also widely known as phase transition in physics. Figure 10.5 shows that for σ less than unity, the transition is smooth. Hence we see that the phenomenon of a critical concentration crucially depends on relatively unfavorable nucleation.

10.3 Macromolecular association

In this section, we study the non-covalent association of macromolecules, say of proteins A and B to form a complex AB:

$$A + B \underset{k_{-1}}{\overset{k_{+1}}{\rightleftharpoons}} AB. \tag{10.14}$$

Such associations are fundamental in protein–protein interaction networks in cell signaling.

Protein–protein networks can be visualized by representing proteins as nodes in a network and potential associations as edges. An example network is shown in Figure 10.6. The figure represents cellular proteins (from yeast) as circles with a line drawn between every pair of proteins observed to associate in the cell. (Loops connecting a circle to itself represent self interactions.) The resulting graph is a rather complicated mess, perhaps more like an abstract painting than a scientifically meaningful diagram.

Here we introduce a theory to help make some sense of protein–protein interactions from statistical perspective. Like our study of the phosphorylation–dephosphorylation cycle in a signaling system in Chapter 5, we analyze a single event that represents a repeating motif in this large network. Specifically, given total amounts of some proteins A and B, we are interested in the expected amount of complex AB, as well as the corresponding probability distribution. Since in a living cell the number of some signaling proteins can be of the order of tens and hundreds, it is important to account not only for average behavior, but also for the inherent variability in systems of relatively small numbers of molecules. Therefore, unlike our analysis of actin polymerization that relied on computing the equilibrium distributions of monomers and polymers in solution, here we may not implicitly invoke the thermodynamic limit by assuming continuous concentrations of large numbers of molecules. Instead, we develop a statistical mechanical approach to computing probability distributions of protein–protein association [185].

10.3.1 A combinatorial theory of macromolecular association

Let us first consider a simple system containing one protein A molecule and one protein B molecule confined in a volume V. Due to intra-molecular interaction there are two possibilities for the system: A and B are separated, or A and B are associated in a complex AB. There are two possible states of the system: A + B and

Figure 10.6 An example network of 1006 proteins, with 948 protein–protein interactions obtained from a two-hybrid experiment to search for interacting pairs of proteins. Vertices (circles in the graph) represent protein types and edges (lines connecting the circles) represent observed interactions. The graph is constructed from data from Uetz *et al.* [200], using the Pajek program [12].

AB, corresponding to the left and right sides of Equation (10.14). If the probability of complex existing (the relative amount of time the system is found in the AB state) is p, then the Boltzmann probability law says that the energy difference between the two possible states is $-k_B T \ln \frac{p}{1-p}$. As we shall see, the probability p is a function of V, the volume of the system. We define $K_a = \frac{p}{1-p}$, which will be shown to be related to (but not equal to) the equilibrium constant for the reaction.

Next, we consider a container with volume held constant containing a number of A and B molecules. We define n_A^0 and n_B^0 to be the total numbers of A and B (including molecules found in complexed and uncomplexed states) in the system and ℓ to be the number of AB complexes. By elementary combinatorics, the number of ways of choosing ℓ A molecules out of the total set of n_A^0 is $\binom{n_A^0}{\ell}$.[5] The number of ways of choosing ℓ B molecules out of the total set of n_B^0 is $\binom{n_B^0}{\ell}$. Since there are $\ell!$ number of ways to pair the ℓ A and B molecules, the total number of ways we can form ℓ complexes is $\binom{n_A^0}{\ell}\binom{n_B^0}{\ell}\ell!$

The energy associated with ℓ AB complexes (compared to the reference state of all in the uncomplexed state) is $-\ell k_B T \ln K_a$. Thus the Boltzmann weight for an individual state with ℓ complexes is

$$e^{-\Delta E / k_B T} = K_a^\ell.$$

If we count all of the independent ways of combining ℓ A and B molecules into ℓ complexes, then the non-normalized probability of ℓ AB complexes existing is

$$\Pr\{n_{AB} = \ell\} \propto \binom{n_A^0}{\ell}\binom{n_B^0}{\ell}\ell! K_a^\ell = \frac{n_A^0! \, n_B^0!}{\ell!(n_A^0 - \ell)!(n_B^0 - \ell)!} K_a^\ell. \qquad (10.15)$$

Normalizing this probability we have

$$p_\ell = \Pr\{n_{AB} = \ell\} = Q^{-1} \frac{n_A^0! \, n_B^0!}{\ell!(n_A^0 - \ell)!(n_B^0 - \ell)!} K_a^\ell \qquad (10.16)$$

where Q is the normalization factor

$$Q = \sum_{\ell=0}^{min(n_A^0, n_B^0)} \frac{n_A^0! \, n_B^0!}{\ell!(n_A^0 - \ell)!(n_B^0 - \ell)!} K_a^\ell. \qquad (10.17)$$

The function $Q(K_a, T)$ is the partition function for the two-molecule association system. The free energy of the system, with the completely non-associated state as the reference, is $-k_B T \ln Q$.

Given K_a, n_A^0, and n_B^0, the average number as well as the most probable number of AB complexes can both be computed from the p_ℓ given in Equation (10.16).

---

[5] The number $\binom{n}{m} = \frac{n!}{m!(n-m)!}$ is the number of ways of choosing m (non-ordered) members from a total set of n molecules.

The most probable number $\hat{\ell}$ can be obtained by setting $p_{\ell+1} - p_\ell = 0$. This is the discrete analog of setting the derivative of a continuous function equal to zero to find the extreme point. By this calculation, $\hat{\ell}$ is not necessarily an integer. Setting

$$\frac{K_a^{\hat{\ell}}}{\hat{\ell}!(n_A^0 - \hat{\ell})!(n_B^0 - \hat{\ell})!} = \frac{K_a^{\hat{\ell}+1}}{(\hat{\ell}+1)!(n_A^0 - \hat{\ell} - 1)!(n_B^0 - \hat{\ell} - 1)!}$$

yields

$$\frac{\hat{\ell} + 1}{(n_A^0 - \hat{\ell})(n_B^0 - \hat{\ell})} = K_a. \tag{10.18}$$

Equation (10.18) is the small-number analog of standard theory of chemical equilibrium in elementary chemistry and biochemistry texts. (See Exercise 3.)

Based on the probability distribution p_ℓ in Equation (10.16), we can also compute the mean and variance of the number of AB complexes. Similar to Equation (10.6), we have

$$\bar{\ell} = \sum_\ell \ell p_\ell = \frac{d \ln Q}{d \ln K_a}, \tag{10.19}$$

$$\sigma_\ell^2 = \overline{(\ell - \bar{\ell})^2} = \overline{\ell^2} - \bar{\ell}^2 = \frac{d\bar{\ell}}{d \ln K_a}. \tag{10.20}$$

For large n_A^0 and n_B^0, we expect the probability distribution for n_{AB} to be narrow and $\bar{\ell} \approx \hat{\ell}$. Using this approximation, and solving for $\hat{\ell}$ in Equation (10.18) we have

$$\bar{\ell} = \frac{1 + K_a n_A^0 + K_a n_B^0 - \sqrt{(1 + K_a n_A^0 + K_a n_B^0)^2 - 4K_a^2 n_A^0 n_B^0}}{2K_a}. \tag{10.21}$$

The variance can be obtained by differentiating both sides of the equation $K_a(n_A^0 - \bar{\ell})(n_B^0 - \bar{\ell}) = \bar{\ell}$ with respect to K_a:

$$\frac{d}{dK_a}\left(K_a(n_A^0 - \bar{\ell})(n_B^0 - \bar{\ell})\right) = \frac{d\bar{\ell}}{dK_a}.$$

Noting that $\bar{\ell}$ is a function of K_a, while n_A^0 and n_B^0 are not,

$$(n_A^0 - \bar{\ell})(n_B^0 - \bar{\ell}) = K_a \frac{d\bar{\ell}}{dK_a}(n_B^0 - \bar{\ell}) + K_a(n_A^0 - \bar{\ell})\frac{d\bar{\ell}}{dK_a} + \frac{d\bar{\ell}}{dK_a},$$

$$\sigma_\ell^2 = \frac{d\bar{\ell}}{d \ln K_a} = K_a \frac{d\bar{\ell}}{dK_a} = \frac{(n_A^0 - \bar{\ell})(n_B^0 - \bar{\ell})}{n_A^0 + n_B^0 + 1/K_a - 2\bar{\ell}}.$$

Substituting Equation (10.18) into the denominator on the right-hand side of the above expression, we have in the limit $\bar{\ell} \gg 1$, $n_A^0 + n_B^0 + 1/K_a - \bar{\ell} = n_A^0 n_B^0/\bar{\ell}$.

Combining these results we have the expression for the relative variance:

$$\text{r.v.} = \frac{\sigma_\ell^2}{\bar{\ell}^2} \approx \frac{1}{K_a(n_A^0 n_B^0 - \bar{\ell}^2)}. \tag{10.22}$$

Thus the relative variance decreases in proportion to the square of the number of molecules in the system. For $n_A^0 = n_B^0 = 100$ and $K_a = 0.5$, Equation (10.21) yields $\bar{\ell} = 86.8$ and r.v. $= 8.1 \times 10^{-4}$. Thus, even for this relatively small number of molecules, the relative variance is small for most practical purposes.

In chemistry and biochemistry, bimolecular associations are characterized by the dissociation constant for the reaction of Equation (10.14), which is defined in terms of equilibrium concentrations: $K_d = [A][B]/[AB] = \frac{k_{-1}}{k_{+1}}$. (The dissociation constant is the inverse of the equilibrium constant.) For our given volume V (when $\bar{\ell} \gg 1$) we have

$$K_a = \left(\frac{n_{AB}}{n_A n_B}\right)_{eq} = \left(\frac{[AB]V}{[A][B]V^2}\right)_{eq} = \frac{1}{K_d V} = \frac{K_{eq}}{V}. \tag{10.23}$$

10.3.2 Statistical thermodynamics of association

The probabilistic model of macromolecular association introduced in the previous section, for the case of large n_A and n_B, may be recast into the formal language in terms of statistical thermodynamics. Recall from Chapter 1 that the chemical potential of a species has two terms, a structural energy (enthalpy) term and a concentration/entropy term:

$$\mu_X = \mu_X^0 + k_B T \ln n_X. \tag{10.24}$$

Here the X can represent A, B, or AB in the current problem. The free energy of the system then is

$$G(\ell) = n_A \mu_A + n_B \mu_B + n_{AB} \mu_{AB} \tag{10.25}$$

where $n_A = n_A^0 - \ell$, $n_B = n_B^0 - \ell$, $n_{AB} = \ell$, and $\mu_A^0 + \mu_B^0 - \mu_{AB}^0 = k_B T \ln K_a$. Hence, we have relative probability

$$\frac{p_{\ell+1}}{p_\ell} = e^{-\frac{G(\ell+1)-G(\ell)}{k_B T}} = \frac{(n_A^0 - \ell)^{n_A^0-\ell}(n_B^0 - \ell)^{n_B^0-\ell}(\ell)^\ell}{(n_A^0 - \ell - 1)^{n_A^0-\ell-1}(n_B^0 - \ell - 1)^{n_B^0-\ell-1}(\ell + 1)^{\ell+1}} K_a, \tag{10.26}$$

which can be obtained directly from Equation (10.15) if one assumes that n_A, n_B, and ℓ are large and uses Sterling's approximation $\ln n! \approx n \ln n - n$.

10.4 A dynamics theory of association

The previous section worked out the relationship between the statistics of association of a collection of A and B molecules as a function of p, the probability of a single pair of A and B molecules to form a complex in a volume V. We saw that the probability p is related to the experimentally determined dissociation constant K_d via

$$K_d = \frac{1-p}{Vp}, \tag{10.27}$$

which in turn is related to kinetic rate constants k_{+1} and k_{-1}

$$K_d = \frac{k_{-1}}{k_{+1}}. \tag{10.28}$$

The kinetic constants k_{+1} and k_{-1}, which are properties of the molecules A and B and their interaction under specific conditions, are more difficult to determine experimentally than their ratio, which does not require resolving rapid transients to measure. One method to measure rate constants is by stopped-flow experiments, in which small reacting volumes are rapidly mixed and reaction progress followed, usually using some spectrophotometric assay.

In this section, we briefly introduce a basic theory of chemical kinetics that provides a relationship between interaction energies (and the interaction-energy landscape) and the kinetics of interaction of two molecules. Here we treat the molecules A and B as spherically symmetric, making the interaction energy landscape one dimensional. (Since the molecules are spherically symmetric, the distance between their centers tells us everything about the relative conformation of the two molecules.) The goal is to understand how one of the rate constants, k_{-1}, and the probability of interaction depend on the molecular energy landscape. Given a measurement of k_{-1}, Equations (10.27) and (10.28) can be combined to provide the estimate of $k_{+1} = \frac{k_{-1}Vp}{1-p}$. (Thus the theory will have to also tell us how the probability of association depends on the volume V.)

Let r be the distance between the centers of mass of the two molecules, and the interaction energy be $U(r)$. The distance r is known as the reaction coordinate for the association process. The potential $U(r)$ could take the form, for example, of Lennard–Jones' 6-12 potential:

$$U(r) = -V_0 \left[2\left(\frac{r_0}{r}\right)^6 - \left(\frac{r_0}{r}\right)^{12} \right], \tag{10.29}$$

which effectively models the non-covalent van der Waals interaction energy. This potential function is illustrated in Figure 10.7A.

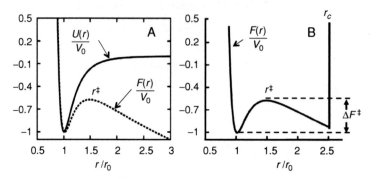

Figure 10.7 A schematic showing the energy and free energy landscapes for the association of simple spherical molecules A and B with the potential defined by Equation (10.29). (A) The solid shows the energy $U(r)$ and the dashed line shows the free energy, which combines the energy with the entropic contribution of the spherical shell volume $4\pi r^2 dr$. The transition state for the dissociation reaction occurs at $r^{\ddagger}$, the location of the free energy maximum. (B) The association–dissociation free energy landscape is shown for the finite concentration case, where $\frac{4}{3}\pi r_c^3[B] = 1$.

As is apparent in the figure, a minimum in energy occurs where the molecules A and B are separated by r_0. Conformations in or near this energy correspond well to the AB complex in Equation (10.14). Values of r that are sufficiently greater than r_0 correspond to the A $+$ B uncomplexed state. To make this definition precise, we note that the probability of the distance between A and B, R_{AB}, is directly related to $U(r)$ following the Boltzmann probability law:

$$\Pr\{r < R_{AB} \le r + dr\} = Q^{-1}e^{-U(r)/k_B T}4\pi r^2 dr, \tag{10.30}$$

for infinitesimally small dr. The term $4\pi r^2 dr$ is the volume in spherical coordinates between r and $r + dr$, accounting for the spherical coordinates of the conformational state space. The normalization factor Q will be determined below.

We can gain important insight by rewriting the right-hand side of Equation (10.30) as $e^{-F(r)/k_B T}dr$ in which

$$F(r) = U(r) - k_B T \ln\left(r^2\right) + \text{const.} \tag{10.31}$$

The function $F(r)$ is called free energy landscape for the spherical A and B association. It has an enthalpic part $U(r)$ and an entropic part $k_B \ln(r^2)$. An important feature of the free energy landscape is that it has a maximum, denoted $r^{\ddagger}$, which is called a transition state. The transition state separates the AB complex, occurring in the neighborhood of $r \sim r_0$, and dissociated state, occurring for $r > r^{\ddagger}$.

The free energy landscape is shown as the dashed line in Figure 10.7A for the Lennard–Jones potential. From observed behavior that $F(r)$ continuously decreases as $-\ln(r^2)$ one might infer that the probability of A and B being dissociated is infinitely greater than that of being associated. In fact that is true in an infinite volume, in which there are no bounds on r. It is not surprising that in an infinitely large volume, the probability of association, p, is zero.

To appropriately take a finite volume V (or equivalently finite concentrations) into account, let us consider a system centered on one A molecule. A fixed concentration of molecule B implies that each single B molecule occupies on average a volume of $1/[B]$. Therefore, the r in Equation (10.30) should not be extended to infinity, but rather bounded by $r \leq r_c$ with $\frac{4}{3}\pi r_c^3 = \frac{1}{[B]}$. Then we have the probability of association computed from Boltzmann statistics

$$p = \frac{\int_0^{r^\ddagger} e^{-U(r)/kT} 4\pi r^2 dr}{\int_0^{r_c} e^{-U(r)/kT} 4\pi r^2 dr}. \tag{10.32}$$

Figure 10.7B shows that when r is bounded, the free energy landscape has two minima, which is typical for a chemical reaction. Each minimum corresponds to one side of the reaction; the transition state occurs at the position $r^\ddagger$ along the reaction coordinate. Note that, given an energy landscape for a reaction coordinate r, we can compute p from $U(r)$ using Equation (10.32). However, it is not possible in general to determine $U(r)$ from an experimental measurement of p.

10.4.1 Transition-state theory and rate constants

The free energy landscape given in Figure 10.7B provides the basis for computing the dissociation and association constants, k_{-1} and k_{+1}, in terms of the height of the transition state, and the curvatures of the energy wells and peak. According to Kramers' theory, the dissociation rate constant is given by [117]

$$k_{-1} = \frac{\sqrt{\nu \nu^\ddagger}}{2\pi \eta} e^{-\Delta F^\ddagger/k_B T}, \tag{10.33}$$

in which ν is the curvature of the energy well where a reactant resides, $\nu^\ddagger$ is the curvature, in absolute value, at the transition state, $\Delta F^\ddagger$ is the free energy barrier height, and η is a frictional coefficient. For a detailed derivation of Equation (10.33), readers should consult the original paper by Kramers [117].

Concluding remarks

The rich subject of the structural and dynamical aspects of macromolecules certainly cannot be contained by a single chapter. Here we have only touched on the subject to illustrate basic concepts and how those concepts fit into the aspects of biochemical systems kinetics and thermodynamics that this text is largely concerned with.

For further reading on the subject, the classic text of Tanford [192] remains an excellent resource, as does the text by Cantor and Schimmel [27]. Recent books by Dill and Bromberg [44] on molecular biophysics and by Schlick [181] on computational chemistry are highly recommended.

Exercises

10.1 For the helix-coil theory of Section 10.1.1, show that $Z = (1 + v)^N$ if $w = v$. This is expected for a chain of independent units, each of which has two states with energy 0 and $-k_B T \ln v$.

10.2 Show that the algorithm in the function `hcontent` in Section 10.1.1 does indeed compute the derivative of the partition function with respect to the parameter w.

10.3 Consider the association reaction in Equation (10.14) with equilibrium association constant $K_{eq} = \frac{k_{+1}}{k_{-1}}$. The equilibrium concentrations for A, B, and AB are

$$\left(\frac{[AB]}{[A][B]} \right)_{eq} = K_{eq}.$$

Use this relation to determine the equilibrium concentration of AB as a function of total numbers of A and B molecules in a finite volume. Compare your result with Equations (10.18) and (10.21).

11

Stochastic biochemical systems and the chemical master equation

Overview

Chemical reactions inside cells are ultimately tied to constant thermal motion that brings reacting species into contact and allows reaction systems to cross the free energy barriers that separate distinct chemical states. From the perspective of a molecule, or a collection of molecules, immersed and interacting in a larger system such as a test tube or living cell, thermal agitation contributes an effectively purely random component to the kinetics of the system. The resulting stochastic nature of biochemical systems has received increasing attention in recent years.

While individual molecules behave stochastically in an aqueous solution in test tubes or inside cells, the dynamics of systems consisting of large numbers of molecules is remarkably non-random, deterministic. The emergence of determinism from randomness is a realization of the same law of large numbers that allows Las Vegas casinos to be confident in their profitability. Consider for example tossing three fair coins in the air and observing how many land heads side up. The probability of observing no heads-up outcomes is 1/8 – not astronomically small by any means. The likelihood that 30 000 tosses will result in no heads-up outcomes is $1/2^{30\,000}$, which is a terrifically small number. Though the chance of any one specific outcome (30 000 tails, for example) is vanishingly small, we can predict with great confidence that nearly one half of the total number of tosses will be heads. In fact, if we were to repeat this experiment many times, the relative variability (standard deviation divided by mean number of heads outcomes) would be approximately 0.006. Thus we can predict the outcome of a single experiment to vary from the predicted outcome of half heads/half tails by less than one percent.

It is, thus, not surprising that when we deal with macroscopic systems with numbers of molecules of the order of the Avogadro constant (6.022×10^{23}), we do not need to concern ourselves with stochastic fluctuations. In a biological cell, however, many reacting species occur with only a few, or a few hundred, copies (see

Figure 4.1). The realization that fluctuations can be important in such a system has led to a surge of interest in stochastic models of biochemical reactions in systems biology.

Just as deterministic models for the kinetics of biochemical reactions are based on the law of mass action, stochastic kinetics are governed by an equivalently fundamental theory. The theory is based on the *chemical master equation* (CME), which in recent years has gained significant visibility due to the influential work of Daniel Gillespie [66, 67, 68]. The CME approach is also known as the Gillespie algorithm. This chapter builds a basic working knowledge of the CME and its application in modeling biochemical reaction systems at the cellular level.

11.1 A brief introduction to the chemical master equation

The mathematical models of reaction systems considered throughout much of this book have relied largely on two fundamental principles from physical chemistry: the Boltzmann probability law (see Section 12.2.1), describing equilibrium probability distributions, and the law of mass action (see Section 3.1.3.1), describing the kinetics of elementary steps in reaction systems. The Boltzmann probability law determines the relative equilibrium probabilities of various states in terms of their energies in an NVT system (or equivalently in terms of enthalpies in an NPT system; see Section 12.3). Kinetic models, on the other hand, require determination of the rate constants associated with the reactions. Reaction energetics and kinetics are not independent. The link between the energy/enthalpy landscape and chemical kinetics was outlined in Section 10.4.1, where it was shown that, at least in principle, it is possible to compute biochemical association reaction rate constants based on biomolecular structures and energy landscapes.

The chemical master equation (CME) for a given system invokes the same rate constants as the associated deterministic kinetic model. Yet the CME is more fundamental than the deterministic kinetic view. Just as Schrödinger's equation is the fundamental equation for modeling motions of atomic and subatomic particle systems, the CME is the fundamental equation for reaction systems. Remember that Schrödinger's equation is not a model for a specific mechanical system. Rather, it is a theoretical framework upon which models for particular systems can be developed. In order to write down a model for an atomic system based on Schrödinger's equation, one needs to know how to write down the Hamiltonian a priori. Similarly, the CME is not a model for a specific biochemical reaction system; it is a theoretical framework. To determine the CME model for a reaction system, one must know what are the possible elementary reactions and the associated rate constants.

As far as we know, the CME first appeared in the work of Max Delbrück in 1940, who studied a small chemical reaction system in terms of a stochastic model

[42].[1] In the same year Hendrik Kramers published his landmark paper [117] on the theory of chemical reaction rates based on thermally activated barrier crossing by Brownian motion [77]. These two papers clearly mark the domains of two related areas of chemical research. Kramers provided the framework for computing the rate constants of chemical reactions based on the molecular structures, energy, and solvent environment. (See Section 10.4.1.) Delbrück's work set the stage for predicting the dynamic behavior of a chemical reaction system, as a function of the presumably known rate constants for each and every reaction in the system.

Deterministic dynamics of biochemical reaction systems can be "visualized" as the trajectory of $(c_1(t), c_2(t), \cdots, c_N(t))$ in a space of concentrations, where $c_i(t)$ is the concentration of ith species changing with time. This mental picture of path traced out in the N-dimensional concentration space by deterministic systems may prove a useful reference when we deal with stochastic chemical dynamics. In stochastic systems, one no longer thinks in terms of definite concentrations at time t; rather, one deals with the probability of the concentrations being $x_1, x_2, \cdots, x_N$ at time t:

$$p(x_1, x_2, \cdots, x_N, t) = \Pr\{c_1(t) = x_1, c_2(t) = x_2, \cdots, c_N(t) = x_N\}. \quad (11.1)$$

The CME is the equation for the probability function p, or equivalently if the system's volume is constant, for the probability function $p(n_1, n_2, \cdots, n_N, t)$ where n_i is the number of molecules of species i. With given concentrations $(c_1, c_2, \cdots, c_N)$ at a time t, deterministic kinetic models give precisely what the concentrations will be at time $t + \delta t$. According to the stochastic CME, however, the concentrations at $t + \delta t$ can take many different values, each with certain probability.

The CME was extensively studied in the 1960s by many people, including Donald McQuarrie, who wrote a comprehensive review [137]. In the 1970s, Thomas Kurtz, a mathematician, proved that in the limit of system's size going to infinity, the solution to the CME is precisely the solution to the corresponding deterministic differential equations based on the law of mass action [121]. Hence, as a mathematical foundation of chemical reaction systems, the CME supersedes the law of mass action, just as Schrödinger's equation supersedes Newton's law of motion since one can prove that the latter occurs as the limit of the former when Planck's constant, $\hbar$, goes to zero. (Figure 11.1 illustrates the analogy between the relationships between the CME and mass-action kinetics and Schrödinger's and Newton's equations.)

Furthermore, the CME framework has been shown to be consistent with the general theory of non-equilibrium thermodynamics [109], and the recently developed

---

[1] In genetics, Delbrück's contribution to stochastic modeling, the Luria–Delbrück distribution, is well known [126]. That theory is also based on a master equation approach.

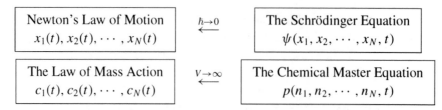

Figure 11.1 A schematic that illustrates the analogy between the theories for mechanical motions and for chemical dynamics. Newton's law of motion, governing a collection of particles with positions $x_1(t), x_2(t), \cdots, x_N(t)$, arises from Schrödinger's equation for the wave function ψ in the limit $\hbar \to 0$. Similarly, the chemical master equation for $p(n_1, n_2, \cdots, n_N, t)$ yields the law of mass action in the limit $V \to \infty$.

statistical physics of open, driven systems with entropy production [46, 164]. Thus, the CME is the mathematical foundation for modeling dynamics of biochemical reactions systems inside cells.

Yet the applications of the CME to biochemical systems are only in their infancy. Because this is a rich area of research at the intersection of cell biology, physical chemistry, and mathematical biology, it is tempting to project future developments by drawing some analogies to three different areas of current chemical and biochemical research.

First, the current state of affairs is remarkably similar to that of the field of computational molecular dynamics 40 years ago. While the basic equations are known in principle (as we shall see), the large number of unknown parameters makes realistic simulations essentially impossible. The parameters in molecular dynamics represent the force field to which Newton's equation is applied; the parameters in the CME are the rate constants. (Accepted sets of parameters for molecular dynamics are based on many years of continuous development and checking predictions with experimental measurements.) In current applications molecular dynamics is used to identify functional conformational states of macromolecules, i.e., free energy minima, from the entire ensemble of possible molecular structures. Similarly, one of the important goals of analyzing the CME is to identify *functional states* of a reaction network from the entire ensemble of potential concentration states. These functional states are associated with the maxima in the steady state probability distribution function $p(n_1, n_2, \cdots, n_N)$. In both the cases of molecular dynamics and the CME applied to non-trivial systems it is rarely feasible to enumerate all possible states to choose the most probable. Instead, simulations are used to intelligently and realistically sample the state space.

Second, with a set of identified functional states, the CME framework may be used to compute the rates of transition between the functional states. This task is similar to the barrier-crossing problem of computing the reaction rate constants based on Kramers' theory. However, the task for the CME involves unique difficulties that arise from the open-chemical nature of living systems [166].

Third, numerically solving the CME and obtaining the probability distribution $p(n_1, n_2, \cdots, n_N, t)$ is a problem mathematically similar to solving the Schrödinger equation for a quantum chemistry problem. In both cases $p(n_1, n_2, \cdots, n_N, t)$ and $\psi(x_1, x_2, \cdots, x_N, t)$ are governed by partial differential, or difference, equations where n can be very large. Perhaps methodologies from quantum chemistry, such as Hartree–Fock approximation [111], may be borrowed for solving the CME.

11.2 Essential materials from probability theory

11.2.1 The law of large numbers

The law of large numbers is fundamental to probabilistic thinking and stochastic modeling. Simply put, if a random variable with several possible outcomes is repeatedly measured, the frequency of a possible outcome approaches its probability as the number of measurements increases. The *weak law of large numbers* states that the average of N identically distributed independent random variables approaches the mean of their distribution.

These laws are the fundamental reason why objects in the macroscopic world behave deterministically while individual atoms and molecules are under constant irregular motion. If there is a sufficient number of atoms and molecules in a system, the stochasticity tends to be canceled out and the system exhibits the average behavior in a deterministic way.

11.2.2 Continuous time Markov chain

As a generic description of a stochastic system, consider a system with N possible states, labeled $1, 2, \cdots, N$. Since the system is stochastic, we cannot define equations that determine the specific state that the system adopts at a specific time. Rather, we look for equations that govern the probability $p_m(t)$ that the system is in state m at time t.

The rate of change in $p_m(t)$ is equal to the rate of transition from other states l to m minus the rate of transition from state m to other states. The simplest model for how the probabilities change with time is the Markov chain, which assumes that the transition probabilities are dependent only on the current state and independent of all past (and future) states. Formally,

$$\frac{dp_m(t)}{dt} = \sum_{\ell \neq m} Q_{lm} p_l - \sum_{\ell \neq m} Q_{ml} p_m \qquad (11.2)$$

where the $Q_{\ell m}$, $\ell \neq m$, is the rate of transition from state ℓ to state m. The $\{Q_{\ell m}\}$ are first-order rate constants with units of $[\text{time}]^{-1}$.

Equation (11.2) can be simplified by defining

$$A_{lm} = Q_{lm} \quad \text{for } \ell \neq m$$

and

$$A_{mm} = -\sum_{\ell \neq m} Q_{ml},$$

yielding

$$\frac{dp_m(t)}{dt} = \sum_{\ell=1}^{N} A_{lm} p_l. \tag{11.3}$$

Equation (11.3) is known as the master equation in the physics literature. Note that this equation must satisfy the property that $\sum_m p_m(t) = 1$ for all t. Therefore

$$\frac{d}{dt} \sum_{m=1}^{N} p_m(t) = \sum_{l=1}^{N} p_l \sum_{m=1}^{N} A_{lm} = 0. \tag{11.4}$$

Since beyond the fact that probabilities are positive and sum to 1, the values of the $\{p_m\}$ are arbitrary, Equation (11.4) requires

$$\sum_{m=1}^{N} A_{lm} = 0.$$

One of the most important consequences of a model based on Equation (11.3) is that, under quite general conditions, the long-time limit of the probability distribution $\{p_m(\infty)\}$ is unique. No matter what is the initial state of the system, the system evolves toward $\{p_m(\infty)\}$. This steady state of the system, however, is characterized not by a single state m, but by a distribution of probabilities of all states. Hence, in general, the steady state of a system governed by a master equation will involve fluctuations.

In the chemical and biochemical literature, Equations (11.2) and (11.3) and similar equations have been used in modeling kinetics of single molecules, from single membrane channel proteins to single enzymes in solution [10, 52, 91]. In those models, the various states represent conformational states of proteins. In the CME approach to chemical reaction systems, the state of a system is the specific set of numbers of copies of all the species in the system. Thus for chemical reaction system applications, the adjective "chemical" is applied. Sometimes these two different applications of master equations are referred to as *state tracking* versus *number tracking* [142]. In the state tracking formalism, the state space of a model has no particular geometry, while in the CME, the state space is an N-dimensional space of non-negative integers.

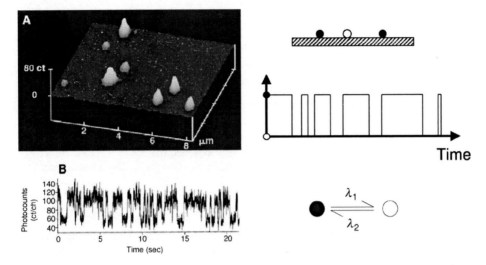

Figure 11.2 The image in the upper left panel shows a snapshot of several individual protein molecules immobilized in a gel. Each protein undergoes conformational fluctuations that can be monitored by a fluorescent probe. The fluorescent signal from a single protein molecule, as a function of time, is recorded in the time trace shown in the lower left panel. On the right, the experimental situation and the fluorescent time trace are idealized as a two-state conformational transition process as given in Equation (11.5), with A representing the darker state and B representing the brighter state. Image and data in left panel obtained from Lu *et al.* [133]. Reprinted with permission from AAAS.

11.3 Single molecules and stochastic models for unimolecular reaction networks

Conformational transitions of single biological macromolecules can be measured in a variety of experimental settings. An important example is that of single-channel recordings of current through ion channels, using patch-clamping techniques established in the 1980s [93]. More recently, optical methods have been used to probe the kinetics of single enzyme molecules in aqueous solution [133]. In these experiments, individual proteins are immobilized on a coverslip and excited by a focused laser. Conformational transitions can be measured using highly sensitive single-photon detection methods. An experiment of this type is illustrated in Figure 11.2.

11.3.1 Rate equations for two-state conformational change

Let us assume that the kinetic scheme for a simple two-state conformational transition is

$$A \underset{\lambda_2}{\overset{\lambda_1}{\rightleftharpoons}} B \qquad (11.5)$$

where λ_1 and λ_2 are two first-order rate constants with dimension [time]$^{-1}$.

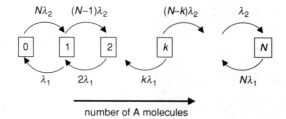

number of A molecules

Figure 11.3 State transition diagram for stochastic discrete model for the single unimolecular reaction of Equation (11.5). It is assumed that there is a total of N molecules in the system; k is the number of molecules in state A. The intrinsic rate constants are $\lambda_{1,2}$.

Unimolecular reactions such as this one involve conformational changes of a single molecule without explicit interaction with other molecules. Implicit interactions with the solvent, of course, are assumed. As for the generic master equation introduced above, rather than asking the question "What is the concentration of species A at time t?," we ask "What is the probability that a single molecule is in state A at time t?"

For this simple two-state transition, the traditional deterministic chemical kinetics (see Chapter 3) is based on rate equations for the concentration of A:

$$\frac{d[A]}{dt} = -\lambda_1[A] + \lambda_2(E_T - [A]) \qquad (11.6)$$

where we have assumed that the total concentration $E_T = [A] + [B]$ remains constant. The concentration [A] is real and non-negative.

If we are dealing with a finite number N of E molecules in a cell, and the number is small enough that fluctuations are expected to be significant, then clearly we need to modify the Equation (11.6) in two crucial ways. First, the species A must now be measured in discrete numbers n_A rather than continuous concentration [A]. Second, we recognize that n_A has a probability distribution

$$p_k(t) = \text{Pr}\{n_A(t) = k\} \qquad (11.7)$$

for every time t. The task of stochastic modeling is to develop the equations for $p_k(t)$.[2]

Figure 11.3 shows how the state of this system, characterized by n_A, can change with time. At the state $n_A = k$ there are k A molecules, each of which transitions to B with rate λ_1. Therefore rate of transition from $n_A = k$ to $n_A = k - 1$ is $k\lambda_1$. Similarly, rate of transition from $n_A = k$ to $n_A = k + 1$ is $(N - k)\lambda_2$.

---

[2] Recall that in Section 10.3 we worked out a detailed theory for the equilibrium distribution for the reaction $A + B \rightleftharpoons C$. Here the task is to determine the governing differential equation (chemical master equation) for the dynamics of the state probabilities in Equation (11.5).

The chemical master equation (CME) for this system accounts for the rate of transition out of state $n_A = k$, as well as the rate of transition from states $n_A = k - 1$ and $n_A = k + 1$ to state $n_A = k$:

$$\frac{dp_k(t)}{dt} = -(k\lambda_1 + (N - k)\lambda_2)p_k + (k + 1)\lambda_1 p_{k+1} + (N - k + 1)\lambda_2 p_{k-1},$$

$$(k = 0, 1, \cdots, N). \tag{11.8}$$

To solve Equation (11.8), we introduce the generating function, which is essentially the Laplace transform for discrete functions. We define

$$G(s, t) = \sum_{k=0}^{N} p_k(t)s^k. \tag{11.9}$$

Multiplying both sides of Equation (11.8) by s^k and summing over k from 0 to N, we obtain

$$\frac{\partial}{\partial t}G(s, t) = [(1 - s)\lambda_1 + s(1 - s)\lambda_2]\frac{\partial}{\partial s}G(s, t) + N(s - 1)\lambda_2 G(s, t). \tag{11.10}$$

Equation (11.10) is a partial differential equation for which the solution is not immediately apparent. Fortunately, we can find a solution by defining $G(s, t) = g^N(s, t)$ and noting that $g(s, t)$ satisfies

$$\frac{\partial}{\partial t}g(s, t) = [(1 - s)\lambda_1 + s(1 - s)\lambda_2]\frac{\partial}{\partial s}g(s, t) + (s - 1)\lambda_2 g(s, t). \tag{11.11}$$

This is a very insightful result: Equation (11.11) is the special case of Equation (11.10) for $N = 1$. It shows that the kinetics for the system of N E molecules is simply related to the kinetics of a single E molecule!

The kinetics for a single enzyme can be obtained from the one-molecule version of Equation (11.6):

$$\frac{dp_A(t)}{dt} = -\lambda_1 p_A + \lambda_2 (1 - p_A) \tag{11.12}$$

in which $p_A = [A]/E_T$ is the probability that the molecule is in state A. Assuming the initial condition that at time zero the molecule is in state A ($p_A(0) = 1$), the solution is

$$p_A(t) = \frac{\lambda_2 + \lambda_1 e^{-(\lambda_1+\lambda_2)t}}{\lambda_1 + \lambda_2}. \tag{11.13}$$

We can now verify that

$$g(s, t) = sp_A(t) + (1 - p_A(t)) = \frac{\lambda_1 + s\lambda_2 + (s - 1)\lambda_1 e^{-(\lambda_1+\lambda_2)t}}{\lambda_1 + \lambda_2}, \tag{11.14}$$

is indeed the solution to Equation (11.10) with $N = 1$. The solution to Equation (11.8) with arbitrary N follows a binomial distribution

$$p_k(t) = \binom{N}{k} (p_A(t))^k (1 - p_A(t))^{N-k} . \tag{11.15}$$

The right-hand side has a clear combinatorial meaning. It is the probability of finding k molecules in state A, among the total N identical and independent molecules, when each molecule has the probability of p_A being in state A and probability $1 - p_A$ being in state B.

Therefore, for N E molecules, the mean number of A at time t is

$$\langle n_A(t) \rangle = N p_A(t), \tag{11.16}$$

and the variance is

$$Var[n_A(t)] = N p_A(t)(1 - p_A(t)). \tag{11.17}$$

What we see is that if N is large, say of the order of 10^4, then the relative variance, i.e., the broadness of the probability distribution for $n_A(t)$, is

$$\frac{\sqrt{Var[n_A]}}{\langle n_A \rangle} \propto \frac{1}{\sqrt{N}} = 0.01. \tag{11.18}$$

Therefore, when N is this large the deterministic model of Equation (11.6) is a very accurate model for the chemical kinetics. There is no need to be concerned with the stochasticity. On the other hand, if N is of the order of tens and hundreds, then the variance can be significant.

11.3.2 Michaelis–Menten kinetics of single enzymes

In Chapter 4 (Section 4.1.1), we derived the Lineweaver–Burk double-reciprocal relation between the steady state flux of an enzyme reaction and its substrate concentrations. (See Equation (4.5).) Furthermore, we showed in Section 4.4.1 that the same equation can be obtained from a stochastic point of view. Recalling this derivation, consider the basic mechanism

$$E + S \underset{k_{-1}}{\overset{k_{+1}}{\rightleftharpoons}} ES \overset{k_{+2}}{\longrightarrow} E + P \tag{11.19}$$

from the perspective of a single enzyme molecule. We wish to determine the average time it takes for the kinetic cycle of Figure 4.2 to turn over, $\langle T_{cycle} \rangle$.

Breaking down the overall reaction of converting one S to one P, the first step of binding of S to E takes an average time $1/(k_1[S])$. The dwell time in state ES is $1/(k_{-1} + k_2)$, after which the ES complex transitions either to E + P or back to

E + S, with corresponding probabilities $\frac{k_2}{k_{-1}+k_2}$ and $\frac{k_{-1}}{k_{-1}+k_2}$. Hence, we have [151]

$$\langle T_{cycle} \rangle = \frac{1}{k_1[S]} + \frac{1}{k_{-1}+k_2} + \left(\frac{k_2}{k_{-1}+k_2} 0 + \frac{k_{-1}}{k_{-1}+k_2} \langle T_{cycle} \rangle \right). \qquad (11.20)$$

Solving for $\langle T_{cycle} \rangle$ we obtain

$$\langle T_{cycle} \rangle = \frac{k_{-1}+k_2}{k_1 k_2 [S]} + \frac{1}{k_2}. \qquad (11.21)$$

Comparing Equations (11.21) and (4.5), we see that $\frac{1}{\langle T_{cycle} \rangle}$ is exactly equal to J in Equation (4.6). The above derivation also indicates that if the enzyme has only one unbound state, then the mean waiting time will always have the double reciprocal form of $a/[S] + b$.

11.4 The CME models for non-linear biochemical reactions with fluctuations

For unimolecular reaction systems described by mass-action kinetics, the kinetic equations are always linear. Hence they are called linear reaction systems. For stochastic unimolecular systems, the key is to understand the multi-state kinetics of a single molecule, as we have seen. If N is the number of states of a molecule, and there are M number of identical, independent molecules, then the probability of having $m_1, m_2, \cdots, m_N$ molecules in the states $1, 2, \cdots, N$ ($m_1 + m_2 + \cdots m_N = M$) will be the multinomial distribution

$$p(m_1, m_2, \cdots, m_N, t) = \frac{M!}{m_1! m_2! \cdots m_N!} p_1^{m_1}(t) p_2^{m_2}(t) \cdots p_N^{m_N}(t), \qquad (11.22)$$

where $p_k(t)$ is the probability of a single molecule being in state k at time t. This result was demonstrated by Terrell Hill in 1971 [89, 92].

For non-linear systems, closed form solutions usually do not exist and it is necessary to simulate their behavior.

11.4.1 Chemical master equation for Michaelis–Menten kinetics

Enzyme-catalyzed reactions involve multi-molecular enzyme-substrate association. Therefore, even when the overall reaction is unimolecular, the enzyme mechanism is generally non-linear. If a system has more than one copy of the enzyme and a small number of the reactant molecules, then one needs the CME framework to represent the stochastic behavior of the system. Note that in cellular regulatory networks, the substrates themselves may be proteins that are present in small numbers of copies. Recall from Section 5.1, for example, that the mitogen-activated protein (MAP) is the substrate of MAP kinase, and the MAPK is the substrate of MAPKK.

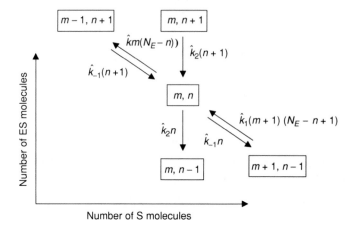

Figure 11.4 Diagrammatic procedure to obtain the CME from a given chemical kinetic scheme. The first step is to translate the standard chemical kinetic scheme, here Equation (11.19), into the *master equation graph* shown above. The traditional kinetic model based on the ordinary differential equations has two independent dynamic species. Correspondingly, the master equation graph is a two-dimensional grid in which each grid point represents the state of the system by specifying the number of molecules for each dynamic species. Each grid point in the graph is connected to several of its neighboring grid points. Each reaction in the kinetic scheme corresponds to an arrow pointing outward from a grid point. The label, i.e., the "rate constant" on an arrow in the diagram, is determined by the product of the number of molecules of reactants multiplied by the corresponding rate constant $\hat{k}_i$. The corresponding CME, Equation (11.23), is then obtained directly from the diagram.

Here to illustrate the procedure for non-linear systems, we work out the chemical master equation for the Michaelis–Menten reaction of Equation (11.19). We let $p(m, n, t)$ be the probability that m, S, and n ES molecules occur in the enzyme reaction system at time t. The probability $p(m, n, t)$ satisfies the CME

$$
\begin{aligned}
\frac{dp(m, n, t)}{dt} = {} & -\left(\hat{k}_1 m(N_E - n) + \hat{k}_{-1} n + \hat{k}_2 n\right) p(m, n, t) \\
& + \hat{k}_1(m + 1)(N_E - n + 1)p(m + 1, n - 1, t) \\
& + \hat{k}_{-1}(n + 1)p(m - 1, n + 1) \\
& + \hat{k}_2(n + 1)p(m, n + 1), \\
& (0 \leq m \leq N_S, \ \ 0 \leq n \leq N_E)
\end{aligned}
\tag{11.23}
$$

in which N_S and N_E are the total numbers of substrate molecules and enzyme molecules. The constants $\{\hat{k}_i\}$ are related to (but not equal to) the constants $\{k_i\}$ in Equation (11.19), as described below.

Figure 11.4 illustrates how the CME is systematically developed from a chemical kinetic scheme such as given in Equation (11.19). The terms in Equation (11.23) correspond to the state transitions that connect state (m, n) with neighboring states in the diagram. Just as the differential equations based on the law of mass action are completely determined from a given kinetic scheme, the CME is completely determined from a given kinetic scheme.

11.4.2 A non-linear biochemical reaction system with concentration fluctuations

The previous example involved a two-dimensional system (involving two independent dynamic species). Thus the CME followed from the two-dimensional reaction diagram. For systems with more species, the dimension of the problem grows accordingly. For a system with three species, say A, B, and C, the CME tracks the three-dimensional probability of ℓ A molecules, m B molecules, and n C molecules present at time t. In general, the mathematical description of an N-dimensional system is the joint probability distribution

$$p(\ell, m, n, \cdots, t) = \Pr\{n_A(t) = \ell, n_B(t) = m, n_C(t) = n, ...\}. \quad (11.24)$$

As for the two-dimensional Michaelis–Menten example, the general procedure to obtain the CME from a given chemical kinetic scheme starts with translating the chemical kinetic scheme into the *master equation* and *master equation graph*. The kinetic schemes of Equations (11.5) and (11.19) correspond to the graphs of Figures 11.3 and 11.4, respectively. Note that the number of independent dynamic species in the chemical reaction system is usually smaller than the total number of chemical species involved since conserved pools of reactants usually exist. For example in the reaction system of Equation (11.5), there are two species but only one independent dynamic species. Similarly, there are four species in the Equation (11.19) but only two independent dynamic species. The master equation graph is an M-dimensional grid, where M is the number of independent species and each grid point represents the state of the system by specifying the numbers $n_1, n_2, \cdots, n_M$, the numbers of molecules of species 1 through M. Each grid point in the graph is connected to several of its neighboring grid points through chemical reactions. Each reaction in the kinetic scheme corresponds to an arrow pointing outward from a grid point. The label on an arrow, i.e., the effective rate constant of a state transition, is given by the product of the number of molecules of all substrates involved in a reaction, multiplied by the corresponding rate constant $\hat{k}$.

As a third example, let us consider a simple non-linear chemical reaction system

$$A + 2X \underset{\alpha_2}{\overset{\alpha_1}{\rightleftharpoons}} 3X$$

$$B + X \underset{\beta_2}{\overset{\beta_1}{\rightleftharpoons}} C \quad (11.25)$$

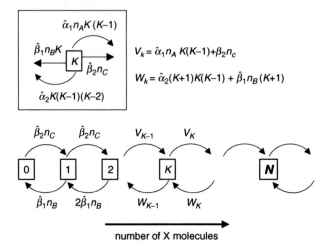

Figure 11.5 The master equation graph for the reaction system shown in Equation (11.25). There are four elementary reactions in the system, which are accounted for by the four arrows pointing outward from state k shown in the upper left. The rates of these reactions are combined to obtain the v_k and w_k, the rates for the corresponding master equation graph, which leads to the CME of Equation (11.27).

in which species A, B, and C are at fixed concentrations a, b, and c, respectively. Therefore, this system involves only one independent dynamic species, X. The macroscopic (deterministic) kinetics of this system evolve according to the law of mass action [146]

$$\frac{dx}{dt} = -\alpha_2 x^3 + \alpha_1 a x^2 - \beta_1 b x + \beta_2 c \qquad (11.26)$$

where x is the concentration of X.

In the CME framework of discrete molecular numbers, X changes one by one stochastically. The stochastic CME model follows the master equation graph of Figure 11.5:

$$\frac{dp_k(t)}{dt} = v_{k-1} p_{k-1}(t) + w_k p_{k+1}(t) - (v_k + w_{k-1}) p_k(t), \qquad (11.27)$$

where $p_k(t) = \Pr\{n_X = k\}$ is the probability of having k X molecules in the system at time t, and

$$v_k = \frac{\alpha_1 a k(k-1)}{V} + \beta_2 c V, \quad w_k = \frac{\alpha_2(k+1)k(k-1)}{V^2} + \beta_1 b(k+1), \qquad (11.28)$$

and V is the volume of the system. For example, $\alpha_1 a x^2$ is the production rate of X, due to reaction $A + 2X \rightarrow 3X$, in units of concentration (number per volume) per time. Hence in pure numbers we have $\hat{\alpha}_1 n_A k(k-1) = V\left(\alpha_1 a x^2\right)$ is the production rate of X in number of molecules per time. The corresponding concentrations are $x = k/V$, $a = n_A/V$, $b = n_B/V$, and $c = n_C/V$. Furthermore, the rate constants of the reaction in pure numbers are $\hat{\alpha}_1 = \alpha/V^2$ in which α_1 is a third-order rate constant. Similarly, $\hat{\alpha}_2 = \alpha_2/V^2$; $\hat{\beta}_1 = \beta_1/V$; $\hat{\beta}_2 = \beta_2$.

It is usually not possible (and never easy!) to solve equations such as Equation (11.27) analytically. So computational simulation of the stochastic trajectories are necessary. The numerical method to obtain stochastic trajectories by Monte Carlo sampling, which we shall discuss in Section 11.4.4, is known as the Gillespie algorithm [68]. However, it happens that the steady state of Equation (11.27) can be obtained in closed form. This is because in steady state, the probability of leaving state 0, $v_0 p_0$ has to exactly balance the probability of entering state 0 from state 1, $w_0 p_1$. Similarly, since $v_0 p_0 = w_0 p_1$, we have $v_1 p_1 = w_1 p_2$, and so on:

$$v_{k-1} p_{k-1} = w_{k-1} p_k. \tag{11.29}$$

Therefore,

$$\frac{p_k}{p_0} = \frac{p_k}{p_{k-1}} \frac{k-1}{p_{k-2}} \cdots \frac{p_1}{p_0} = \prod_{\ell=0}^{k-1} \frac{v_\ell}{w_\ell}, \tag{11.30}$$

in which p_0 is determined through normalization:

$$p_0 = \left(1 + \sum_{k=1}^{\infty} \prod_{\ell=0}^{k-1} \frac{v_\ell}{w_\ell}\right)^{-1}. \tag{11.31}$$

Figure 11.6 shows the resulting probability distribution for this model.

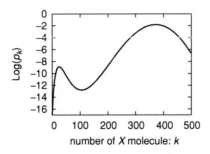

Figure 11.6 The steady state probability distribution for the number of molecule n_X. Parameters used: $\alpha_1 = \alpha_2 = 10^{-6}$, $\beta_1 = \beta_2 = 0.05$, $V = 1$; $a = 500$, $b = 1$, and $c = 20$. The steady state distribution shows two peaks, corresponding to two stable states of the non-linear biochemical reaction system.

11.4.3 Bistability and non-equilibrium steady state

The probability distribution in Figure 11.6 indicates that there are two stable states for the chemical reaction system of Equation (11.25). Since the system is open to species A, B, and C, these states are non-equilibrium steady states (NESS). A more careful discussion of the terminology is in order here. The concept of an NESS has different meanings depending on whether we are considering a macroscopic or a microscopic view. This difference is best understood in comparison to the term *chemical equilibrium*. From a macroscopic standpoint, an equilibrium simply means that the concentrations of all the chemical species are constant, and all the reactions have no net flux. However, from a microscopic standpoint, all the concentrations are fluctuating.

The concentrations fluctuate in a non-equilibrium steady state as well. In fact, the concentrations may fluctuate around multiple probability peaks, as illustrated in Figure 11.6. This system tends to fluctuate around one state, and then occasionally jump to the other. The situation is quite analogous to the transitions between two conformational states of a protein and the local fluctuations within the conformational states.

11.4.4 Stochastic simulation of the CME

To use a computer to simulate a stochastic trajectory of the chemical master equation such as described in Figure 11.4, one must establish the rules of how to move the system from one grid point to its neighboring points. The essential idea is to draw random moves from the appropriate distribution and to assign random times (also drawn from the appropriate distribution) to each move. Thus each simulation step in the simulation involves two random numbers, one to determine the associated time step and one to determine the grid move.

The two random numbers, denoted r_1 and r_2, are randomly sampled from a uniform distribution over the interval [0, 1]. The time T, as dictated by any first-order decay process, is exponentially distributed:

$$f_T(t) = qe^{-qt} \quad (t \geq 0) \tag{11.32}$$

where q is the sum of all the outward rate constants, and $\frac{1}{q}$ is the mean lifetime of the state. For example for the state (m, n) in Figure 11.4,

$$q = \hat{k}_1 m(N_E - n) + \hat{k}_{-1} n + \hat{k}_2 n.$$

The upper panel in Figure 11.7 shows a state with four outward steps and the corresponding rate constants are q_1, q_2, q_3, and q_4. Hence $q = q_1 + q_2 + q_3 + q_4$. To convert the uniformly distributed (pseudo)random number r_1 obtained from the

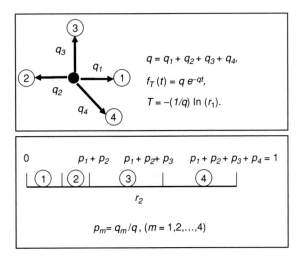

Figure 11.7 A schematic illustration of the Monte Carlo simulation method for computing the stochastic trajectories of a chemical reaction system following the CME. Two random numbers, r_1 and r_2, are sampled from a uniform distribution to simulate each stochastic step: r_1 determines when to move, and r_2 determines where to move. For a given state of a master equation graph shown in the upper panel, there are four outward reactions, labeled 1–4, each with their corresponding rate constants q_i ($i = 1, 2, \cdots, 4$). The upper and lower panels illustrate, respectively, the calculation of the random time T associated with a stochastic move, and the probability p_m of moving to state m.

computer to the random time T given by Equation (11.32), we use

$$T = -\frac{1}{q} \ln r_1.$$ (11.33)

Using Equations (11.32) and (11.33), it is straightforward to verify that the random time T follows the exponential probability distribution if r_1 follows a uniform probability distribution:

$$
\begin{aligned}
\Pr\{x < r_1 \leq x + dx\} &= \Pr\{x < e^{-qT} \leq x + dx\} \\
&= \Pr\{\ln x < -qT \leq \ln x + dx/x\} \\
&= \Pr\left\{-\frac{1}{q}\ln x > T \geq -\frac{1}{q}\ln x - \frac{dx}{qx}\right\} \\
&= q e^{q(\ln x)/q}\frac{dx}{qx} = dx \quad (0 \leq x \leq 1).
\end{aligned}
$$

Once we have determined when to "jump" out of a state, we must next determine where to jump. For the four possible destinations in Figure 11.7, the corresponding probabilities are $p_m = q_m/q, (1 \leq m \leq 4)$. Obtaining a second uniformly

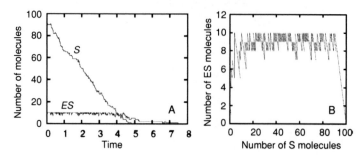

Figure 11.8 An example of the stochastic trajectory from Monte Carlo simulation according to the Gillespie algorithm for reaction system given in Equation (11.19) and corresponding master equation graph given in Figure 11.4. Here we set $N_S = 100$ and $N_{ES} = 0$ at time zero and total enzyme number $N_E = 10$. (A) The fluctuating numbers of S and ES molecules as functions of time. (B) The stochastic trajectory in the phase space of (m, n).

distributed random number from the computer, $r_2 \in [0, 1]$, the destination state is chosen as follows. If r_2 is in $[0, p_1)$, then the system jumps to state 1; if r_2 is in $[p_1, p_1 + p_2)$, then the system jumps to state 2, etc. The procedure is shown in the lower panel of Figure 11.7.

After completing a step (a jump in state) the corresponding outward rate constants for the new state are now all different. To continue the simulation, we draw another two random numbers, make another move, and so on. A stochastic trajectory is thus obtained. One notes that the trajectory has randomly variable time steps, a feature indicative of the Gillespie algorithm.

Figure 11.8 illustrates an example of a stochastic simulation. It is based on the model given in Section 11.4.1 with $\hat{k}_1 = 1.0$, $\hat{k}_{-1} = 0.5$, $\hat{k}_2 = 2$, and $N_E = 10$. In the simulation, the initial values for m and n, the numbers of S and ES, are set at 100 and 0. Since the enzyme reaction is irreversible, both numbers of S and ES eventually go to zero.

11.5 The CME model for protein synthesis in a single cell

The stochastic behavior of biochemical reactions has been observed in single living cells as well as in in vitro experiments. Sunney Xie and colleagues have observed a stochastic, randomly timed bursting behavior in the production of proteins, i.e., translation of an mRNA [26, 217]. The observed number of the protein molecules produced in each burst, n_P, follows a geometric distribution,

$$p_{n_P}(\ell) = \Pr\{n_P = \ell\} = (1 - \theta)\theta^\ell, \quad \ell = 1, 2, \cdots, \tag{11.34}$$

and the duration between two consecutive bursts, T, follows an exponential probability distribution,

$$f_T(t) = \lambda e^{-\lambda t}. \tag{11.35}$$

Based on the CME framework it is possible to demonstrate that these observations are exactly consistent with a very simple kinetic model for translation. In this model we let R be a ribosome that carries out the protein synthesis. We further assume that a ribosome and an mRNA can form a complex that continues to synthesize copies of the protein until the complex dissociates. The kinetic scheme is

$$R + mRNA \underset{k_2}{\overset{k_1^o}{\rightleftharpoons}} mRNA \cdot R,$$

$$mRNA \cdot R + n\, AA \overset{k_3^o}{\longrightarrow} mRNA \cdot R + P, \tag{11.36}$$

in which AA represents a constant source of amino acid-loaded tRNA, n is the number of amino acids in the protein, and P represents synthesized protein.

We define pseudo-first order rate constants $k_1 = k_1^o[mRNA]$, $k_3 = k_3^o[AA]^n$. In addition, for simplicity, we assume there is only a single ribosome and that the mRNA concentration remains fixed. The CME for the kinetic in Equation (11.36) then is

$$\frac{dp_0(\ell, t)}{dt} = k_1 p_1(\ell) - k_2 p_0(\ell) - k_3 p_0(\ell) + k_3 p_0(\ell - 1), \tag{11.37a}$$

$$\frac{dp_1(\ell, t)}{dt} = -k_1 p_1(\ell) + k_2 p_0(\ell), \tag{11.37b}$$

where $p_m(\ell, t) = \Pr\{n_P(t) = \ell, n_R(t) = m\}$, $m = 0, 1$.

The solution to Equation (11.37) can be obtained in two parts. From the two-state transition $R \rightleftharpoons mRNA \cdot R$ it follows that the duration between two bursts of protein synthesis is exponentially distributed with mean time $1/k_1$. The duration of each burst is also exponentially distributed with mean time $1/k_2$. Thus the λ in Equation (11.35) is related to the kinetic constant: $\lambda = k_1$.

For each burst, we can obtain the probability distribution for the number of proteins produced from Equation (11.37) with $k_1^o = 0$:

$$\frac{dp_0(\ell, t)}{dt} = -(k_2 + k_3)\, p_0(\ell) + k_3 p_0(\ell - 1), \tag{11.38a}$$

$$\frac{dp_1(\ell, t)}{dt} = k_2 p_0(\ell), \tag{11.38b}$$

$$p_{n_P}(\ell) = p_1(\ell, \infty), \tag{11.38c}$$

with initial condition $p_0(\ell, 0) = \delta_{\ell 0}$, $p_1(\ell, 0) = 0$.

Equation (11.38a) can be solved to obtain

$$p_0(\ell, t) = \frac{(k_3 t)^\ell}{\ell!} e^{-(k_2 + k_3)t}.$$ (11.39)

Therefore, Equations (11.38b) and (11.38c) together give

$$p_{n_p}(\ell) = k_2 \int_0^\infty p_0(\ell, t) dt = \frac{k_2 k_3^\ell}{(k_2 + k_3)^{\ell+1}}.$$ (11.40)

Comparing Equation (11.40) with Equation (11.34), we have $\theta = k_3/(k_2 + k_3)$. Equation (11.40), which has been confirmed in recent experiments [26, 217], was first predicted theoretically by Berg [22] based on a model similar to that of Equation (11.36). The assumption we made here about weak ribosome–mRNA association with multiple translation possible while each complex exists can be replaced by assuming that mRNA complexes with multiple ribosomes, as suggested by Cai *et al.* [26].

Recalling the definitions of the pseudo-first order rate constants, we have

$$\theta = \frac{k_2}{k_2 + k_3^o[AA]^n}, \quad \lambda = k_1^o[\text{mRNA}].$$ (11.41)

The CME model for stochastic protein production provides specific predictions: the λ increase with the level of mRNA in a cell, and θ is a function of the concentrations of the amino acids charged tRNA, and the size of the protein.

Concluding remarks

It is widely appreciated that chemical and biochemical reactions in the condensed phase are stochastic. It has been more than 60 years since Delbrück studied a stochastic chemical reaction system in terms of the chemical master equation. Kramers' theory, which connects the rate of a chemical reaction with the molecular structures and energies of the reactants, is established as a central component of theoretical chemistry [77]. Yet study of the dynamics of chemical and biochemical reaction systems, in terms of either deterministic differential equations or the stochastic CME, is not the exclusive domain of chemists. Recent developments in the simulation of reaction systems are the work of many sorts of scientists, ranging from control engineers to microbiologists, all interested in the dynamic behavior of biochemical reaction systems [199, 210].

Stochastic models for biochemical reaction systems in terms of the CME are not an alternative to the differential equation approach, but a more general theoretical framework that deserves further investigation. In particular, the relation between the dynamic CME and the general theory of statistical thermodynamics of closed and open systems requires further elucidation. We suggest that the importance of the CME to small biochemical reaction systems is on a par with the Boltzmann equation for gases and the Navier–Stokes equation for fluids. This is a big claim; whether it is justified remains to be tested through more studies along the lines presented in this chapter to simple as well as ever more complex biochemical reaction systems. It is also worth pointing out that one of the most promising laboratory methods for testing predictions from the CME is fluorescence correlation spectroscopy (FCS), which can be used to measure the concentration fluctuations in a small non-linear reaction system [173], beyond the conformational transition of single molecules.

Exercises

11.1 Show that Equations (11.10) and (11.11) are related by $G(s, t) = g^N(s, t)$.

11.2 To find the deterministic steady state of reaction system of Equation (11.25), set the right-hand side of Equation (11.26) to zero and find the root(s) to the equation. Using the parameters given in Figure 11.6 ($\alpha_1 = \alpha_2 = 10^{-6}$, $\beta_1 = \beta_2 = 0.05$, $a = 500$, $b = 1$, and $c = 20$) numerically find the steady states. How many steady states are there? Compare these results with Figure 11.6. What happens if all the parameters are the same as above but $a = 100$?

11.3 Show that the concentrations a, b, and c in the system of Equation (11.25) satisfy $c/(ab) = \alpha_1\beta_1/(\alpha_2\beta_2)$ in equilibrium. Show that if this equation holds true then it is impossible to have bistability. Rather, the distribution is Poissonian.

11.4 Perform a computer simulation of the CME for the system of Equation (11.25), using the parameters given in the legend of Figure 11.6. (Assume that A, B, and C are held at fixed numbers.) From the simulated stochastic trajectory, can you reproduce the steady state probability distribution shown in Figure 11.6?

12

Appendix: the statistical basis of thermodynamics

Overview

To truly appreciate how thermodynamic principles apply to chemical systems, it is of great value to see how these principles arise from a statistical treatment of how microscopic behavior is reflected on the macroscopic scale. While this appendix by no means provides a complete introduction to the subject, it may provide a view of thermodynamics that is refreshing and exciting for readers not familiar with the deep roots of thermodynamics in statistical physics. The primary goal here is to provide rigorous derivations for the probability laws used in Chapter 1 to introduce thermodynamic quantities such as entropy and free energies.

12.1 The NVE ensemble

Thermodynamic principles arise from a statistical treatment of matter by studying different idealized ensembles of particles that represent different thermodynamic systems. The first ensemble that we study is that of an isolated system: a collection of N particles confined to a volume V, with total internal energy E. A system of this sort is referred to as an NVE system or *ensemble*, as N, V, and E are the three thermodynamic variables that are held constant. N, V, and E are *extensive* variables. That is, their values are proportional to the size of the system. If we combine NVE subsystems into a larger system, then the total N, V, and E are computed as the sums of N, V, and E of the subsystems. Temperature, pressure, and chemical potential are *intensive* variables, for which values do not depend on the size of the system.

An NVE system is also referred to as a *microcanonical ensemble* of particles. In addition to the NVE system, we will encounter NVT (*canonical*) and NPT (*isobaric*) systems. Sticking for now to the NVE system, let us imagine that for any given thermodynamic state, or *macrostate*, the many particles making up the system may adopt a number of possible configurations, or *microstates*. Thus each macrostate (defined by N, V, and E) has associated with it a number of microstates for which the N particles confined to the fixed volume V have total energy adding up to E. According to the principles of quantum mechanics there is a finite number of microscopic states that may be adopted by such a system. We denote this number by $\Omega(N, V, E)$ – the number $\Omega(N, V, E)$ can be obtained by counting the number of independent solutions to the Schrödinger equation that the system can adopt for a given eigenvalue E of the Hamiltonian [156]. The quantity Ω is commonly referred to as the *microcanonical partition function*, a partition function being a statistically weighted sum over the possible states of a system. The microcanonical partition function Ω is a non-biased enumeration of the microstates.

For classical systems the microstates are not discrete and the number of possible states for a fixed NVE ensemble is in general not finite. To see this imagine a system of a single particle ($N = 1$) traveling in an otherwise empty box of volume V. There are no external force fields acting on the particle so its total energy is $E = \frac{1}{2}mv^2$. The particle could be found in any location within the box, and its velocity could be directed in any direction without changing the thermodynamic macrostate defined by the fixed values of N, V, and E. To apply ensemble theory to classical systems $\Omega(N, V, E)$ is defined as the (appropriately scaled) total volume accessible by the state variables of position and momentum accessible by the particles in the system.

Applications in statistical mechanics are based on constructing expressions for $\Omega(N, V, E)$ (and other partition functions for various ensembles) based on the nature of the interactions of the particles in a given system. To understand how thermodynamic principles arise from statistics, however, it is not necessary to worry about how one might go about computing $\Omega(N, V, E)$, or how Ω might depend on N, V, and E for particular systems (classical or quantum mechanical). It is necessary simply to appreciate that the quantity $\Omega(N, V, E)$ exists for an NVE system.

Pathria [156] begins his treatment of the subject with the following thought experiment. Consider two systems, denoted system 1 and system 2, having macrostates defined by (N_1, V_1, E_1) and (N_2, V_2, E_2), respectively. Imagine that these two systems are in thermal contact (see Figure 12.1). By *thermal contact* we mean that the systems are allowed to exchange energy, but nothing else. That is E_1 and E_2 may change, but N_1, N_2, V_1, and V_2 remain fixed. Of course the total energy remains fixed as well. That is,

$$E_0 = E_1 + E_2 = \text{constant}, \qquad (12.1)$$

if the two systems interact with only one another.

| N_1, V_1, E_1 | N_2, V_2, E_2 |

Figure 12.1 Two NVE systems in thermal contact.

Now we introduce a fundamental postulate of statistical thermodynamics: at a given N_1, V_1, and E_1, system 1 is equally likely to be in any one of its Ω_1 microstates; similarly system 2 is equally likely to be in any one of its Ω_2 microstates (more on this assumption later). The combined system, consisting of systems 1 and 2, has associated with it a total partition function $\Omega_0(E_1, E_2)$, which represents the total number of possible microstates. The number $\Omega_0(E_1, E_2)$ may be expressed as the multiplication:

$$\Omega_0(E_1, E_2) = \Omega_1(E_1)\Omega_2(E_2). \tag{12.2}$$

To seek the statistical consequences of thermal equilibrium, we look for the distribution of internal energy (values of E_1 and E_2) for which the number of microstates $\Omega_0(E_1, E_2)$ achieves its maximum value. We will call this achievement equilibrium, or more specifically thermal equilibrium, the assumption here being that physical systems naturally move from less probable macrostates to more probable macrostates.[1] Due to the large numbers with which we deal on the macro level ($N \sim 10^{23}$), the most probable macrostate is orders of magnitude more probable than even closely related macrostates [156]. That means that for equilibrium we must maximize $\Omega_0(E_1, E_2)$ under the constraint that the sum $E_0 = E_1 + E_2$ remains constant.

At the maximum $\partial \Omega_0/\partial E_1 = 0$, or

$$\frac{\partial \left[\Omega_1(E_1)\Omega_2(E_2)\right]}{\partial E_1} = \left[\frac{\partial \Omega_1}{\partial E_1}\Omega_2 + \Omega_1\frac{\partial E_2}{\partial E_1} \cdot \frac{\partial \Omega_2}{\partial E_2}\right]_{E_1=E_1^*, E_2=E_2^*} = 0, \tag{12.3}$$

where (E_1^*, E_2^*) denote the maximum point. Since $\partial E_2/\partial E_1 = -1$ from Equation (12.1), Equation (12.3) reduces to:

$$\frac{1}{\Omega_1} \cdot \frac{\partial \Omega_1}{\partial E_1}(E_1^*) = \frac{1}{\Omega_2} \cdot \frac{\partial \Omega_2}{\partial E_2}(E_2^*), \tag{12.4}$$

which is equivalent to

$$\frac{\partial}{\partial E_1} \ln \Omega_1(E_1^*) = \frac{\partial}{\partial E_2} \ln \Omega_2(E_2^*). \tag{12.5}$$

---

[1] Again, the term *macrostate* refers to the thermodynamic state of the composite system, defined by the variables N_1, V_1, E_1, and N_2, V_2, E_2. A more probable macrostate will be one that corresponds to more possible microstates than a less probable macrostate.

To generalize, for any number of systems in equilibrium thermal contact,

$$\frac{\partial}{\partial E} \ln \Omega = \beta = \text{constant} \tag{12.6}$$

for each system.

Next consider that systems 1 and 2 are not only in thermal contact, but also their volumes are allowed to change in such a way that both the total energy E_o and the total volume $V_0 = V_1 + V_2$ remain constant. For this example imagine a flexible wall separates the two chambers – the wall flexes to allow pressure to equilibrate between the chambers, but the particles are not allowed to pass. Thus N_1 and N_2 remain fixed. For such a system we find that maximizing $\Omega_0(V_1, V_2)$ yields

$$\frac{\partial [\Omega_1(V_1)\Omega_2(V_2)]}{\partial V_1} = \left[\frac{\partial \Omega_1}{\partial V_1} \Omega_2 + \Omega_1 \frac{\partial V_2}{\partial V_1} \cdot \frac{\partial \Omega_2}{\partial V_2} \right]_{V_1 = V_1^*, V_2 = V_2^*} = 0, \tag{12.7}$$

or

$$\frac{1}{\Omega_1} \cdot \frac{\partial \Omega_1}{\partial V_1}(V_1^*) = \frac{1}{\Omega_2} \cdot \frac{\partial \Omega_2}{\partial V_2}(V_2^*), \tag{12.8}$$

or

$$\frac{\partial}{\partial V_1} \ln \Omega_1(V_1^*) = \frac{\partial}{\partial V_2} \ln \Omega_2(V_2^*), \tag{12.9}$$

or

$$\frac{\partial}{\partial V} \ln \Omega = \eta = \text{constant.} \tag{12.10}$$

We shall see that the variables β and η are related to the intensive thermodynamic quantities temperature and pressure, respectively. But before completing the picture of how macroscopic thermodynamics emerges from the NVE ensemble, we first have one more equilibration to consider – concentration equilibration. For this case, imagine that the partition between the chambers is perforated and particles are permitted to freely travel from one system to the next. The equilibrium statement for this system is

$$\frac{\partial}{\partial N} \ln \Omega = \zeta = \text{constant.} \tag{12.11}$$

The derivation of the above equation for concentration equilibration follows the same approach as the derivations for thermal and pressure equilibration.[2]

---

[2] See exercise 1.

To summarize, we have the following:

For constant N and V (thermal equilibrium) we arrive at

$$\left(\frac{\partial \ln \Omega}{\partial E}\right)_{N,V} = \beta \quad (d \ln \Omega)_{N,V} = \beta dE. \tag{12.12}$$

For constant E and N (pressure equilibrium) we arrive at

$$\left(\frac{\partial \ln \Omega}{\partial V}\right)_{E,N} = \eta \quad (d \ln \Omega)_{E,N} = \eta dV. \tag{12.13}$$

For constant E and V (concentration equilibrium) we arrive at

$$\left(\frac{\partial \ln \Omega}{\partial N}\right)_{E,V} = \zeta \quad (d \ln \Omega)_{E,V} = \zeta dN. \tag{12.14}$$

From the above expressions we write

$$d \ln \Omega = (d \ln \Omega)_{N,V} + (d \ln \Omega)_{E,N} + (d \ln \Omega)_{E,V}$$
$$= \beta dE + \eta dV + \zeta dN, \tag{12.15}$$

which relates changes in $\Omega(N, V, E)$ to changes in N, V, and E in a given system. The variables β, η, and ζ are clearly intensive thermodynamic variables. How are they related to the intensive thermodynamic parameters familiar to our everyday experience – temperature, pressure, and chemical potential? Remember that the variable β is constant for systems in thermal equilibrium – i.e., systems having the same temperature. Pressure equilibrium implies constant η, and concentration equilibrium implies constant ζ. The remaining key is in the formula $S = k_B \ln \Omega$, which was introduced by Boltzmann. As Pathria puts it, this formula "provides a bridge between the microscopic and the macroscopic" [156]. With this key, we can assign physical meaning to the variables in Equation (12.15) through comparison to Equation (1.1). From this comparison, we arrive at the following.

$$\beta = 1/k_B T$$
$$\eta = P/k_B T$$
$$\zeta = -\mu/k_B T. \tag{12.16}$$

To summarize, we have shown that a specific physical interpretation of the intensive variables governed by Equation (1.1) – temperature, pressure, and chemical potential – arises from the assumption that systems move to thermodynamic macrostates that maximize the number of accessible microstates. This is our first application of the famous second law of thermodynamics, which, as is implicit in the above derivations, is stated as *the entropy of a closed system never decreases*. It is worth noting that our interpretation of the intensive thermodynamic variables is not arbitrary. For example, the temperature of the NVE ensemble is expressed as

$$T = \frac{1}{k_B} \left(\frac{\partial E}{\partial \ln \Omega(N, V, E)} \right)_{N,V} = \left(\frac{\partial E}{\partial S} \right)_{N,V}. \tag{12.17}$$

This expression serves as a precise mathematical definition of temperature. It is interesting to note that temperature, a variable with which we have intuitive and sensory familiarity, is defined based on entropy, one with which we may be less familiar. In fact, we shall see that entropy and temperature are intimately related in the concept of free energy, in which temperature determines the relative importances of energy and entropy in driving thermodynamic processes.

12.2 The NVT ensemble

While the NVE (microcanonical) ensemble theory is sound and useful, the NVT (canonical) ensemble (which fixes the number of particles, volume, and temperature while allowing the energy to vary) proves more convenient than the NVE for numerous applications.

12.2.1 Boltzmann statistics and the canonical partition function: a derivation

Our study of the NVT ensemble begins by treating a large heat reservoir thermally coupled to a smaller system using the NVE approach. The energy of the heat reservoir is denoted E_r and the energy of the smaller subsystem, E_i. The composite system is assumed closed and the total energy is fixed: $E_r + E_i = E_o = $ constant. The composite system is assumed to be a closed NVE system and the subsystem is assumed to have constant N and V.

This system is illustrated in Figure 12.2. For a given microstate i and corresponding energy E_i of the subsystem, the reservoir can obtain $\Omega_r(E_o - E_i)$ microstates, where Ω_r is the microcanonical partition function of the heat reservoir. For each state i obtained by the subsystem, the total number of states available to the composite system is enumerated by Ω_r, the partition function for the reservoir. According to our standard assumption that the probability of a state is proportional to the number of microstates available to the system:

$$P_i \sim \Omega_r(E_o - E_i). \tag{12.18}$$

In addition, from Equation (12.6) we know that $\ln \Omega_r$ is equal to $\beta E_r + a$, where the constant a does not depend on the energy of the system. From this relationship we have

$$\Omega_r(E_o - E_i) \sim e^{\beta(E_o - E_i)} \tag{12.19}$$

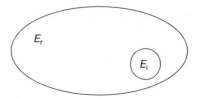

Figure 12.2 A system with energy E_i thermally coupled to a large heat reservoir with energy E_r.

and therefore

$$P_i \sim e^{-\beta E_i}, \tag{12.20}$$

where $\beta = 1/k_B T$ has been defined previously. Equation (12.20) is called the Boltzmann probability law; it is the central result in NVT ensemble theory. It tells us how the probability of a given state of a system in a constant-temperature heat bath depends on its energy.

It should be noted that in our derivation (as is the usual case in derivations of the Boltzmann probability law) we have invoked the idea that a subsystem is embedded in a much larger heat reservoir. However, this idea that one system in Figure 12.2 is much larger than the other is not applied or needed in this derivation. The important things are that the composite system is large enough to be thought of as having a constant temperature and that the subsystem is held at fixed N and V. (Constant N and V allow us to use the relationship $\ln \Omega_r = \beta E_r = a$.)

12.2.2 Another derivation

A second approach to the NVT ensemble found in Feynman's lecture notes on statistical mechanics [55] is also based on the central idea from NVE ensemble theory that the probability of a microstate is proportional to the number of microstates available to the system. Thus

$$\frac{P(E_1)}{P(E_2)} = \frac{\Omega(E_o - E_1)}{\Omega(E_o - E_2)} \tag{12.21}$$

where again E_o is the total energy of the system and a heat reservoir to which the system is coupled. The energies E_1 and E_2 are possible energies of the system and Ω is the microcanonical partition function for the reservoir. (The subscript r has been dropped.)

Next Feynman makes use of the fact that energy is defined only up to an additive constant. In other words, there is no absolute energy value, and we can always add a constant, say ϵ, so long as we add the same ϵ to all relevant values of energy.

Without changing its physical meaning Equation (12.21) can be modified:

$$\frac{P(E_1)}{P(E_2)} = \frac{\Omega(E_o - E_1 + \epsilon)}{\Omega(E_o - E_2 + \epsilon)} . \tag{12.22}$$

Next we define the function $g(x) = \Omega(E_o - E_2 + x)$. Equating the right-hand sides of Equations (12.21) and (12.22) results in

$$\Omega(E_o - E_1)\Omega(E_o - E_2 + \epsilon) = \Omega(E_o - E_2)\Omega(E_o - E_1 + \epsilon) \tag{12.23}$$

or

$$g(E_2 - E_1)g(\epsilon) = g(0)g(\epsilon + E_2 - E_1) . \tag{12.24}$$

Equation (12.24) is uniquely solved by:

$$g(\epsilon) = g(0)e^{\beta\epsilon}, \tag{12.25}$$

where β is some constant.[3] Therefore the probability of a given energy E is proportional to $e^{-\beta E}$, which is the result from Section 12.2.1. To take the analysis one step further we can normalize the probability:

$$P(E) = \frac{e^{-\beta E}}{Q}, \tag{12.26}$$

where

$$Q = \sum_i e^{-\beta E_i} \tag{12.27}$$

is called the canonical partition function and Equation (12.26) defines the NVT probability distribution function. (Feynman does not go on to say why $\beta = 1/k_B T$, as we saw is the case in the previous derivation.) The quantity Q is called the canonical partition function. Summation in Equation (12.27) is over all possible microstates. Equation (12.27) is Equation #1 on the first page of Feynman's notes on statistical mechanics [55]. Feynman calls this equation the "summit of statistical mechanics" and notes that "the entire subject is either the slide-down from this summit ... or the climb-up." The climb took us a little bit longer than it takes Feynman, but we got here just the same.

12.2.3 One more derivation

Since the NVT probability distribution function is the summit, it may be instructive to scale the peak once more via a different route. In particular we seek a derivation that stands on its own and does not rely on the NVE theory introduced earlier. The

---

[3] See exercise 4.

following derivation follows from those found in Section 3.2 of Pathria [156] and Chapter 1 of Hill [86].

Consider a collection of M identical systems, which are thermally coupled and thus share energy at a constant temperature. If we label the possible states of the system $i = 1, 2, \ldots$ and denote the energy of these obtainable microstates as E_i, then the total number of subsystems in the system is equal to the summation,

$$\sum_i n_i = M, \qquad (12.28)$$

where n_i are the number of systems that correspond to microstate i. The total energy of the ensemble can be computed as

$$\sum_i n_i E_i = MU \qquad (12.29)$$

where U is the average internal energy of the systems in the ensemble.

Equations (12.28) and (12.29) represent constraints on the ways microstates can be distributed among the members of the ensemble. Analogous to our study of NVE statistics, here we assume that the probability of obtaining a given set $\{n_i\}$ of numbers of systems in each microstate is proportional to the number of ways this set can be obtained. Imagine the numbers n_i to represent bins counting the number of systems at a given state. Since the systems are identical, they can be shuffled about the bins as long as the numbers n_i remain fixed. The number of possible ways to shuffle the states about the bins is given by:

$$W(\{n_i\}) = \frac{M!}{n_1! n_2! \ldots}. \qquad (12.30)$$

One way to arrive at the canonical distribution is via maximizing the number W under the constraints imposed by Equations (12.28) and (12.29). At the maximum value,

$$\nabla W \cdot \vec{\delta} = 0, \qquad (12.31)$$

where the gradient operator is $\nabla = \{\frac{\partial}{\partial n_1}, \frac{\partial}{\partial n_2}, \ldots\}$, and $\vec{\delta}$ is a vector which represents a direction allowed by the constraints.[4]

We can maximize the number W by using the method of Lagrange multipliers. Again, it is convenient to work with the logarithm of the number W, which allows

---

[4] The occasional mathematician will point out the hazards of taking the derivative of a discontinuous function with respect to a discontinuous variable. Easy-going types will be satisfied with the explanation that for astronomically large numbers of possible states, the function W and the variables $\{n_i\}$ are effectively continuous. Sticklers for mathematical rigor will have to find satisfaction elsewhere.

us to apply Stirling's approximation, $\ln x! \approx x \ln x - x$. Thus

$$\ln W = M \ln(M) - \sum_i n_i \ln(n_i). \qquad (12.32)$$

This equation is maximized by setting

$$\nabla \ln W - \alpha \nabla \sum_i n_i - \beta \nabla \sum_i n_i E_i = 0, \qquad (12.33)$$

where α and β are the unknown Lagrange multipliers. The second two terms in this equation are the gradients of the constraint functions. Evaluating Equation (12.33) results in:

$$-\ln n_i - 1 - \alpha - \beta E_i = 0 \qquad (12.34)$$

for each of the independent entries of the gradients in Equation (12.33). Thus Equation (12.34) gives us a straightforward expression for the optimal n_i:

$$n_i = e^{-(a + \beta E_i)} \qquad (12.35)$$

where $a = \alpha + 1$ and the unknown constants a and β can be obtained by returning to the constraints.

The probability of a given state j can be computed from $P_j = n_j/M = (e^{-a}/M)e^{-\beta E_j}$. Since $\sum_i P_i = 1$, $a = \ln(M/\sum_i e^{-\beta E_i})$, and

$$P_j = \frac{e^{-\beta E_j}}{\sum_i e^{-\beta E_i}} \qquad (12.36)$$

which is by now familiar as the NVT probability distribution function. As you might guess, the parameter β once again turns out to be $1/k_B T$ when we examine the thermodynamics of the NVT ensemble.

Note that the above derivation assumed that the numbers of states $\{n_i\}$ assumes the most probable distribution, e.g., maximizes W. For a more rigorous approach, which evaluates the expected values of n_i, see Section 3.2 of Pathria [156].

12.2.4 Equipartition

The equipartition theorem, which describes the correlation structure of the variables of a Hamiltonian system in the NVT ensemble, is a central component of the field of statistical mechanics. Although the intent of this chapter is to introduce aspects of statistical thermodynamics essential for the remainder of this book – and not to be a complete text on statistical mechanics – the equipartition theorem provides an interpretation of the intrinsic variable T that is useful in guiding our intuition about temperature in chemical reaction systems.

Here we derive the theorem for classical systems. The classical laws of motion can be formulated in terms of the Hamiltonian function for the particles in a system, which is defined in terms of the particle positions $\{q_i\}$ and momenta $\{p_i = m_i \dot{q}_i\}$. Let the scalar quantities p_i and q_i be the entries of the vectors $\mathbf{p}$ and $\mathbf{q}$. For a collection of N particles $\mathbf{p} \in \mathfrak{R}^{3N}$ and $\mathbf{q} \in \mathfrak{R}^{3N}$ are the collective positions and momenta vectors listing all $3N$ entries. The Hamiltonian function is an expression of the total energy of a system:

$$\mathcal{H} = \sum_{i=1}^{3N} \frac{p_i^2}{2m_i} + U(q_1, q_2, \ldots, q_{3N}). \tag{12.37}$$

The equations of motion are written as,

$$\dot{q}_i = \frac{\partial \mathcal{H}}{\partial p_i} \tag{12.38}$$

$$\dot{p}_i = -\frac{\partial \mathcal{H}}{\partial q_i}, \tag{12.39}$$

which are equivalent to Newton's second law:

$$\dot{q}_i = p_i/m_i, \quad \dot{p}_i = -\partial U/\partial q_i = f_i. \tag{12.40}$$

To derive the equipartition theorem we denote the $6N$ independent momentum and position coordinates by x_i and seek to evaluate the ensemble average:

$$\left\langle x_i \frac{\partial \mathcal{H}}{\partial x_j} \right\rangle = \frac{\int \left(x_i \frac{\partial \mathcal{H}}{\partial x_j} \right) e^{-\beta \mathcal{H}} d^{6N}\mathbf{x}}{\int e^{-\beta \mathcal{H}} d^{6N}\mathbf{x}}, \tag{12.41}$$

where the integration $d^{6N}\mathbf{x}$ is over all possible values of the $6N$ x coordinates. The Hamiltonian $\mathcal{H}$ depends on the internal coordinates although the dependence is not explicitly stated in Equation (12.41).

Using integration by parts in the numerator to carry out the integration over the x_j coordinate produces:

$$\left\langle x_i \frac{\partial \mathcal{H}}{\partial x_j} \right\rangle = \frac{\int \left[\left(-\frac{x_i}{\beta} e^{-\beta \mathcal{H}} \right) \Big|_{x_j^-}^{x_j^+} + \int \frac{1}{\beta} \left(\frac{\partial x_i}{\partial x_j} \right) e^{-\beta \mathcal{H}} dx_j \right] d^{6N-1}\mathbf{x}}{\int e^{-\beta \mathcal{H}} d^{6N}\mathbf{x}} \tag{12.42}$$

where the integration over $d^{6N-1}\mathbf{x}$ indicates integration over all x coordinates excluding x_j. The notation x_j^- and x_j^+ indicates the extreme values accessible to the coordinate x_j. Thus for a momentum coordinate these extreme values would be $\pm\infty$, while for a position coordinate the extreme values would come from the boundaries of the container. In either case, the first term of the numerator in

Equation (12.42) vanishes because the Hamiltonian is expected to become infinite at the extreme values of the coordinates.

Equation (12.42) can be further simplified by noting that since the coordinates are independent, $\partial x_i / \partial x_j = \delta_{ij}$, where δ_{ij} is the usual Kronecker delta function. ($\delta_{ij} = 1$ for $i = j$; $\delta_{ij} = 0$ for $i \neq j$.) After simplification we are left with

$$\left\langle x_i \frac{\partial \mathcal{H}}{\partial x_j} \right\rangle = k_B T \delta_{ij}, \tag{12.43}$$

which is the general form of the equipartition theorem for classical systems. It should be noted that this theorem is only valid when all coordinates of the system can be freely and independently excited, which may not always be the case for certain systems at low temperatures. So we should keep in mind that the equipartition theorem is rigorously true only in the limit of high temperature.

Equipartition tells us that for any coordinate $\langle x \frac{\partial \mathcal{H}}{\partial x} \rangle = k_B T$. Applying this theorem to a momentum coordinate, p_i, we find,

$$\left\langle p_i \frac{\partial \mathcal{H}}{\partial p_i} \right\rangle = \langle p_i \dot{q}_i \rangle = k_B T. \tag{12.44}$$

Similarly,

$$\langle q_i \dot{p}_i \rangle = -k_B T. \tag{12.45}$$

From Equation (12.44), we see that the average kinetic energy associated with the ith coordinate is $\langle m v_i^2 / 2 \rangle = k_B T / 2$. For a three-dimensional system, the average kinetic energy of each particle is specified by $3 k_B T / 2$. If the potential energy of the Hamiltonian is a quadratic function of the coordinates, then each degree of freedom will contribute $k_B T / 2$ energy, on average, to the internal energy of the system.

12.3 The NPT ensemble

The NPT ensemble can be analyzed by returning to derivation of the NVT partition function in Section 12.2.3, which determined how energy is statistically distributed among a collection of systems in thermal equilibrium. In that case the assumption that each system maintained a constant volume was implicit in the fact that while the energy was allowed to exchange between the subsystems, the subsystems were treated as identical in every other way. Holding the mean energy constant at the value U provides an important constraint on how the total energy may be distributed among the subsystems.

To obtain the statistical distribution for the NPT case (constant pressure, constant temperature), we hold the mean volume of each subsystem fixed at V and allow the volume of the subsystems to vary under the constraint

$$\sum_i n_i V_i = MV. \tag{12.46}$$

Adding this constraint to those of Section 12.2.3 and maximizing the number of ways to shuffle the system, we have

$$\nabla \ln W - \alpha \nabla \sum_i n_i - \beta \nabla \sum_i n_i E_i - \gamma \nabla \sum_i n_i V_i = 0, \tag{12.47}$$

where γ is the additional Lagrange multiplier associated with Equation (12.46). Evaluating this equation leads to:

$$-\ln n_i - 1 - \alpha - \beta E_i - \gamma V_i = 0 \tag{12.48}$$

for each of the independent entries of the gradients in Equation (12.33).

Defining the thermodynamic quantity *enthalpy* $H = E + PV$ and recognizing that $\gamma = P/k_B T$, we have

$$n_i \sim e^{-\beta(E_i + PV_i)} \tag{12.49}$$

and the NPT probability law

$$P_i = e^{-H_i/k_B T}/Z$$

$$Z = \sum_i e^{-H_i/k_B T}. \tag{12.50}$$

The above formula for Z, the NPT partition function, was first reported by Guggenheim [74], who wrote the expression down by analogy rather than based on a detailed derivation. While this form of the partition function is thought to be broadly valid and is widely applied (for example in molecular dynamics simulation [6]), it introduces the conceptual difficulty that the meaning of the discrete volumes $\{V_i\}$ is not clear. Discrete energy states arise naturally from quantum statistics. Yet it is not necessarily obvious what discrete volumes to sum over in Equation (12.50). In fact for most applications it makes sense to replace the discrete sum with a continuous volume integral. Yet doing so results in a partition function that has units of volume, which is inappropriate for a partition function that formally should be unitless.

Hill discusses how to make reasonable choices for the discrete volumes in the sum [85] and more recent authors have developed a complete theory of the NPT ensemble where continuous volume integrals are made unitless by the proper normalization [37, 38, 115].

Exercises

12.1 Derive Equation (12.11) from the assumptions stated in Section 12.1.

12.2 Show that the result of Equation (12.20) can be obtained by first taking the logarithm of both sides of Equation (12.18) and expanding $\ln \Omega_r(E_o - E_i)$ about $\ln \Omega_r(E_o)$.

12.3 For a system of non-interacting monatomic particles (an ideal gas) the microcanonical partition function is proportional to V^N. Based on $\Omega \sim V^N$, we can derive the state equation known as the ideal gas law:

$$\frac{P}{k_B T} = \frac{\partial \ln \Omega}{\partial V} = \frac{N}{V}.$$

Consider a gas of particles that do not interact in any way except that each particle occupies a finite volume v_o, which cannot be overlapped by other particles. What consequences does this imply for the gas law? [Hint: use the relationship $\Omega(V) \sim \int d\mathbf{x}$. You might try assuming that each particle is a solid sphere.]

12.4 Show that Equation (12.25) is the unique solution to Equation (12.24).

Bibliography

[1] B. Alberts, A. Johnson, J. Lewis, M. Raff, K. Roberts, and P. Walter. *Molecular Biology of the Cell*. Garland Publishing, Inc., New York, NY, fourth edition, 2002.

[2] R. A. Alberty. Equilibrium compositions of solutions of biochemical species and heats of biochemical reactions. *Proc. Natl. Acad. Sci. USA*, 88:3268–3271, 1991.

[3] R. A. Alberty. Standard transformed formation properties of carbon dioxide in aqueous solutions at specified pH. *J. Phys. Chem.*, 99:11028–11034, 1995.

[4] R. A. Alberty. *Thermodynamics of Biochemical Reactions*. Wiley-Interscience, Hoboken, NJ, 2003.

[5] R. A. Alberty. Thermodynamic properties of weak acids involved in enzyme-catalyzed reactions. *J. Phys. Chem. B*, 110:5012–5016, 2006.

[6] H. C. Anderson. Molecular dynamics simulations at constant pressure and/or temperature. *J. Phys. Chem.*, 72:2384–2393, 1980.

[7] J. S. Bader, A. Chaudhuri, J. M. Rothberg, and J. Chant. Gaining confidence in high-throughput protein interaction networks. *Nat. Biotechnol.*, 22:78–85, 2004.

[8] J. E. Bailey. Toward a science of metabolic engineering. *Science*, 252:1668–1675, 1991.

[9] D. Baker, G. Church, J. Collins *et al.* Engineering life: building a FAB for biology. *Scientific American*, 294:44–51, June 2006.

[10] F. G. Ball and J. A. Rice. Stochastic models for ion channels: introduction and bibliography. *Math. Biosci.*, 112:189–206, 1992.

[11] J. B. Bassingthwaighte and C. A. Goresky. Modeling in the analysis of solute and water exchange in the microvasculature. In E. M. Renkin and C. C. Michel, editors, *Handbook of Physiology. Section 2, The Cardiovascular System, Volume IV, The Microcirculation*, pages 549–626. American Physiological Society, Bethesda, MD, 1984.

[12] V. Batagelj and A. Mrvar. Pajek – Analysis and visualization of large networks. In M. Jüunger and P. Mutzel, editors, *Graph Drawing Software*, pages 77–103. Springer, Berlin, 2003.

[13] D. A. Beard. A biophysical model of the mitochondrial respiratory system and oxidative phosphorylation. *PloS Comput. Biol.*, 1:e36, 2005.

[14] D. A. Beard. Modeling of oxygen transport and cellular energetics explains observations on in vivo cardiac energy metabolism. *PloS Comput. Biol.*, 2:e107, 2006.

[15] D. A. Beard, J. B. Bassingthwaighte, and A. S. Greene. Computational modeling of physiological systems. *Physiol. Genomics*, 23:1–3, 2005.

[16] D. A. Beard, E. Babson, E. E. Curtis, and H. Qian. Thermodynamic constraints for biochemical networks. *J. Theor. Biol.*, 228:327–333, 2004.

[17] D. A. Beard, S. D. Liang, and H. Qian. Energy balance for analysis of complex metabolic networks. *Biophys. J.*, 83:79–86, 2002.

[18] D. A. Beard and H. Qian. Thermodynamic-based computational profiling of cellular regulatory control in hepatocyte metabolism. *Am. J. Physiol.*, 288:E633–E644, 2005.

[19] D. A. Beard and H. Qian. Relationship between thermodynamic driving force and one-way fluxes in reversible processes. *PLoS One*, 2:e144, 2007.

[20] A. Becskei and L. Serrano. Engineering stability in gene networks by autoregulation. *Nature*, 405:590–593, 2000.

[21] J. M. Berg, J. L. Tymoczko, and L. Stryer. *Biochemistry*. W.H. Freeman and Co., New York, NY, fifth edition, 2002.

[22] O. G. Berg. A model for the statistical fluctuations of protein numbers in a microbial population. *J. Theoret. Biol.*, 71:587–603, 1978.

[23] R. B. Bird, W. E. Stewart, and E. N. Lightfoot. *Transport Phenomena*. John Wiley & Sons, Inc., New York, NY, second edition, 2001.

[24] G. E. Briggs and J. B. S. Haldane. A note on the kinetics of enzyme action. *Biochem. J.*, 19:338–339, 1925.

[25] E. Buchner. Cell-free fermentation. Nobel Lecture, December 11, 1907.

[26] L. Cai, N. Friedman, and X. S. Xie. Stochastic protein expression in individual cells at the single molecule level. *Nature*, 440:358–362, 2006.

[27] C. R. Cantor and P. R. Schimmel. *Biophysical Chemistry Part III: The Behavior of Biological Macromolecules*. W. H. Freeman and Company, New York, NY, 1980.

[28] S. Carnot. *Réflections sur la Puissance Motrice du Feu et sur Les Machines Propres a Développer cete Puissance*. Paris, France, 1824.

[29] S. J. Chapman. *MATLAB Programming for Engineers*. Brooks/Cole, Pacific Grove, CA, second edition, 2002.

[30] R. J. Charbeneau. *Groundwater Hydraulics and Pollutant Transport*. Waveland Press Inc., Long Grove, IL, 2006.

[31] B. L. Clarke. Stability of complex reaction networks. *Adv. Chem. Phys.*, 43:1–215, 1980.

[32] B. L. Clarke. Stoichiometric network analysis. *Cell Biophys.*, 12:237–253, 1989.

[33] E. C. W. Clarke and D. N. Glew. Evaluation of Debye–Hückel limiting slopes for water between 0 and 150°C. *J. Chem. Soc., Faraday Trans. 1*, 76:1911–1916, 1980.

[34] W. W. Cleland. The kinetics of enzyme-catalyzed reactions with two or more substrates or products. I. Nomenclature and rate equations. *Biochim. Biophys. Acta*, 67:104–137, 1963.

[35] A. Cornish-Bowden. *Fundamentals of Enzyme Kinetics*. Portland Press, London, UK, third edition, 2004.

[36] A. Cornish-Bowden and M. L. Cárdenas. Information transfer in metabolic pathways. Effects of irreversible steps. *Eur. J. Biochem.*, 268:6616–6624, 2001.

[37] D. S. Corti. Isothermal-isobaric ensemble for small systems. *Phys. Rev. E*, 64:016128, 2001.

[38] D. S. Corti and G. Soto-Campos. Deriving the isothermal-isobaric ensemble: the requirement of a "shell" molecule applicability to small systems. *J. Chem. Phys.*, 108:7959–7966, 1998.

[39] J. Crank. *The Mathematics of Diffusion*. Oxford University Press, Oxford, UK, second edition, 1980.

[40] C. Crone. The permeability of capillaries in various organs as determined by use of the 'indicator diffusion' method. *Acta Physiol. Scand.*, 58:292–305, 1963.

[41] G. E. Crooks. Entropy production fluctuation theorem and the nonequilibrium work relation for free energy differences. *Phys. Rev. E*, 60:2721–2726, 1999.

[42] M. Delbrück. Statistical fluctuations in autocatalytic reactions. *J. Chem. Phys.*, 8:120–124, 1940.

[43] A. Deussen and J. B. Bassingthwaigthe. Modeling [15]oxygen tracer data for estimating oxygen consumption. *Am. J. Physiol.*, 270:H1115–H1130, 1996.

[44] K. A. Dill and S. Bromberg. *Molecular Driving Forces: Statistical Thermodynamics in Chemistry and Biology*. Garland Publishing, Inc., New York, NY, 2002.

[45] M. L. L. Dolar, P. Suarez, P. J. Ponganis, and G. L. Kooyman. Myoglobin in pelagic small cetaceans. *J. Exp. Biol.*, 202:227–236, 1999.

[46] J. R. Dorfman. *An Introduction to Chaos in Nonequilibrium Statistical Mechanics*. Cambridge University Press, London, 1999.

[47] J. T. Edsall. Hemoglobin and the origin of the concept of allosterism. *Fed. Proc.*, 39:226–235, 1980.

[48] J. S. Edwards and B. O. Palsson. The *Escherichia coli* mg1655 in silico metabolic genotype: its definition, characteristics, and capabilities. *Proc. Natl. Acad. Sci. USA*, 97:5528–5533, 2000.

[49] H. Eisenberg. Focal contributions to molecular biophysics and structural biology. *Biophys. Chem.*, 100:33–48, 2003.

[50] M. B. Elowitz and S. Leibler. A synthetic oscillatory network of transcriptional regulators. *Nature*, 403:335–338, 2000.

[51] G. Enciso and E. D. Sontag. On the stability of a model of testosterone dynamics. *J. Math. Biol.*, 49:627–634, 2004.

[52] B. P. English, W. Min, A. M. van Oijen *et al.* Ever-fluctuating single enzyme molecules: Michaelis–Menten equation revisited. *Nature Chem. Biol.*, 2:87–94, 2006.

[53] D. A. Fell. *Understanding the Control of Metabolism*. Portland Press, London, UK, 1996.

[54] D. A. Fell and H. M. Sauro. Metabolic control and its analysis. *Eur. J. Biochem.*, 148:555–561, 1985.

[55] R. P. Feynman. *Statistical Mechanics. A Set of Lectures*. Addison-Wesley, Reading, MA, 1998.

[56] A. Fick. Uber diffusion. *Ann. der Physik*, 94:59–86, 1855.

[57] E. H. Fischer, H. Charbonneau, and N. K. Tonks. Protein tyrosine phosphatases: a diverse family of intracellular and transmembrane enzymes. *Science*, 253:401–406, 1991.

[58] E. H. Fischer and E. G. Krebs. Conversion of phosphorylase b to phosphorylase a in muscle extracts. *J. Biol. Chem.*, 216:121–132, 1955.

[59] J. Forster, I. Famili, P. Fu, B. O. Palsson, and J. Nielsen. Genome-scale reconstruction of the *Saccharomyces cerevisiae* metabolic network. *Genome Res.*, 13:244–253, 2003.

[60] C. Frieden. Slow transitions and hysteretic behavior in enzymes. *Ann. Rev. Biochem.*, 48:471–489, 1979.

[61] C. Frieden and R. A. Alberty. The effect of pH on fumarase activity in acetate buffer. *J. Biol. Chem.*, 212:859–868, 1955.

[62] T. S. Gardner, C. R. Cantor, and J. J. Collins. Construction of a genetic toggle switch in *Escherichia coli*. *Nature*, 403:339–342, 2000.

[63] P. B. Garland and D. Shepherd. The kinetic properties of citrate synthase from rat liver mitochondria. *Biochem. J.*, 114:597–610, 1969.

[64] S. Ghaemmaghami, W.-K. Huh, K. Bower *et al.* Global analysis of protein expression in yeast. *Nature*, 425:737–741, 2003.

[65] C. Giersch. Control analysis and metabolic networks. *Eur. J. Biochem.*, 174:509–519, 1988.

[66] D. T. Gillespie. General method for numerically simulating stochastic time evolution of coupled chemical reactions. *J. Comp. Phys.*, 22:403–434, 1976.

[67] D. T. Gillespie. Exact stochastic simulation of coupled chemical reactions. *J. Phys. Chem.*, 81:2340–2361, 1977.

[68] D. T. Gillespie. Stochastic simulation of chemical kinetics. *Ann. Rev. Phys. Chem.*, 58:35–55, 2007.

[69] A. Goldbeter, G. Dupont, and M. J. Berridge. Minimal model for signal-induced Ca^{2+} oscillations and for their frequency encoding through protein phosphorylation. *Proc. Natl. Acad. Sci. USA*, 87:1461–1465, 1990.

[70] A. Goldbeter and D. E. Koshland. An amplified sensitivity arising from covalent modification in biological systems. *Proc. Natl. Acad. Sci. USA*, 78:6840–6844, 1981.

[71] B. C. Goodwin. Oscillatory behavior in enzymatic control processes. *Adv. Enzym. Regul.*, 3:425–438, 1965.

[72] M. W. Gorman, J. B. Bassingthwaighte, R. A. Olsson, and H. V. Sparks. Endothelial cell uptake of adenosine in canine skeletal muscle. *Am. J. Physiol.*, 250:H482–H489, 1986.

[73] A. Y. Grosberg and A. R. Khokhlov. *Statistical Physics of Macromolecules.* AIP Press, Woodbury, NY, 1994.

[74] E. A. Guggenheim. Grand partition functions and so-called "thermodynamic probability". *J. Chem. Phys.*, 7:103–107, 1939.

[75] A. C. Guyton and J. E. Hall. *Textbook of Medical Physiology.* W. B. Saunders Company, Philadelphia, PA, ninth edition, 1996.

[76] R. D. Hamer, S. C. Nicholas, D. Tranchina, P. A. Liebman, and T. D. Lamb. Multiple steps of phosphorylation of activated rhodopsin can account for the reproducibility of vertebrate rod single-photon responses. *J. Gen. Physiol.*, 122:419–444, 2003.

[77] P. Hänggi, P. Talkner, and M. Borkovec. Reaction-rate theory: fifty years after Kramers. *Rev. Mod. Phys.*, 62:251–341, 1990.

[78] L. H. Hartwell, J. J. Hopfield, S. Leibler, and A. W. Murray. From molecular to modular cell biology. *Nature*, 402:C47–C52, 1999.

[79] S. E. Heffron, R. Moeller, and F. Jurnak. Solving the structure of *Escherichia coli* elongation factor Tu using a twinned data set. *Acta Crystallogr. D Biol. Crystallogr.*, 62:433–438, 2006.

[80] V. Henri. *Comptes Rendus Hebdomadaires des Séances de l'Academie des Sciences, Paris*, 135:916–919, 1902.

[81] V. Henri. *Lois Générales de l'Action des Diastases.* Hermann, Paris, 1903.

[82] B. Hess and A. Boiteux. Oscillatory phenomena in biochemistry. *Annu. Rev. Biochem.*, 40:237–258, 1971.

[83] W. J. Heuett and H. Qian. Combining flux and energy balance analysis to model large-scale biochemical networks. *J. Bioinform. Comput. Biol.*, 4:1227–1243, 2006.

[84] R. Hilgenfeld, J. R. Mesters, and T. Hogg. Insights into the GTPase mechanism of EF-Tu from structural studies. In R. Garrett, S. R. Douthwaite, A. Liljas, A. T. Matheson, P. B. Moore, and H. F. Noller, editors, *The Ribosome: Structure, Function, Antibiotics, and Cellular Interactions*, chapter 28, pages 347–357, "American Society for Microbiology, Washington DC." 2000.

[85] T. L. Hill. *Statistical Mechanics*. McGraw-Hill, New York, NY, 1956.

[86] T. L. Hill. *An Introduction to Statistical Thermodynamics*. Addison-Wesley Publishing Company, Inc., Reading, MA, 1960.

[87] T. L. Hill. Studies in irreversible thermodynamics IV. Diagrammatic representation of steady state fluxes for unimolecular systems. *J. Theor. Biol.*, 10:442–459, 1966.

[88] T. L. Hill. *Thermodynamics for Chemists and Biologists*. Addison-Welsley, Reading, MA, 1968.

[89] T. L. Hill. Approach of certain systems, including membranes, to steady-state. *J. Chem. Phys.*, 54:34–35, 1971.

[90] T. L. Hill. Discrete-time random walks on diagrams (graphs) with cycles. *Proc. Natl. Acad. Sci. USA*, 85:5345–5349, 1988.

[91] T. L. Hill. *Free Energy Transduction and Biochemical Cycle Kinetics*. Dover Publications, Inc., New York, NY, 2004.

[92] T. L. Hill and I. W. Plesner. Studies in irreversible thermodynamics II. A simple class of lattice models for open systems. *J. Chem. Phys.*, 43:267–285, 1965.

[93] B. Hille. *Ion Channels of Excitable Membranes*. Sinauer Associates, Inc., Sunderland, MA, third edition, 2001.

[94] A. L. Hodgkin and A. F. Huxley. A quantitative description of membrane current and its application to conduction and excitation in nerve. *J. Physiol.*, 117:500–544, 1952.

[95] A. L. Hodgkin and A. F. Huxley. Currents carried by sodium and potassium ions through the membrane of the giant axon of Loligo. *J. Physiol.*, 116:449–472, 1952.

[96] A. L. Hodgkin and A. F. Huxley. The components of membrane conductance in the giant axon of Loligo. *J. Physiol.*, 116:473–496, 1952.

[97] A. L. Hodgkin and A. F. Huxley. The dual effect of membrane potential on sodium conductance in the giant axon of Loligo. *J. Physiol.*, 116:497–506, 1952.

[98] A. L. Hodgkin, A. F. Huxley, and B. Katz. Measurement of current–voltage relations in the membrane of the giant axon of Loligo. *J. Physiol.*, 116:424–448, 1952.

[99] S. Hoops, S. Sahle, R. Gauges *et al.* COPASI a COmplex PAthway SImulator. *Bioinformatics*, 22:3067–3074, 2006.

[100] J. J. Hopfield. Kinetic proofreading: a new mechanism for reducing errors in biosynthetic processes requiring high specificity. *Proc. Natl. Acad. Sci. USA*, 71:4135–4139, 1974.

[101] J. J. Hopfield. The energy relay: a proofreading scheme based on dynamic cooperativity and lacking all characteristic symptoms of kinetic proofreading in DNA replication and protein synthesis. *Proc. Natl. Acad. Sci. USA*, 77:5248–5252, 1980.

[102] V. R. T. Hsu. *Ion Transport Through Biological Cell Membranes: From Electro-diffusion to Hodgkin–Huxley via a Quasi Steady-State Approach*. University of Washington, Seattle, WA, Ph.D. dissertation edition, 2004.

[103] E. M. Izhikevich. *Dynamical Systems in Neuroscience: The Geometry of Excitability and Bursting*. The MIT Press, Cambridge, MA, 2007.

[104] E. T. Jaynes. Gibbs vs. Boltzmann entropies. *Am. J. Phys.*, 33:391–398, 1965.

[105] W. P. Jencks, M. Gresser, M. S. Valenzuela, and F. C. Huneeus. Acetyl coenzyme a: arylamine acetyltransferase. Measurement of the steady state concentration of the acetyl-enzyme intermediate. *J. Biol. Chem.*, 247:3756–3760, 1972.

[106] J. A. Jeneson, H. V. Westerhoff, T. R. Brown, C. J. Van Echteld, and R. Berger. Quasi-linear relationship between Gibbs free energy of ATP hydrolysis and power output in human forearm muscle. *Am. J. Physiol.*, 268:C1474–1484, 1995.

[107] A. R. Joyce and B. Ø. Palsson. Toward whole cell modeling and simulation: comprehensive functional genomics through the constraint-based approach. *Prog. Drug Res.*, 64:266–309, 2007.

[108] J. Keener and J. Sneyd. *Mathematical Physiology*. Springer, New York, NY, 1998.

[109] J. Keizer. *Statistical Thermodynamics of Nonequilibrium Processes*. Springer, New York, NY, 1987.

[110] J. Kevorkian and J. D. Cole. *Multiple Scale and Singular Perturbation Methods*. Springer-Verlag, New York, NY, 1996.

[111] K. Y. Kim and J. Wang. Potential energy landscape and robustness of a gene regulatory network: toggle switch. *PLoS Comp. Biol.*, 3:e60, 2007.

[112] E. L. King and C. Altman. A schematic method of deriving the rate laws for enzyme-catalyzed reactions. *J. Phys. Chem.*, 60:1375–1378, 1956.

[113] J. Knight. Bridging the culture gap. *Nature*, 419:244–246, 2002.

[114] M. C. Kohn, M. J. Achs, and D. Garfinkel. Computer simulation of metabolism in pyruvate-perfused rat heart. III. Pyruvate dehydrogenase. *Am. J. Physiol.*, 237:R167–R173, 1979.

[115] G. J. M. Koper and H. Reiss. Length scale for the constant pressure ensemble: application to small systems and relation to Einstein fluctuation theory. *J. Phys. Chem.*, 100:422–432, 1996.

[116] D. E. Koshland, G. Némethy, and D. Filmer. Comparison of experimental binding data and theoretical models in proteins containing subunits. *Biochemistry*, 5:365–385, 1966.

[117] H. A. Kramers. Brownian motion in a field of force and the diffusion model of chemical reactions. *Physica (Utrecht)*, 7:284–304, 1940.

[118] A. Krogh. The number and distribution of capillaries in muscle with calculations of the oxygen pressure head necessary for supplying the tissue. *J. Physiol.*, 52:409–415, 1919.

[119] A. Krogh. Progress of physiology. *Am. J. Physiol.*, 90:243–251, 1929.

[120] K. Kruse and F. Jülicher. Oscillations in cell biology. *Curr. Opin. Cell Biol.*, 17:20–26, 2005.

[121] T. G. Kurtz. The relationship between stochastic and deterministic models for chemical reactions. *J. Chem. Phys.*, 57:2976–2978, 1972.

[122] M. J. Kushmerick. Multiple equilibria of cations with metabolites in muscle bioenergetics. *Am. J. Physiol.*, 272:C1739–C1747, 1997.

[123] N. Lakshminarayanaiah. *Transport Phenomena in Membranes*. Academic Press, New York, NY, 1969.

[124] K. LaNoue, W. J. Nicklas, and J. R. Williamson. Control of citric acid cycle activity in rat heart mitochondria. *J. Biol. Chem.*, 245:102–111, 1970.

[125] E. H. Larsen. Hans H. Ussing – Scientific work: contemporary significance and perspectives. *Biochim. Biophys. Acta*, 1566:2–15, 2002.

[126] S. E. Luria and M. Delbrück. Mutations of bacteria from virus sensitivity to virus resistance. *Genetics*, 28:491–511, 1943.

[127] G. N. Lewis. A new principle of equilibrium. *Proc. Natl. Acad. Sci. USA*, 11:179–183, 1925.

[128] G.-P. Li and H. Qian. Kinetic timing: a novel mechanism for improving the accuracy of GTPase timers in endosome fusion and other biological processes. *Traffic*, 3:249–255, 2002.

[129] S. Lifson and A. Roig. On the theory of helix-coil transition in polypeptides. *J. Chem. Phys.*, 34:1963–1974, 1961.

[130] C. C. Lin and L. A. Segel. *Mathematics Applied to Deterministic Problems in the Natural Sciences*. SIAM, Philadelphia, PA, 1988.

[131] P.-C. Lin, U. Kreutzer, and T. Jue. Anisotropy and temperature dependence of myoglobin translational diffusion in myocardium: implication for oxygen transport and cellular architecture. *Biophys. J.*, 92:2608–2620, 2007.

[132] H. Lineweaver and D. Burk. The determination of enzyme dissociation constants. *J. Am. Chem. Soc.*, 56:658–666, 1934.

[133] H. P. Lu, L. Xun, and X. S. Xie. Single-molecule enzymatic dynamics. *Science*, 282:1877–1882, 1998.

[134] A. E. Martell and R. M. Smith. NIST Standard Reference Database 46 Version 7.0: NIST Critically Selected Stability Constants of Metal Complexes, 2003. NIST Standard Reference Data.

[135] Y. Matsuoka and A. Srere. Kinetic studies of citrate synthase from rat kidney and rat brain. *J. Biol. Chem.*, 248:8022–30, 1973.

[136] B. J. McGuire and T. W. Secomb. Estimation of capillary density in human skeletal muscle based on maximal oxygen consumption rates. *Am. J. Physiol.*, 285:H2382–H2391, 2003.

[137] D. A. McQuarrie. Stochastic approach to chemical kinetics. *J. Appl. Prob.*, 4:413–478, 1998.

[138] D. A. McQuarrie and J. D. Simon. *Physical Chemistry, a Molecular Approach.* University Science Books, Sausalito, CA, 1997.

[139] T. Meyer and L. Stryer. Calcium spiking. *Annu. Rev. Biophys. Biophys. Chem.*, 20:153–174, 1991.

[140] L. Michaelis and M. Menten. Die kinetik der invertinwirkung. *Biochem. Z.*, 49:333–369, 1913.

[141] S. Middelman. *Transport Phenomena in the Cardiovascular System.* Wiley-Interscience, New York, NY, 1972.

[142] H. Mino, J. T. Rubinstein, and J. A. White. Comparison of algorithms for the simulation of action potentials with stochastic sodium channels. *Ann. Biomed. Engr.*, 30:578–587, 2002.

[143] J. Monod, J. Wyman, and J.-P. Changeux. On the nature of allosteric transition: a plausible model. *J. Mol. Biol.*, 12:88–118, 1965.

[144] J. D. Murray. On the molecular mechanism of facilitated oxygen diffusion by haemoglobin and myoglobin. *Proc. R. Soc. Lond. B. Biol. Sci.*, 178:95–110, 1971.

[145] J. D. Murray. On the role of myoglobin in muscle respiration. *J. Theor. Biol.*, 47:115–126, 1974.

[146] J. D. Murray. *Mathematical Biology.* Springer-Verlag, Berlin, second edition, 1993.

[147] D. L. Nelson and M. M. Cox. *Lehninger Principles of Biochemistry.* W. H. Freeman and Company, New York, NY, fourth edition, 2005.

[148] D. D. Nguyen, X. Huang, D. W. Greve, and M. M. Domach. Fibroblast growth and h-7 protein kinase inhibitor response monitored in microimpedance sensor arrays. *Biotechnol. Bioeng.*, 87:138–144, 2004.

[149] D. G. Nicholls and S. J. Ferguson. *Bioenergetics 3.* Academic Press, London, 2002.

[150] J. Ninio. Kinetic amplification of enzyme discrimination. *Biochimie*, 57:587–595, 1975.

[151] J. Ninio. Alternative to the steady-state methods: derivation of reaction rates from first-passage times and pathway probabilities. *Proc. Natl. Acad. Sci. USA*, 84:663–667, 1987.

[152] D. Noble. From the Hodgkin–Huxley axon to the virtual heart. *J. Physiol.*, 580:15–22, 2007.

[153] F. W. J. Olver. Bessel functions of integer order. In M. Abramowitz and I. A. Stegun, editors, *Handbook of Mathematical Functions with Formulas, Graphs, and*

Mathematical Tables, chapter 9, pages 355–436. National Bureau of Standards, Washington, DC, 1964.

[154] L. Onsager. Reciprocal relations in irreversible processes. I. *Phys. Rev.*, 37:405–426, 1931.

[155] L. Onsager. Reciprocal relations in irreversible processes. II. *Phys. Rev.*, 38:2265–2279, 1931.

[156] R. K. Pathria. *Statistical Mechanics*. Butterworth-Heinemann, Oxford, UK, second edition, 1996.

[157] R. N. Pittman. Oxygen transport and exchange in the microcirculation. *Microcirculation*, 12:59–70, 2005.

[158] J. R. Platt. Strong inference. *Science*, 146:347–353, 1964.

[159] D. Poland. On the stability of mass action reactions in open systems. *J. Chem. Phys.*, 94:4427–4439, 1991.

[160] D. Poland and H. A. Scheraga. *Theory of Helix-Coil Transitions in Biopolymers: Statistical Mechanical Theory of Order – Disorder Transitions of Biological Macromolecules*. Academic Press, New York, NY, 1970.

[161] A. S. Popel. Theory of oxygen transport to tissue. *Crit. Rev. Biomed. Eng.*, 17:257–321, 1989.

[162] N. D. Price, J. L. Reed, and B. Ø. Palsson. Genome-scale models of microbial cells: evaluating the consequences of constraints. *Nat. Rev. Microbiol.*, 2:886–897, 2004.

[163] H. Qian. Thermodynamic and kinetic analysis of sensitivity amplification in biological signal transduction. *Biophys. Chem.*, 105:585–593, 2003.

[164] H. Qian. Open-system nonequilibrium steady state: statistical thermodynamics, fluctuations and chemical oscillations. *J. Phys. Chem. B.*, 110:15063–15074, 2006.

[165] H. Qian. Reducing intrinsic biochemical noise in cells and its thermodynamic limit. *J. Mol. Biol.*, 362:387–392, 2006.

[166] H. Qian. Phosphorylation energy hypothesis: open chemical systems and their biological functions. *Annu. Rev. Phys. Chem.*, 58:113–142, 2007.

[167] H. Qian and D. A. Beard. Thermodynamics of stoichiometric biochemical networks in living systems far from equilibrium. *Biophys. Chem.*, 114:213–220, 2005.

[168] H. Qian, D. A. Beard, and S. D. Liang. Stoichiometric network theory for nonequilibrium biochemical systems. *Eur. J. Biochem.*, 270:415–421, 2003.

[169] H. Qian, S. L. Mayo, and A. Morton. Protein hydrogen exchange in denaturant: quantitative analysis a two-process model. *Biochemistry*, 33:8167–8171, 1994.

[170] H. Qian and J. A. Schellman. Helix-coil theories: a comparative study for finite length polypeptides. *J. Phys. Chem.*, 96:3987–3994, 1992.

[171] J. L. Reed, T. D. Vo, C. H. Schilling, and B. Ø. Palsson. An expanded genome-scale model of *Escherichia coli* K-12 (iJR904 GSM/GPR). *Genome Biol.*, 4:R54, 2003.

[172] E. M. Renkin. Exchangeability of tissue potassium in skeletal muscle. *Am. J. Physiol.*, 197:1211–1215, 1959.

[173] R. Rigler and E. L. Elson (Eds.). *Fluorescence Correlation Spectroscopy: Theory and Applications*. Springer-Verlag, Berlin, Heidelberg, 2001.

[174] H. Risken. *The Fokker–Planck Equation. Methods of Solution and Applications*. Springer, New York, NY, 1996.

[175] M. Rodbell, L. Birnbaumer, S. L. Pohl, and H. M. J. Krans. The glucagon-sensitive adenyl cyclase system in plasma membranes of rat liver. V. An obligatory role of guanylnucleotides in glucagon action. *J. Biol. Chem.*, 246:1877–1882, 1971.

[176] E. M. Ross and A. G. Gilman. Resolution of some components of adenylate cyclase necessary for catalytic activity. *J. Biol. Chem.*, 252:6966–6969, 1977.

[177] G. R. Sambrano. Developing a navigation and visualization system for signaling pathways, 2003. Alliance for Cellular Signaling, http://afcs.lbl.gov/reports/v1/DA0009.pdf.

[178] W. C. Sangren and C. W. Sheppard. Mathematical derivation of the exchange of a labeled substance between a liquid flowing in a vessel and an external compartment. *Bull. Math. Biophys.*, 15:387–394, 1953.

[179] H. M. Sauro. Jarnac: a system for interactive metabolic analysis. In J.-H. Hofmeyr, J. M. Rohwer, and J. Snoep, editors, *Animating the Cellular Map. 9th International BioThermoKinetics Meeting*, chapter 33, pages 221–228. Stellenbosch University Press, Stellenbosch, 2000.

[180] J. A. Schellman. The stability of hydrogen-bonded peptide structures in aqueous solution. *C. R. Trav. Lab Carlsberg Ser. Chim.*, 29:230–259, 1955.

[181] T. Schlick. *Molecular Modeling and Simulation*. Springer, New York, NY, 2002.

[182] L. M. Schwartz, T. R. Bukowski, J. H. Revkin, and J. B. Bassingthwaighte. Cardiac endothelial transport and metabolism of adenosine and inosine. *Am. J. Physiol.*, 277:H1241–H1251, 1999.

[183] I. H. Segel. *Enzyme Kinetics: Behavior and Analysis of Rapid Equilibrium and Steady-State Enzyme Systems*. John Wiley and Sons Interscience, New York, NY, 1975.

[184] D. Segrè, D. Vitkup, and G. M. Church. Analysis of optimality in natural and perturbed metabolic networks. *Proc. Natl. Acad. Sci. USA*, 99:15112–15117, 2002.

[185] Y. Y. Shi, G. A. Miller, O. Denisenko, H. Qian, and K. Bomsztyk. Quantitative model for binary measurements of protein–protein interactions. *J. Comput. Biol.*, 14:1011–1023, 2007.

[186] D. Shore, J. Langowski, and R. L. Baldwin. DNA flexibility studied by covalent closure of short DNA fragments into circles. *Proc. Natl. Acad. Sci. USA*, 78:4833–4837, 1981.

[187] B. M. Slepchenko, J. C. Schaff, I. Macara, and L. M. Loew. Quantitative cell biology with the virtual cell. *Trends Cell. Biol.*, 13:570–576, 2003.

[188] C. M. Smith and J. R. Williamson. Inhibition of citrate synthase by succinyl-CoA and other metabolites. *FEBS Lett.*, 18:35–38, 1971.

[189] G. Stephanopoulos. Metabolic engineering. *Curr. Opin. Biotechnol.*, 5:196–200, 1994.

[190] G. Strang. *Introduction to Applied Mathematics*. Wellesley-Cambridge Press, Wellesley, MA, 1986.

[191] S. H. Strogatz. *Nonlinear Dynamics and Chaos: With Applications to Physics, Biology, Chemistry and Engineering*. Perseus Books Group, Cambridge, MA, 2001.

[192] C. Tanford. *Physical Chemistry of Macromolecules*. John Wiley & Sons, New York, NY, 1961.

[193] Y. Termonia and J. Ross. Oscillations and control features in glycolysis: numerical analysis of a comprehensive model. *Proc. Natl. Acad. Sci. USA*, 78:2952–2956, 1981.

[194] B. Teusink, J. Passarge, C. A. Reijenga *et al.* Can yeast glycolysis be understood in terms of in vitro kinetics of the constituent enzymes? Testing biochemistry. *Eur. J. Biochem.*, 267:5313–5329, 2000.

[195] I. Thiele, T. D. Bo, N. D. Price, and B. O. Palsson. An expanded metabolic reconstruction of *Helicobacter pylori* (iIT341 GSM/GPR): an in silico genome-scale characterization of single and double deletion mutants. *J. Bacteriol.*, 187:5818–5830, 2005.

[196] N. M. Tsoukias, D. Goldman, A. Vadapalli, R. N. Pittman, and A. S. Popel. A computational model of oxygen delivery by hemoglobin-based oxygen carriers in three-dimensional microvascular networks. *J. Theor. Biol.*, 2007.

[197] J. J. Tyson. Biochemical oscillations. In C. P. Fall, E. Marlang, J. Wagner, and J. J. Tyson, editors, *Computational Cell Biology*, chapter 3, pages 230–260. Springer, New York, NY, 2002.

[198] J. J. Tyson, K. Chen, and B. Novak. Network dynamics and cell physiology. *Nature Rev. Mol. Cell Biol.*, 2:908–916, 2001.

[199] J. J. Tyson. Bringing cartoons to life. *Nature*, 445:823–823, 2007.

[200] P. Uetz, L. Giot, G. Cagney *et al.* A comprehensive analysis of protein-protein interactions in Saccharomyces cerevisiae. *Nature*, 403:623–627, 2000.

[201] H. H. Ussing. The distinction by means of tracers between active transport and diffusion. The transfer of iodide across the isolated frog skin. *Acta Physiol. Scand.*, 19:43–56, 1949.

[202] S. J. van Dien and M. E. Lidstrom. Stoichiometric model for evaluating the metabolic capabilities of the facultative methylotroph methylobacterium extorquens am1, with application to reconstruction of c(3) and c(4) metabolism. *Biotechnol. Bioeng.*, 78:296–312, 2002.

[203] D. D. Van Slyke and G. E. Cullen. The mode of action of urease and of enzymes in general. *J. Biol. Chem*, 19:141–180, 1914.

[204] A. Varma and B. Ø. Palsson. Metabolic capabilities of *Escherichia coli*: I. Synthesis of biosynthetic precursors and cofactors. *J. Theor. Biol.*, 165:477–502, 1993.

[205] K. C. Vinnakota, M. L. Kemp, and M. J. Kushmerick. Dynamics of muscle glycogenolysis modeled with pH time-course computation and pH-dependent reaction equilibria and enzyme kinetics. *Biophys. J.*, 91:1264–1287, 2006.

[206] P. H. von Hippel. From "simple" DNA-protein interactions to the macromolecular machines of gene expression. *Ann. Rev. Biophys. Biomol. Struct.*, 36:79–105, 2007.

[207] J. D. Watson and F. H. C. Crick. Molecular structure of nucleic acids: a structure for deoxyribose nucleic acid. *Nature*, 171:737–738, 1953.

[208] H. V. Westerhoff and Y.-D. Chen. How do enzyme activities control metabolite concentrations? An additional theorem in the theory of metabolic control. *Eur. J. Biochem.*, 142:425–430, 1994.

[209] H. V. Westerhoff and K. van Dam. *Thermodynamics and Control of Biological Free-Energy Transduction*. Elsevier, Amsterdam, 1987.

[210] D. J. Wilkinson. *Stochastic Modelling for Systems Biology*. Chapman & Hall/CRC, New York, NY, 2006.

[211] J. B. Wittenberg. Myoglobin-facilitated oxygen diffusion: role of myoglobin in oxygen entry into muscle. *Physiol. Rev.*, 50:559–636, 1970.

[212] F. Wu, J. A. Jeneson, and D. A. Beard. Oxidative ATP synthesis in skeletal muscle is controlled by substrate feedback. *Am. J. Physiol.*, 292:C115–C124, 2007.

[213] F. Wu, F. Yang, K. C. Vinnakota, and D. A. Beard. Computer modeling of mitochondrial tricarboxylic acid cycle, oxidative phosphorylation, metabolite transport, and electrophysiology. *J. Biol. Chem.*, 282:24525–24537, 2007.

[214] J. Wyman. Facilitated diffusion and the possible role of myoglobin as a transport mechanism. *J. Biol. Chem.*, 241:115–121, 1966.

[215] J. Wyman and S. J. Gill. *Binding and Linkage: Functional Chemistry of Biological Macromolecules*. University Science Books, Mill Valley, CA, 1990.

[216] F. Yang, H. Qian, and D. A. Beard. Ab initio prediction of thermodynamically feasible reaction directions from biochemical network stoichiometry. *Metab. Eng.*, 7:251–259, 2005.

[217] J. Yu, J. Xiao, X. J. Ren, K. Q. Lao, and X. S. Xie. Probing gene expression in live cells, one protein molecule at a time. *Science*, 311:1600–1603, 2006.

[218] J. Zhang, K. Ugurbil, A. H. From, and R. J. Bache. Myocardial oxygenation and high-energy phosphate levels during graded coronary hypoperfusion. *Am. J. Physiol.*, 280:H318–H326, 2001.

Index

Lightning Source UK Ltd.
Milton Keynes UK
18 February 2011
167773UK00001B/9/P